AF757954

GEOMETRIC REPRESENTATIONS OF PERCEPTUAL PHENOMENA

Papers in Honor of Tarow Indow on His 70th Birthday

GEOMETRIC REPRESENTATIONS OF PERCEPTUAL PHENOMENA

Papers in Honor of Tarow Indow on His 70th Birthday

Edited by

R. Duncan Luce
Michael D'Zmura
Donald Hoffman
Geoffrey J. Iverson
A. Kimball Romney
University of California, Irvine

LEA LAWRENCE ERLBAUM ASSOCIATES, PUBLISHERS
1995 Mahwah, New Jersey Hove, UK

Copyright © 1995 by Lawrence Erlbaum Associates, Inc.
All rights reserved. No part of this book may be reproduced in any form, by photostat, microform, retrieval system, or any other means, without the prior written permission of the publisher.

Lawrence Erlbaum Associates, Inc., Publishers
10 Industrial Avenue
Mahwah, New Jersey 07430

Library of Congress Cataloging-in-Publication Data

Geometric representations of perceptual phenomena : papers in honor of Tarow Indow on his 70th birthday / edited by R. Duncan Luce . . . [et al.].
p. cm.
Papers originally presented at a conference held at the University of California, Irvine, July 22–28, 1993.
Includes bibliographical references and indexes.
ISBN 0–8058–1686–0 (alk. paper)
1. Psychophysics—Congresses. 2. Psychometrics—Congresses. 3. Space perception–Congresses. 4. Color vision—Congresses. I. Indow, Tarow, 1923– . II. Luce, R. Duncan (Robert Duncan) III. University of California, Irvine.
BF237.G39 1995
152.14—dc20 94–24199
CIP

Books published by Lawrence Erlbaum Associates are printed on acid-free paper, and their bindings are chosen for strength and durability.

Printed in the United States of America

10 9 8 7 6 5 4 3 2 1

Contents

Preface

This volume arose from a conference held July 22–28, 1993 at the University of California, Irvine (UCI) on the topic that provides its title: geometric representations of perceptual phenomena. The conference was run jointly by the Department of Cognitive Sciences and the Institute for Mathematical Behavioral Sciences, and it was supported by funds provided the Institute by UCI, by a National Science Foundation Research and Training Grant, and by funds from UCI's Committee on Research and Graduate Studies obtained by the chair of the Department of Cognitive Sciences, Professor Mary-Louise Kean, for this purpose.

The conference was held, and the volume prepared, in honor of Professor Tarow Indow on his 70th birthday, which occurred during his 16th year at UCI. The social climax of the conference was a dinner held to pay tribute to Professor Indow and to his lovely wife, Minako. Various people who have known them over the years provided warm reminiscences and toasts. Everyone agreed that the high point of the evening was a delightful slide show presented by Professor John I. Yellott, Jr. (assisted by Mrs. Indow in assembling the slides) that covered many significant events of Indow's life from childhood to the present. Some of the faculty appeared in slides of earlier eras and, to their dismay, were not always recognized by their colleagues until identified by Professor Yellott.

The volume, following the conference, is organized into three major topics concerning the use of geometry in perception: space, color, and scaling. The first topic refers to attempts to represent the subjective space within which we locate ourselves and perceive objects to reside. The second topic concerns attempts to represent the structure of color percepts as revealed by various experimental procedures. The third has as its goal the organization of various bodies of data (in this case perceptual) through scaling techniques, primarily multidimensional ones. These topics provide a natural organization of the work in the field, as well as one that corresponds to the major aspects of Professor Indow's contributions. He has participated in a seminal fashion in the development of each of these areas. The magnitude of these contributions are reflected in his, the first, chapter and his bibliography in the Appendix to that chapter. We comment briefly on his work on each of the topics.

Indow's work on perceptual space is widely agreed to be classic. He was the first to devise a clear-cut procedure for discriminating among hyperbolic, Euclidean, and elliptic models for perception. Moreover, his continued work on this problem based on "alleys" of lights in different planes relative to the observer has raised very serious doubts about the homogeneity of visual space, which means that all of the classical geometries are ruled out as a global description of perceptual space. This raises a major mathematical challenge that has yet to be met.

He has contributed significantly to our understanding of the constraints that exist in the perception of color. Once again, these constraints are most simply described as a cone in a geometric color space. The earliest work was on aperture colors and, in particular, the Munsell chips used to represent them. Later, and current work, is focused on the more complex domain of surface colors that lead to a pattern of percepts that is rather different from that of aperture colors.

Finally, his work on scaling has exhibited two striking aspects. First, he was one of the earliest actually to use the, then new, multidimensional scaling techniques, which were computationally laborious at the time. Second, some of these uses involved highly applied problems of direct relevance to a Japanese industry. These are superb examples of a person doing excellent science in the context of a socially significant application.

The goal of the book is both to provide the reader with some overview of the issues in each of the areas and to present some current results. Introductions by Donald Hoffman, Michael D'Zmura and Geoffrey Iverson, and A. Kimball Romney are provided for each part.

We take this opportunity to thank the authors for their prompt preparation of manuscripts and for providing useful comments on each other's chapters.

R. Duncan Luce

List of Contributors

Marc K. Albert Department of Information and Computer Science, University of California, Irvine, Irvine, CA 92717.

John C. Baird Department of Psychology, Dartmouth College, Hanover, NH 03755.

William H. Batchelder University of California, Irvine, Irvine, CA 92717.

Tim Brazill University of California, Irvine, Irvine, CA 92717.

J. Douglas Carroll Graduate School of Management, Rutgers University, 92 New Street, Newark, NJ 07102.

Anil Chaturvedi AT&T Bell Laboratories, 600 Mountain Ave., Murray Hill, NJ 07974.

Jan Drösler Institut for Psychologie, Universit Regensburg, 8400 Regensburg, Regensburg, Germany.

Ehtibar Dzhafarov Department of Psychology, University of Illinois at Urbana-Champaign, 603 E. Daniel Street, Champaign, IL 61820.

Michael D'Zmura Department of Cognitive Sciences and Institute for Mathematical Behavioral Sciences, University of California, Irvine, Irvine, CA.

Hannes Eisler Department of Psychology, University of Stockholm, Stockholm S-106 91, Sweden.

Donald D. Hoffman Department of Cognitive Science, University of California, Irvine, CA 92717.

G. Iverson Department of Cognitive Science and Institute for Mathematical Behavioral Sciences, University of California, Irvine, Irvine, CA 92717.

Chingiz A. Izmailov Department of Psychology, Moscow University, Marx Prospect 18/5, Moscow K9, Russia.

Karl Christoph Klauer Universitaet Heidelberg, Psychologisches Institut, Hauptstrasse 47-51, 69117 Heidelberg, FR Germany.

John Krauskopf Department of Psychology, New York University, Washington Square, New York, NY 10003.

Vengu Lakshminarayanan School of Optometry, University of Missouri-St. Louis, 8001 Natural Bridge Road, St. Louis, MO 63121-4499.

R. Duncan Luce Institute for Mathematical Behavioral Sciences, University of California, Irvine, Social Science Tower, Irvine, CA 92714.

Donald Macleod Univ. of CA at San Diego, Dept. of Psychology, 5121 McGill Hall C-009, San Diego, CA 92093.

Laurence Maloney Department of Psychology, New York University, Washington Square, New York, New York 10003.

Lawrence Marks John B. Pierce Laboratory, 290 Congress Avenue, New Haven, CT 06519.

A. Kimball Romney University of California, Irvine, Irvine, CA 92717.

T. S. Santhanam Department of Science and Mathematics, Parker College of St. Louis University, St. Louis, MO 62206.

Benjamin Singer Department of Cognitive Science and Institute for Mathematical Behavioral Sciences, University of California, Irvine, Irvine, CA 92717.

Patrick Suppes Lucie Stern Professor of Philosophy, Ventura Hall, Stanford University, Stanford, CA 94305.

J. Douglas Willen Univ. of CA at San Diego, Dept. of Psychology, 5121 McGill Hall C-009, San Diego, CA 92093.

Sophie M. Wuerger Department of Psychology, New York University, Washington Square, New York, NY 10003.

1 Psychophysical Scaling: Scientific and Practical Applications

Tarow Indow
University of California, Irvine

ABSTRACT

Two approaches, (a) without scaling and (b) with scaling, are contrasted in studies of the geometry of visual space (VS) and color perception. The discrepancy between parallel and equidistance alleys in VS illustrates type (a) and MDS scaling of perceptual distances illustrates type (b). The latter yielded finer-grain information about the geometry of VS. Standard Munsell color chips, which were selected so that colors change with perceptually equal steps in each of the three attributes, illustrates type (a). Scaling uni- and multiattribute color differences makes clear that the Munsell solid as a whole can be regarded as a structure having a perceptually meaningful metric (b). Quantitative representation of sensation is useful in some industries and technologies. As an example, the scaling of four qualities of taste is discussed.

1. PSYCHOPHYSICAL SCALES

Experiments in which subjects make judgments about stimuli are structured as in Figure 1.1. Such judgments yield one or both of the following types of results. One is a set of stimuli arising from the subject's judgments, such as just-noticeable differences (JND), and points of subjective equality (PSE). The other is a scale representing some aspect of perception caused by the stimuli, such as loudnesses of tones. Stevens (1975) often used ϕ and ψ to denote the physical stimulus magnitude and the sensation magnitude, respectively. In contrast, here Greek letters are reserved to represent latent variables within the subject (treated

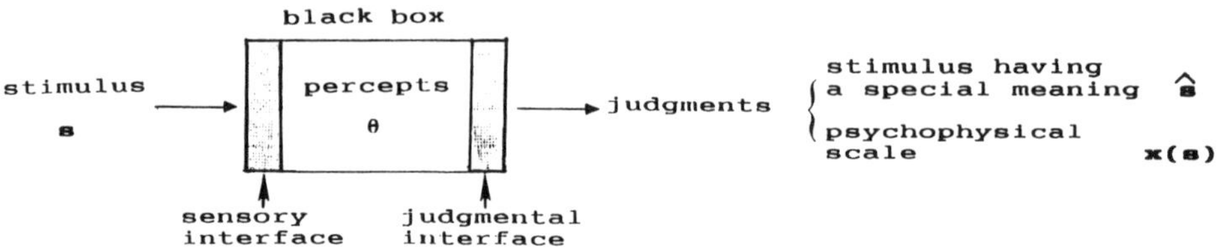

FIG. 1.1. The structure of psychophysical experiment to elicit judgments on stimulus.

as a black box), and **s** and **x** are used as the general terms for stimuli and scale values. For example, the loudness that a subject perceives for a stimulus **s** is denoted by $\theta(\mathbf{s})$. Then $\mathbf{x}(\mathbf{s}_1) > \mathbf{x}(\mathbf{s}_2)$ corresponds to the subject judging that "$\theta(\mathbf{s}_1)$ is louder than $\theta(\mathbf{s}_2)$," symbolized by $\theta(\mathbf{s}_1) > \theta(\mathbf{s}_2)$. When we can conclude from **x** only the order relationship in θ, **x** is called an ordinal scale. According to the situation, different symbols are used in lieu of **s** and **x**, namely e_{jk} for physical distances between two stimuli (**s**) and d_{jk} for scaled values (**x**) of perceptual distances δ_{jk} (θ). From d_{jk} we can tell something more than the order among δ_{jk}. Even when the same sensory magnitude is involved, different symbols are used, as necessary, for different judgments.

For present purposes, *psychophysical scaling* refers to scales **x**(**s**) stronger than ordinal. We discuss the types of scales **x**(**s**) that can be defined and what information is obtained by their use in three areas: visual space, the color system, and the relative strengths of four qualities of taste. The first two are scientific applications of scaling, and the use and nonuse of scales are contrasted. The third example illustrates a practical use of psychophysical scaling. In each case, **s** can be specified physically, and hence **x**(**s**) denotes a functional relationship between two continuous variables. Just as the sensory interface or transducer exists on the input side of the black box, so the judgmental interface exists on the output side. To describe the processing involved in this interface, a number of models were proposed: the Thurstonian model (Falmagne, 1985; Thurstone, 1927; Torgerson, 1958) and detection theory (e.g., Green and Swets, 1966) for binary judgments, and models of Krantz (1972) and Shepard (1978) for numerical judgments such as magnitude estimation.

The main objective of this chapter is to provide experimental results shedding light on the nature of the judgmental process in the three concrete cases.

2. THE GLOBAL STRUCTURE OF VISUAL SPACE

2.1. Visual Space

Aside from those who are visually handicapped, people always perceive an environment extending in the three directions in front of us. This perceived space is called visual space (VS) in order to distinguish it from the physical space (X) from which light stimuli come. VS is the final product of a long series of processes from physical objects to the retina (a physical process) and from the retina to the brain (a physiological process). Under ordinary conditions, VS consists of individual objects, their backgrounds, and the perceived self. Just as perceived objects are different from physical ones, the perceived self (θ) must also be distinguished from one's physical body (Köhler, 1929; Schiler, 1950). Except for the visually perceived hands, arms, etc., the perception of the self is due to proprioceptive stimulation from the body. Nonetheless, the self is localized as a percept in VS, and the direction in VS is usually determined from this point. If the eyes are exposed to homogeneous light of sufficiently low intensity, one experiences the phenomenon called "Ganzfeld" in which one sees an unstructured mist of light around one's self (Metzger, 1930). In order to have a structured VS, some heterogeneity in retinal stimulation is necessary. Ordinarily this requirement is satisfied and one sees, for example, a surface in front of one's self. In most studies of visual perception, this articulation of VS is taken for granted, and a local phenomenon in VS, such as perception on a frontoparallel surface, is the matter of concern. In the present context, the more global structure of VS *per se* is discussed.

(VS 1) According to VS, we are able to guide our bodies appropriately through the physical space X so as to reach, manipulate, or avoid physical objects. Hence, at least in the neighborhood of the self, VS must be structured so as to make such movement possible. Beyond this range, the correspondence between VS and X is not completely "isomorphic." For example, X does not include a physical entity that has the same form as the perceptual sky in VS (Indow, 1991). VS is closed and continuous at the boundary in the sense that we do not see anything as being at an infinite distance and all physical objects beyond a certain distance are seen as at the same finite distance. At that boundary, we perceive something at the end of the line of sight. Indoors, the boundary of VS consists of walls; and outdoors, the terrain, horizon, or sky delimits VS. VS is dynamic. It is not like a solid container into which various percepts are placed; rather it is more like a balloon, and perceived distance to the boundary is very sensitive to the stimulating conditions. When visible, the horizon always appears at the level of the eye in VS, and the perceptual distance to the horizon changes according to condition.

(VS 2) Under a fixed stimulus configuration, VS is stable and independent of the direction of the eyes. As sight is redirected, different parts of VS are seen as remaining still. In other words, the whole VS is a product of multiple glances. There is a hierarchy among the percepts in VS, with one level acting as the *framework* for another. In a room, the self lies within the framework of perceived walls and floor. Hence, like the induced movement of a point in a framework on a frontoparallel surface (Duncker, 1929), it is possible to create the induced movement of self with respect to the walls of a room. Furthermore, localizing sounds is related to VS. There must be some interaction between visual processes and those of other modalities. No one knows at present how the sequences of physical and physiological processes mentioned earlier generate VS as a coherent self-organized dynamic complex, somewhat independent of the pattern of stimulation on the retina. However, for VS as a perceptual phenomena, it seems worthwhile to raise the following question.

(VS 3) We perceive geometrical properties—curves, straight lines, intersections, angles, congruences, and parallelness, etc.—in VS. That being so, we may ask: what geometry best describes the structure of VS? Although we can safely treat X as a 3-D Euclidean space, no a priori reason exists why VS has also to be structured as Euclidean geometry.

Perhaps, most people think that VS is not structured according to any geometry, at least, not to one of the mathematically well-established ones. However, Luneburg (1947, 1950), a mathematician, postulated that VS is a Riemannian space (R) of constant curvature, which position was reiterated by another mathematician, Blank (1958, 1959) (see also Suppes, Krantz, Luce, & Tversky, 1989). The basic motive for this postulate can be briefly summarized as follows (Indow, 1991).

(R1) If one can perceive distance δ between any two points in VS (taken to be finitely compact and convex) and if δ satisfies Fréchet's conditions ($\delta_{ik} \sim \delta_{kj} > 0$, $\delta_{jj} \sim 0$, $\delta_{ij} \oplus \delta_{jk} \geqslant \delta_{ik}$) then VS is a metric space.

(R2) If VS is also locally Euclidean, then VS is Riemannian, R. In this case, the property of space around a point is characterized by the Gaussian total curvature K at that point.

(R3) If VS exhibits the Desarguesian property and free mobility, then K is a constant. In this case, three possibilities exist: elliptic ($K > 0$), Euclidean ($K = 0$), and hyperbolic ($K < 0$).

Some experimental data and discussion related to these three conditions were stated in an earlier paper (Indow, 1991). Roughly speaking, the last requirement means that a figure can be rotated at any location of VS without changing its shape and, for a given figure, one can construct a congruent figure at any other location in VS. Congruent—being of the same form and size—means perceptual identity in VS, not identity between stimulus patterns in X. In other words, it is postulated that the subject can adjust the stimulus figure in X so that its percept satisfies these conditions.

2.2. Parallel and Distance Alleys: Traditional Experiments

The following experimental finding, which was first systematically observed by Blumenfeld (1913), allows us to determine the sign and even the value of K. Figure 1.2 shows positions of two series of pairs of light points $\{Q_{Li}, Q_{Ri}\}$ as adjusted by a subject. All these light points were presented in a dark space (X)

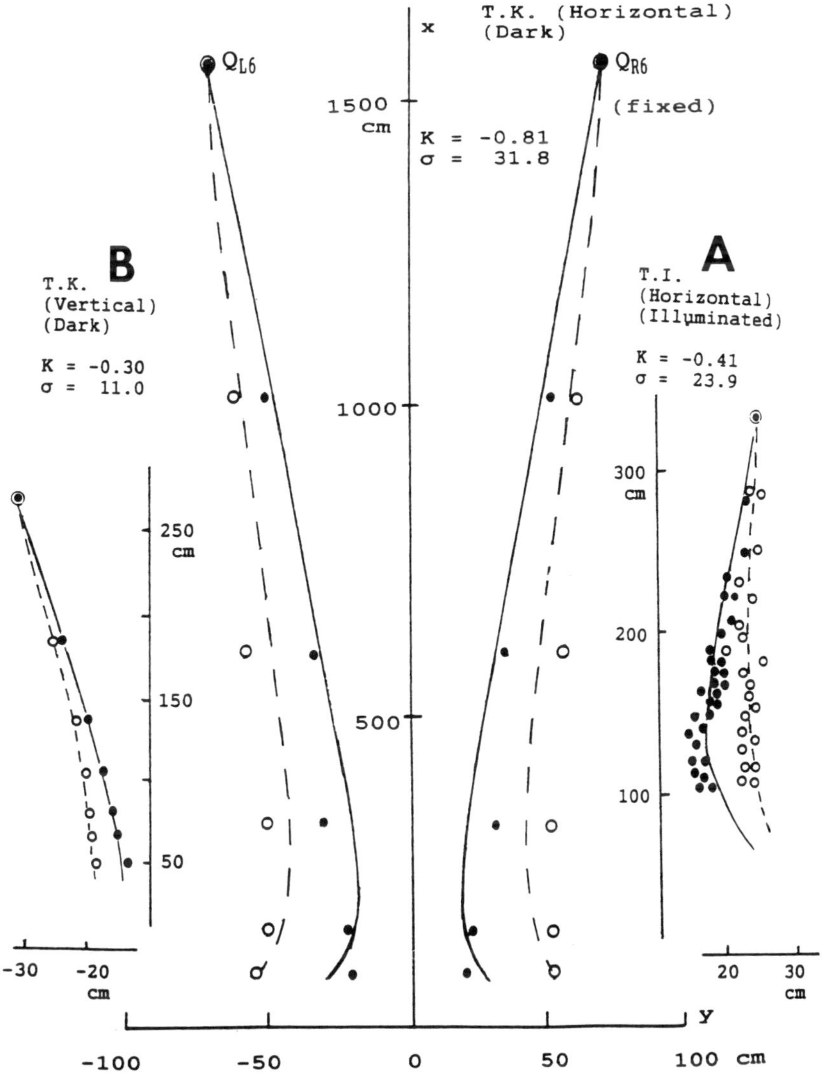

FIG. 1.2. Three examples of parallel and distance alleys. (Inset A was taken by permission from T. Indow [1982], *Journal of Mathematical Psychology, 26,* Fig. 5 on p. 214.)

and lay on the horizontal plane at eye level (HZ). The location of the furthest pair, $\{Q_{L6}, Q_{R6}\}$ in this case, was fixed. The subject was asked to adjust the other Q's along the y-axis in two ways: once so that they were seen as two straight and parallel series in VS (solid points in Figure 1.2), and once so that all pairs were equally separated and had the same lateral distances in VS (open points in Figure

1.2). The former is called a *parallel alley* (P) and the latter a *distance alley* (D). Because the two alleys differ, one concludes that VS is not Euclidean. The points of the D alley always lie outside those of the P alley, which for Riemannian geometries implies that $K < 0$.

A Riemannian space (R) can be embedded in a Euclidean space (E) in various ways. Poincaré's model represents R^3 of constant curvature in E^3 (e.g., Berger, 1987; Busemann, 1955; Spivak, 1979). In this model, straight lines (geodesics) are represented by arcs satisfying a condition and their lengths must be measured in a specific way (not isometric); however, angles are preserved (conformal). Let us call this representation of VS its *Euclidean map* (EM) (Indow, 1979, 1991). VS is represented within a sphere whose radius is related to K. Luneburg gave nice representations of P- and D-alleys in EM and showed that, within the sphere, the set of arcs for D lies outside those for P when and only when $K < 0$. Hence, if the correspondence between X and EM is monotone, the above-stated interpretation for the discrepancy between the two alleys holds. Let us call the correspondence between these two spaces *mapping functions.* In order to define theoretical curves of P- and D-alleys in X, Luneburg assumed a particular form for the mapping functions and mapped the sets of arcs in EM to X. We fitted these theoretical curves to experimental data by optimizing values of two parameters involved, K and σ, where σ, which is a parameter in the mapping functions, tells how the distance from the one's body to a stimulus point in X is mapped to the perceived distance from the self to the point in VS. Figure 1.2 presents the fitted curves and the parameter values. This is a case where $\{Q_i\}$ covers a large area using a few points (Indow, Inoue, & Matsushima, 1963). In the inset A on the right, $(Q_{Ri}\}$ for P and D are shown from an experiment with small black points under illumination and the number of points is extraordinarily large (Indow, 1982).

The discrepancy between the P- and D-alleys in X does not necessarily lead to the conclusion that the two curves differ in VS. When constructing a P-alley, the subject usually scans $\{Q_{Li}, Q_{Ri}\}$ along the x-axis to see whether they appear as two straight and parallel lines. When constructing a D-alley, the main direction of scanning is along the y-axis. Hence, if correspondence between VS and X changes according to the direction of scanning, even if the subject perceives the same series under two instructions, the series $\{Q_{Li}, Q_{Ri}\}$ for P and D will differ in X.

This possibility was checked in an experiment in which a single pair of stimulus points Q_R and Q_L was moved. The subject adjusted the trajectories so they appeared straight and parallel (P) or so as to maintain the same interval (D). In this case, the scan direction was controlled by the two directions of movement, e.g., Q_R toward and Q_L away from the subject simultaneously. Irrespective of the scan direction, exactly the same results were obtained as with the stationary configuration of points $\{Q_{Li}, Q_{Ri}\}$ (Indow and Watanabe, 1984a). The inset B in Figure 1.2 shows an example of $\{Q_{Li}\}$ of P- and D-alleys constructed in the vertical direction. The subject looked upward and adjusted the positions of Q's that were suspended from the ceiling (Indow, Inoue, & Matsushima, 1962). So for alleys on the plane extending from the subject, no matter whether in the horizontal or vertical direction, the same discrepancy between P- and D-alleys is found, and these subspaces in VS can be interpreted as R^2 with $K < 0$.

If Luneburg's set of mapping functions are assumed, we can define the value of K. From these functions, the sphere representing the boundary of VS is defined in EM. It is convenient to take the radius of this sphere as the unit in terms of which the length of geodesic is measured in EM. When that radius is defined to be 2, then $-1 < K < 1$ in terms of this unit. The values given in Figure 1.2 are defined in this way. Of course, there are individual differences in the estimated values of K and σ and, even for the same subject, both change systematically according to the size of the configuration of points $\{Q_{Li}, Q_{Ri}\}$ (the dynamic property of VS) and also whether X is dark or illuminated (Table 1 in Indow and Watanabe, 1984a; Indow, 1991). In order to estimate K numerically, we need to specify a set of mapping functions. Luneburg's set was egocentric and very simple in form. Directional angles from the subject in X are preserved in VS, and the boundary of VS is represented by a sphere, which implies that VS is isotropic. It seems unlikely that such simple egocentric mapping functions can be correct when we perceive the configuration of points within a definite framework. This is the reason that most alley experiments were performed with light points in a dark room. Under the illuminated condition, small black points were presented on a white top table at the level of the eyes and the table was surrounded by the white sheet. The subject could not see the edges of the table as well as the walls.

2.3. Psychophysical Scaling of Perceptual Distances

Thus far, the results have been sets of stimulus points adjusted by the subject. The approach can be extended and improved in two ways by introducing psychophysical scaling on perceptual distances between points. First, it is not necessary to use the a priori assumed Luneburg's mapping functions in which all angles in VS are solely determined by the corresponding angles in X irrespective of the form of $\{P_j\}$. The correspondence can depend on the framework and $\{P_j\}$; hence we can extend the study to VS under more natural conditions. Second, a finer-grain analysis becomes possible, which includes geometrical properties such as distances and angles. This mode of analysis is schematized in Figure 1.3. The subject makes judgments on perceived distances (δ) with a configuration of n stimulus points $\{Q_j\}$ to obtain a matrix of scaled distances $\mathbf{D} = (d_{jk})$. If a value of K is assumed, d can be converted to the length of a geodesic, ρ, in EM under K. Then, the configuration of points in EM, $\{P_j\}$, can be defined from (ρ_{jk}) using multidimensional scaling (MDS). Denote by $(\hat{\rho}_{jk})$ a matrix of interpoint distances

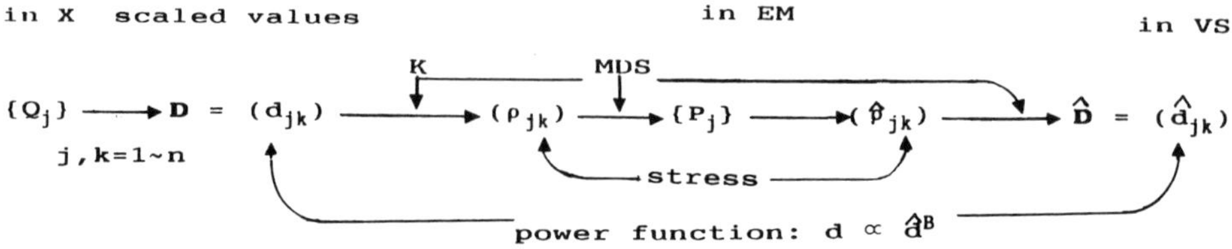

FIG. 1.3. Direct mapping through powered Riemaniann distances (DMPRD).

in $\{P_j\}$. These distances in EM are converted through K to distances in VS and we have $\hat{\mathbf{D}} = (\hat{d}_{jk})$. The program adjusts $\{P_j\}$ and K until closest fits are obtained between $\{\rho_{jk}\}$ and $\{\hat{\rho}_{jk}\}$ as well as between (d_{jk}) and $(\hat{d}_{jk})$. The relationship between data d and theoretical distances $\hat{d}$ can be any monotone function, but the optimization is easier when some additional constraints are imposed. The program Direct Mapping through Powered Riemannian Distances (DMPRD) assumes that $d \propto \hat{d}^B$, and it treats B as a free parameter. The initial value for B was set to 1 and B turned out ultimately to be unity in most cases and slightly larger in the remainder. The data $\mathbf{D}$ can be incomplete. If we use as $\{Q_j\}$ a configuration having an internal structure, e.g., $\{Q_{Rj}, Q_{Lj}\}$ of an alley experiment for the subject, then $\{P_j\}$ can be compared with the theoretical equations in EM and that degree of agreement can be another criterion. By comparing $\{P_j\}$ and $\{Q_j\}$, we have an a posteriori definition of the mapping functions (Indow, 1982, 1991).

The scaling to obtain $\mathbf{D}$ was based on judgments of ratios of two perceptual distances from a common point i, δ_{ij} versus δ_{ik}. The subject assigned a value $r_{i.jk}$ for each of appropriately selected triplets, Q_i, Q_j, Q_k, and then d_{jk} can be defined, with a common unit, from the matrix $(r_{i.jk})$. If judgments are limited to such triplets for which the subject assigns $r_{i.jk}$ with confidence, ratios of obtained scales, d_{ij}/d_{ik}, numerically reproduce the data $r_{i.jk}$ very closely. It means that, if δ is regarded as a quantity,

$$d = \alpha\delta^{\beta}, \qquad \alpha, \beta > 0. \tag{1.1}$$

The final result $\hat{d}$ is proportional to $d^{1/B}$, and hence $\hat{d} \propto \delta^{\beta/B}$. Notice that $B = 1$ in most cases. The "self" must be included among $\{P_j\}$, say as a point P_0, and values of all d_{0j}, $j = 1, 2, \ldots, n$, need to be given in $\mathbf{D}$.

Such psychophysical scaling was provided by each subject on his or her $\{Q_j\}$ for P and D alleys in which $\{Q_{Li}, Q_{Ri}\}$ were also adjusted to appear on a frontoparallel plane (horopter). According to whether Euclidean or Riemannian geometry was at issue, either B alone or K and B were optimized. Because no mapping functions were assumed in this approach, the boundary sphere of VS in EM, which is only defined via a mapping function, is necessarily indeterminate in advance. Hence, though its sign is defined, a value of K cannot be specified by the above-stated radius. Denote by P_n the furthest point from P_0. Then K is defined in terms of the radial distance ρ_{0n} in EM. K estimated in this way always turned out to be negative. In a scatter diagram, d lies very close to $\hat{d}$ in most cases and $B \approx 1$. If there is an asymptote for an increasing sequence of radial distances ρ_{0j}, then the boundary sphere of VS can be determined, and its length can be different, in theory, according to direction. However, estimating the asymptote was not easy, even when EM was assumed to be isotropic, and the agreement of $\{P_j\}$ with the theoretical curves in EM was not quite satisfactory (Indow, 1982, 1991). This finding casts some doubt on the validity of the hyperbolic interpretation of the discrepancy between P- and D-alleys.

Three subjects, looking at the night sky, scaled perceptual distances between 10 stars that are clearly distinguishable from the other stars. Since the "self" was included, $\{P_j\}$ consisting of 11 points was constructed in EM^3 with curvature K. Representing them in three different spaces, $K <, =, >0$, gave almost the same results, and so the geometry of this VS is indeterminate from these data (Indow, 1968, 1991). We do not have any theoretical curves in this case. If $K = 0$ in this VS, $\{P_j\}$ in E^3 is the undistorted representation of the night sky being perceived by the subject. Most subjects admitted that the shape of $\{P_j\}$ was close to his or her perception of the night sky. The observation of stars was made at a long seashore, and the subjects were unanimous in saying that the sky appeared closer in the direction where nothing was visible than in the direction where lights of a town were in the sight. This appearance of the sky is reflected in $\{P_j\}$. The boundary of VS is always determined by what we see in the direction of sight (the dynamic property of VS).

2.4 The Geometry of a Plane in Front of the Subject

In many natural situations, we are concerned with figures on a plane in front of the self. Hence, constructing alleys and using psychophysical scaling were extended to this subspace of VS. The subject adjusted positions of stimulus point Q's in the three directions so that all appeared on a frontoparallel plane and constructed two different configurations of three horizontal series of five points $\{Q_{\alpha j}\}$, where α = *t*op, *m*iddle, *b*ottom, and j = 1–5. Once three sets of five points, $\{Q_{tj}\}$, $\{Q_{mj}\}$, $\{Q_{bj}\}$ and j = 1–5, were adjusted so as to appear three horizontal straight lines parallel with each other (P alley). Once five sets of three points $\{Q_{\alpha 1}\}$–$\{Q_{\alpha 5}\}$ and α = t, m, b, were adjusted so as to appear to have a constant interval in the vertical direction (D alley). Then a scaling experiment on perceptual distances between points was performed with his or her own $\{Q_{\alpha j}\}$'s. Indow (1988a) provided theoretical equations for the horopter plane and these alleys. Two sets of these experiments gave unequivocal results (Indow and Watanabe, 1984b, 1988). The two alleys, P and D, were in complete agreement (K = 0), and the estimated values of K in the scaling experiment were also very close to 0. Hence, frontoparallel planes in VS can be regarded as being Euclidean.

When the subject looks at $\{Q\text{'s}\}$ of the D-alley on the frontoparallel plane, the series of points appear to be both straight and parallel as in the P-alley. This contrasts to the case of D alleys when the subspace of VS extends from the subject (the second subsection of this section). In the traditional procedure of constructing a D alley, while a pair $\{Q_{Li}, Q_{Ri}\}$ is being adjusted, usually in the order $i = n - 1, n - 2, \ldots, 1$, the subject sees only this pair and the fixed one, $\{Q_{Ln}, Q_{Rn}\}$. When the n pairs $\{Q_L, Q_R\}$ are presented, after all the adjustments had been completed, the subject noticed that the two series, Q_{Li} and Q_{Ri}, diverged toward him or her. Most subjects also realized that each of the two series appeared slightly curved. We therefore introduced a different procedure for D-alleys. While a pair $\{Q_{Li}, Q_{Ri}\}$ was being adjusted, all pairs that had been already adjusted, $\{Q_{Lj}, Q_{Rj}\}$, $j > i$, remained visible in VS. The same discrepancy between P and D-alleys occurred and the same impressions of the D alley were noticed (Indow and Watanabe, 1984a).

2.5 Discussion

In $\{P_j\}$ constructed, there are several series of P_i, P_j, P_k, etc., that are guaranteed to be collinear in VS and scaled values of d's between them are given in **D**. Hence, it is easy to test the additivity of distances: $d_{ik} = d_{ij} + d_{jk}$. This requirement was shown to hold very well with d's scaled in the procedure stated in the third subsection (Indow, 1991, Figure 13). The finding suggests that β in Equation 1.1 is equal to 1, and $d \propto \delta$. In other words, from the scale d, we can define for the black box of Figure 1.1 the perceptual distance δ that behaves as a quantity and satisfies the additivity between collinear points in perception. Furthermore, $B = 1$ and $\hat{d} \propto \delta$ in most cases. In these cases where $\hat{d} \propto \delta$, $\{P_j\}$ is regarded as a quantitative representation of perceptual distances in the black box. For the results related to δ, a Riemannian representation of VS (including the case $K = 0$) is persuasive. In order to think of the geometrical structure of VS, we need one more step. When such a $\{Q_j\}$ that has an internal structure is used, the corresponding $\{P_j\}$ constructed in EM can be compared with the theoretical curves in EM under a geometry. If the fit is satisfactory, we can say that VS is really structured according to this geometry on the basis of fitting both distances and angles. This conclusion is independent of any assumption about the mapping correspondence between X and EM. As mentioned before, however, the agreement between $\{P_j\}$ and the equations is not quite satisfactory. It was true even in horizontal alleys on the frontoparallel subspace (2.4). It might well be possible that, when the subject pays attention to a triplet, Q_i, Q_j, Q_k, to assess δ's, the geometrical property of that part of VS is perburbed. Hence, in order to make explicit the real geometrical structure of VS, we have to invent a scaling procedure in which the possibility of interference of this kind is minimized.

3. SURFACE COLOR SPACE

3.1 Color Spaces

The number of colors that we can discriminate one from another is about 7 million. Hence, to organize such a variety of stimuli, we have few options other than some sort of spatial representation. It has been well established that, except for colors attributed to gloss of the surface (e.g., gold, silver, etc.), all colors can be specified as points in a 3-D space, X^3. This is called a *color space,* and according to the principle by which colors are represented, we have various color spaces. These are classified into two large groups; one is exemplified by the *CIE* (Commission Internationale de l'Eclarge) (x, y, Y) space, and the other by the *Munsell* system. Perceived color can be in several different modes of appearance. The two most important are *aperture* and *surface* color modes. The former is the color of light filling an area (e.g., the blue of a clear sky), and its brightness varies from dark to bright. The latter is the color of a perceived surface (e.g., the blue of a cloth) which varies from blackish to whitish.

A color stimulus, s, is a pattern of distribution of radiant energy $P(\lambda)$ where λ denotes wavelength in the visible part Ω of spectrum. Correspondence between perceived color (χ) and $P(\lambda)$ is one to (infinitely) many. Those $P(\lambda)$ corresponding to a single color are called *metameric* to each other. On the basis of color-matching experiments, a system has been established so that all $P(\lambda)$'s which are metameric under standard observing conditions are represented by the same point F. A point $F(x, y, Y)$ uniquely specifies a color in such a way that hue and saturation are determined by (x, y) and luminance by Y. The spectral composition of light emitted from a TV display is very different from that of the light entering the TV camera. In the CIE system, the two are represented by points that are close to each other. A color receiver uses all the information (x, y, Y), whereas a black-and-white receiver uses Y only. The plane (x, y) of a fixed luminance level Y is called a *chromaticity diagram.* Define

$$X \propto \int P(\lambda)\,\bar{x}(\lambda)\,d\lambda, \quad Y \propto \int P(\lambda)\bar{y}(\lambda)\,d\lambda, \quad Z \propto \int P(\lambda)\bar{z}(\lambda)\,d\lambda, \tag{1.2}$$

$$x = X/S, \quad y = Y/S, \quad z = Z/S, \quad S = X + Y + Z,$$

where $\bar{x}(\lambda)$, $\bar{y}(\lambda)$, $\bar{z}(\lambda)$ are variables defined from color-matching experiments. In particular, $\bar{y}(\lambda)$ is often written as $V(\lambda)$ with an appropriate normalization. $V(\lambda)$ is called the *spectral luminous efficiency function,* because it represents the effects of light of wavelength λ upon the eye in producing brightness (χ). Notice that $P(\lambda)$ is a pure physical process outside the black box, whereas neither (X, Y, Z) nor (x, y, Y) is purely physical. The specification of color stimulus incorporates the function of the sensory interface in Figure 1.1.

The Munsell system consists of painted standard color chips that are arranged according to cylindrical coordinates $(H, V/C)$, where the vertical axis in the center represents lightness (black to white) (V), and the polar angle and distance represent hue (H) and saturation (C), respectively. In the Munsell notation, the change of lightness is called *Value V*. The series of achromatic chips representing V is called the Munsell gray or Value scale ($V = 0$ and 10 for ideal black and ideal white, respectively). *Hue H* is denoted in terms of five Munsell principal hues, *R* (red), *Y* (yellow), *G* (green), *B* (blue), *P* (purple), and their combinations (e.g., *YR* meaning orange) with a number between 0 to 10 as the prefix. In each, the most representative hue is denoted by 5*H*. *Chroma* C changes from 0 (achromatic) to from 8 to 14 (the most saturated color of a given *H*). By visual comparison with this standard system, a surface color is specified such as (6*R*, 5.5/14), meaning the most saturated color (/14) of the near medium lightness level (5.5/) of a slightly yellowish red (6*R*). Standard chips were selected so that two neighboring chips have the same magnitude of perceptual difference, $\delta = \chi_i \ominus \chi_{i+1}$, along each axis. This uniformity of steps in each attribute is needed to make visual interpolation easier in visual color specification. Selecting standard chips according to this principle is called *color spacing*.

The system was originally developed by A. H. Munsell, an art educator (the first publication was in 1903). The present version, Renotation Munsell, is based upon the very extensive visual examination of spacing by a subcommittee of the Colorimetric Committee of the Optical Society of America (OSA). The "color-producing effect" of a surface, (X, Y, Z), is determined by its spectral reflectance function $R(\lambda)$ and the illuminating light. Denote the illuminating light by (X_{ill}, Y_{ill}, Z_{ill}) and the reflected light by (X_{ref}, Y_{ref}, Z_{ref}). Then $X = X_{ref}/Y_{ill}$, $Y = Y_{ref}/Y_{ill}$, and $Z = Z_{ref}/Y_{ill}$, and (x, y, Y) are defined as in Equation 1.2. Each Munsell standard chip, when illuminated by the CIE standard light source (C or D65), is specified by $F(x, y, Y\%)$, where $Y\% = Y \times 100$, and tabulated in the CIE 1931 equivalents of Munsell renotations (Newhall, Nickerson, & Judd, 1943; Wyszecki and Stiles, 1982).

The CIE (x, y, Y) space relied upon equality judgments ($\sim$) between the two halves of a bipartite target in which lights $P_R(\lambda)$ and $P_L(\lambda)$ are presented. Usually the surround is darker than the target, and the subject sees colors filling the area without any texture (aperture mode). The subject adjusts one $P(\lambda)$ until two colors become identical (the method of adjustments). We have two CIE standards, one based on two sets of matching experiments with a 2° target (1931) and the other on two sets of matching experiments with a 10° target (1961). From the original matching results, which were highly stable and consistent, the coordinate axes for (X, Y, Z) were obtained by an affine transformation. The chromaticity diagram (x, y) was obtained by a projective transformation from (X, Y, Z). The particular forms of these transformations were determined so that the colorimetric calculations of $F(x, y, Y)$ from $P(\lambda)$, as in Equation 1.2, would be as easy as possible. Because three numerical integrations are involved, these calculations were very laborious before the computer age.

According to Grassmann's laws (Krantz, 1975a; Wyszecki and Stiles, 1982), these transformations do not alter the two properties inherent in the color-matching experiment: the unique specification of color [if $F_1 = F_2$, $\chi(F_1) \sim \chi(F_2)$], and the prediction of color mixtures [if F_1 and F_2 are additively mixed, F of the resulting color is a weighted mean of F_1 and F_2, and three are collinear in the order F_1, F, F_2 in $(x, y\ Y)$]. Obviously we feel $\chi(F)$ is more similar to $\chi(F_1)$ than $\chi(F_2)$ is. Hence, the (x, y, Y) space is a topological representation of the perceptual relationships among colors, but it lacks a metric to represent the size of perceived color differences.

This last fact comes as no surprise because the coordinates (x, y, Y) were selected mainly for the convenience of colorimetric calculations; no attempt was made to incorporate other perceptual phenomena (e.g., color differences, unique hues, etc.) or underlying physiological processes. Color scientists used either of the following two criteria to convert the (x, y, Y) space to a space with uniform color scales (UCS) that are more isometric to perceived color differences, δ. One may be called a *global criterion,* i.e., whether the Munsell system is represented without distortion. It will be discussed in the next section whether the Munsell space deserves to be used in this way. The other may be called a *local critrion,* i.e., whether JNDs are represented, around any point and in any direction, by segments of the same length. This problem will be discussed in the next subsection.

3.2 Multidimensional Studies of Munsell Color Solid

The Munsell system is based on visual selection of chips having the same attributes, H, V, and C, and judgments of equality ($\sim$) of perceptual differences between two neighboring chips ($\delta_{i,i+1} \sim \delta_{i+1,i+2}$) in each of these attributes taken separately. The procedure actually used by the OSA subcommittee is described in Indow (1988b). The Munsell Value series, unidimensional change from black to white of achromatic color, were first scaled by equisections of 14 subjects (Munsell, Sloan, & Godlove, 1933). When plotted against Y of gray chips, the V curve is convex upward. The curve adopted by the OSA subcommittee differs slightly from this result. The main part of $V(Y\%)$, say $1.0 < V < 9.0$, is well represented by the simple equation

$$V = 2.5(Y\%)^{1/3} - 1.7, \tag{1.3}$$

in which $V = 5$ ($Y\% = 19.77$) is the middle gray. Uniformities of steps in Munsell chips in the remaining two unidimensional changes, H and C, were checked by a less systematic procedure. However, even if Munsell chips satisfy the requirement $\delta_{i,i+1} \sim \delta_{i+1,i+2}$ in the respective attributes, it is not sufficient to adopt the Munsell system as the global criterion, because no information is given on multiattribute color differences.

Most color differences we see between two objects in VS are of multiattributes. Actually, when we see two colors, their perceptual relationship comes first, and then, if asked, we can say something about their hues, levels of lightness, etc. In other words, perceptual color difference is not consciously composed from their differences in several attributes. Because of the practical need to control the colors of products, color practitioners proposed various formulae to define numerical values, d, for multiattribute differences from distances either in (x, y, Y) or in $(H, V/C)$ for small color differences, δ, not greatly in excess of one JND. The macroscopic approach to use the Munsell system as the global criterion is equivalent to look for a space in which uni- and multiattribute color differences of larger sizes are directly represented as distances.

When we look at the Munsell color solid displayed in a physical space, E^3, we see more than the uniformity along each dimension. We see multiattribute color differences between chips in various directions and a gradual change of color as a whole on the hue circle. Since Torgerson's (1952, 1958) first scaling algorithm, we have several psychophysical procedures appropriate to examining the global structure of the Munsell solid. There are now various forms of multidimensional scaling (MDS), but I begin with the traditional form of MDS that embeds n objects as a configuration of points $\{P_j\}$ in a Euclidean space of appropriate dimensionality m, E^m. It is essentially the same as DMPRD in Figure 1.3 except that K is fixed to be 0. In some methods, m is determined through the analysis and, in some others, m must be specified prior to the analysis. In this case, $\{Q_j\}$ is a configuration of n Munsell chips used in an experiment. The most comprehensive Munsell system consists of 1,928 standard chips. Denote by n the number of colors used in an experiment. In the earlier studies, $n < 25$. In more recent experiments, relying on greatly increased computer power, $n = 120$–178, and the results became more realistic (Table 1 in Indow, 1988b). Various scaling methods were tested to obtain data $\mathbf{D} = (d_{jk})$, but only one procedure, which was used in these more comprehensive studies, will be explained. On the background of middle gray, a pair of color chips, Q_j and Q_k, and a series of Munsell grays of fine steps $\{V_X\}$ were presented with a standard gray (V_A). The subject was asked to select from the gray series the one, V_B, for which the perceptual difference in lightness from the standard gray, δ_{AB}, matches in magnitude the perceptual color difference δ_{jk} ($\sim$). If necessary, the subject could interpolate between two neighboring grays in the series. Assuming that V is an interval scale for gray differences, we can define the scale for δ_{jk} by $d_{jk} = |V_A - V_B|$. In one run, the standard gray V_A was chosen from the blacker side; in a second, from the whiter side; and the mean result was used. Thus, color differences were converted to lightness differences in terms of Munsell V. Only a portion of the possible pairs in $\{Q_j\}$ were scaled in this way, and hence $\mathbf{D}$, $n \times n$, was incomplete, which was necessary not only to reduce judgments to a manageable number but also to meet the limits inherent in human capacity for evaluating color differences. When two colors are too different (e.g., red and green), we feel they are "entirely different" and so cannot determine a difference in size. Hence, the pairs of colors presented were limited to a range within which size of color difference is intuitively clear to the subjects: $d_{jk} < 3.5V$. In some of the earlier studies, the MDS analyses were carried out individually in the sense that $\{P_j\}$ was obtained for each subject. Because these studies demonstrated no qualitative individual differences, we averaged d_{jk} over individuals before using DMPRD and so one $\{P_j\}$ was defined for all (usually five) subjects.

Because the Munsell color solid is displayed in E^3, it is natural to ask whether the solid is embeddable in E^3 as $\{P_j\}$. Hence, $K = 0$ and the intermediate step (ρ_{jk}) can be skipped. Under this constraint, the following $\{P_j\}$ was obtained in E^3. Denote by $\hat{\mathbf{D}}$ the matrix of interpoint distances in $\{P_j\}$. Then $\hat{\mathbf{D}}$ reproduced the data $\mathbf{D}$ very well with the exponent B (Figure 1.3) being close to 1 and the root-mean-square (RMS) value of the discrepancies ($d_{jk} - \hat{d}_{jk}$) was about 0.20–0.26 of the V unit. Since JNDs for lightness discrimination in the main part of the V scale are about $0.07V$, this numerical coincidence between d_{jk} and $\hat{d}_{jk}$ is only about three JDNs. Results based on $\{P_j\}$ and plots of d_{jk} against $\hat{d}_{jk}$ were provided in detail in Indow (1988b).

A consistent finding is that P_j's for colors at the highest level of C (the outermost hue circle) are located too close to P_j's for colors of the next level of C compared with separations between other adjacent levels of C. Two possible reasons may be pointed out: (1) According to the Euclidean structure, the distance $\hat{d}$ between two vectors representing hues increases linearly as a function of the distance, C, from the central axis representing achromatic colors ($C = 0$). It is clear that, for lower levels of C, the difference δ between two colors appears more enhanced when the colors are more saturated. If there is an asymptotic level for this effect, then the cylindrical form in E^3 is intrinsically inappropriate to represent the perceptual differences quantitatively. (2) For points P_j located on the outermost hue circle, all P_k's yielding data d_{jk} are necessarily colors within that outer hue circle. Thus, nothing exists to pull P_j toward the outside. This situation may create an artifact of the form described.

No matter what the reason may be, the distortion is the same as those we have when concentric circles on an sphere are vertically projected on E^2, and hence we can expect that when the $\{P_j\}$ are embedded in an elliptic space with $K > 0$, the same step size will be maintained between successive levels of C (Indow, 1988b). Figure 1.4a gives an example of such a representation computed using DMPRD, as outlined in Figure 1.3; it shows the projection of the $\{P_j\}$ on a plane of constant V level. Panel B plots the data d_{jk} against interpoint distances $\hat{d}_{jk}$, lengths of elliptic geodesics between points in $\{P_j\}$. The optimization, including the value of K and the exponent B in the power function between d and $\hat{d}$, was obtained by minimizing RMS in this plot. The relationship between d and $\hat{d}$ turned out to be proportional ($B = 1.02$), as in other cases, and the value of RMS was again about three JNDs in the V scale.

Minimizing RMS of $d_{jk} - \hat{d}_{jk}$ does not, by itself, lead to $\{P_j\}$ that faithfully represents the perceptual structure. Noise in the data $\mathbf{D}$ and lack of balance in

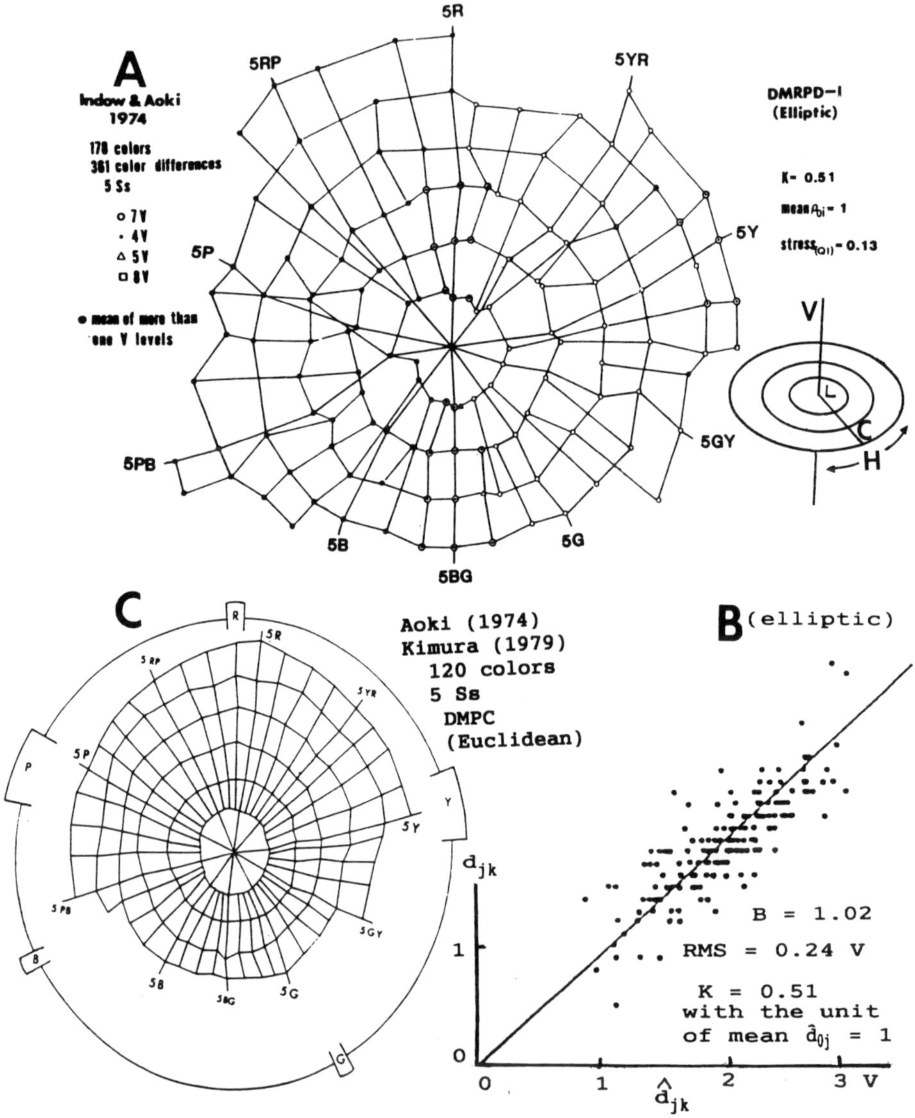

FIG. 1.4. Embedding of Munsell colors in an elliptic space. (A) The configuration $\{P_j\}$ of 178 colors of various lightness projected to a plane of hue and saturation. (B) Reproducibility of scaled color differences d_{jk} by interpoint distances $\hat{d}_{jk}$ of $\{P_j\}$. (C) The configuration $\{P_j\}$ and five bundles of individual principal hue vectors based on two sets of data. (A and C are taken by permission from T. Indow [1988b], *Psychological Review, 95,* Fig. 8 on p. 467, and from T. Indow [1987], *Die Farbe, 34,* Fig. 2 on p. 256.)

$\{Q_j\}$ may introduce some unwanted irregularities in $\{P_j\}$. It is, therefore, desirable to add some other criterion that has a substantive meaning. The use of $\{Q_j\}$ having an internal structure in the subsection "Psychophysical Scaling of Perceptual Distances" was such an example. Here we can use the principal hue components, ξ_α, that we see in each color χ. An orange color appears reddish, ξ_R, and yellowish, ξ_Y. Figure 1.4c shows $\{P_j\}$ based on scaling both color differences, δ, and the principal hue components, ξ_α. Individual chips, Q_j, were presented one at a time, and each subject assessed the degree of principal hue components ξ_α, $\alpha = R, Y, G, B$, and P. Because there may be individual differences in how to interpret pure red, pure yellow, etc., individual vectors, $f_{i\alpha}$, hues α for subject i, were defined in the same space with $\{P_j\}$. The algorithm is called Direct Mapping through Powered Components (DMPC) because it assumes that data ξ_j of subject i are represented by contravariant components of P_j on the vector $f_{i\alpha}$ in the form of power function with an individual exponent (Indow, 1980, 1987). It was also assumed that the same $\{P_j\}$ holds for all the subjects, because the algorithm did not work otherwise. In Figure 1.4c, the range of bundle of five individual vectors $f_{i\alpha}$, for each principal hue α, is indicated along the circle. Because I am not fully satisfied with the method used to scale ξ_α and the algorithm DMPC by which $f_{i\alpha}$ were defined, as to $f_{i\alpha}$, $i\alpha = 1$ to 5×5, the following three simple comments suffice. (1) In contrast to other principal hues, the B bundle is not in the direction of 5B Munsell chips, the most representative blue in the Munsell notation. The bundle is in the direction of colors of 5PB. This inadequacy of the Munsell blue notation is also well established by other MDS studies, and was recently emphasized by a color scientist (McCamy, 1993). (2) In two analyses, bundles for R, G, B, turned out considerably narrower than those for Y and P. In the retina, there are three kinds of cones, with maximal sensitivities at long, medium, and short wavelengths. Physiological processes underlying perception of red, green, and blue have their origins in excitations of the corresponding cones. Yellow is perceptually unique, but it is supported by a physiological process due to excitations of both red and green cones. Purple is not perceptually unique, and everyone points out that it contains redness and blueness. Hence, large individual differences in Y and P may stem from the fact that each has multiple sources in the retina and individual differences in respective sources are superimposed. However, the above-stated result in the two analyses must be taken with some reservation because of the following recent finding. When the algorithm is changed, the P bundle becomes narrower and the B bundle wider (Indow, 1993a). (3) Most color vision theories now accept two opponent processes, R–G and Y–B (Jameson and Hurvich, 1955; Krantz, 1975b; Suppes et al., 1989). The subjects have no difficulty in assessing principal hue components in each color in terms of just these four hue names (Indow, 1987, McCamy, 1993). The reason why Munsell included purple among his principal hues was discussed in Indow (1988b).

In the Munsell notation, the V-axis is defined to be orthogonal to the plane of

H and C. MDS studies have always given the results, such as in Figure 1.4., that endorese this definition. For aperture color mode, however, the following Helmholtz-Kohlrausch effect is known: when two light stimuli of the same luminance Y but different in spectral purity are presented side by side, the one that appears more saturated tends also to appear brighter. The orthogonality between V- and C-axes obtained in MDS studies implies that the effect is not too conspicuous in surface colors. By definition, Munsell divided the hue circle into five equal sectors. This definition did not undergo rigorous examination in the study of the OSA subcommittee. Hence, it is surprising that such a $\{P_j\}$ was obtained by MDS that is structured as expected from the Munsell notation. If the direction of 2.5*PB* is defined to be 5*B* and some minor amendments are made with color spacing, the hue circle in $\{P_j\}$ at the middle level of V is almost in agreement with the Munsell system and also with the relationship between complementary colors. The Munsell system does not have a unified scale for the three axes. Color practitioners have long regarded that one step in V is perceptually equivalent to two steps in C. The configuration $\{P_j\}$ is constructed with a unified unit that agrees with this view: $\hat{d}$ between V and $V + 1$ turned out to correspond, approximately, to $\hat{d}$ between C and $C + 2$ (Indow, 1974, 1980).

The discussion up to this point is based on the relation between d and $\hat{d}$ for uni- and multiattribute color differences beyond JND levels. However, the judged color differences were limited to those that are intuitively clear to the subjects, so **D** was incomplete. Hence, what we can say from these studies is not that the Munsell solid as a whole is Euclidean or elliptic in nature but that it can be embedded in a 3-D manifold that has either locally Euclidean or elliptic metric $\hat{d}$. Suppose that perceved color difference δ can be regarded as a quantity. If judgments of the subject give such a scale d that is proportional to δ, then, because the exponent $B \approx 1$ and hence $d \propto \hat{d}$, we can say that $\{P_j\}$ is a compact and coherent quantitative summary of all perceptual relationships in surface colors. However, contrary to the scale d for perceptual distances δ in VS (see the subsection 2.5) additivity ($d_{ik} = d_{ij} + d_{jk}$) was found not to hold for collinear triplets P_i, P_j, P_k in $\{P_j\}$. Always, $d_{ik} < d_{ij} + d_{jk}$, and this subadditivity exhibits a regularity. Perhaps, we can assume Equation 1.1, but with the exponent $\beta < 1$. The implication of $\{P_j\}$ as a representation of perceptual structure of colors has to be understood with this reservation.

3.3 JNDs in (x, y, Y) Space and the Munsell Solid

As stated earlier in this section, the (x, y, Y) space does not have a perceptually meaningful metric. Now it has become clear that the Munsell system can be used as the global criterion for UCS in the sense stated above. When values of the CIE equivalents of Munsell Renotation are plotted, Munsell colors of the same V do not exhibit concentric circles in the chromaticity diagram (x, y), i.e., violation of the global criterion for UCS. The (x, y, Y) fails to meet the local criterion also.

When JNDs from a color stimulus $F(x, y, Y)$ are measured in various directions, they form an ellipsoid and, when Y is kept constant, an ellipse on (x, y) (Brown and MacAdam, 1949; MacAdam, 1942; Wyszecki and Fielder, 1971). If radii are regarded as Riemannian line elements, the chromaticity diagram becomes an extremely complicated Riemannian surface in which K changes its value and sign from point to point.

It has been traditional in color science to define JNDs by the method of adjustments, the procedure by which the color-matching data for (x, y, Y) were obtained. A light F_0 is fixed and the subject adjusts the other light F so that the two appear the same, $\chi(F) \sim \chi(F_0)$. F is a random variable, and if we take the area around F_0 that embraces matches F with probabilities equal to and larger than a value π $(0 < \pi < 1)$, we have an ellipsoid. The radius of this *color discrimination ellipsoid* is defined to be the JND in that direction. The ellipsoidal form of distribution of matches strongly suggests that random noise inherent in the process underlying perception of an aperture color (χ) follows a 3-D Gaussian distribution. This procedure cannot be used to define JNDs of painted colors because they cannot be continuously changed while being observed. In most studies, paired comparisons are made between two painted samples, but the algorithm to define JNDs from percentages of discrimination has not been standardized.

A procedure, which is an extension of the traditional method of constant stimuli for measuring differential threshold in unidimensional stimulus continuum, was proposed and has been successfully applied to three sets of experimental data using painted samples (Indow and Morrison, 1991). An experiment was performed to compare ellipsoids generated by this method with those generated by the method of adjustments (Indow, Robertson, von Gruneau, & Fielder, 1992). Two stimuli, F_0 and F were presented on a color display, for 0.5 sec. Depending on whether the luminance of the surround is less or greater than these stimuli, the perception is of the aperture mode (A) or of a color similar to the surface mode. The latter do not appear exactly the same as painted surfaces, S, because they lack texture, and so they are called simulated surface colors mode (S′). Pairs (F_0, F) were repeatedly presented until the subject came to a judgment. In the method of adjustment, the subject adjusted F to match F_0, and the same subject also made paired comparison judgments in the method of constant stimuli. In the latter, a set of comparison stimuli were presented as F, one by one, and the subject judged whether F_0 and F were distinguishable. In the matching experiment, M, the subject could change F in seven directions through F_0, and matches were repeated 16 times in each direction. In the constant method, C, $[(7 \times 10) + 1]$ comparison stimuli were paired with F_0—10 different distances from F_0 in each of the seven directions and one identical pair (F_0, F_0)—and each pair was judged 30 times. Six F_0 giving achromatic, reddish, yellowish, greenish, bluish, and brownish colors were used. With each F_0, four ellipsoids, two modes (A, S′) by two methods (C, M), were obtained from the same two subjects.

Ellipsoids by the M method were defined on the assumption of 3-D Gaussian distribution of matches, and the data fit this assumption very well. For this reason, sigmoidal psychometric curve was used in each of the seven directions (W) to relate the percentage of discrimination in the C method, $P_W(F)$, to the distance between F_0 and F, $d(F_0, F)_W$. The sigmoid varied according to direction W. Then, such an ellipsoid was sought for. When plotted against $d(F_0, F)_W$ divided by the radius in the direction W, $P_W(F)$ for all directions form a single sigmoid. In both methods, JNDs were defined with $\pi = 0.5$.

The main findings were as follows: (1) The M method always gave the same ellipsoids for two modes of appearance, A and S′. (2) For S′-mode colors, the C method gave almost the same ellipsoids as in (1), except for yellow. For A-mode colors, however, ellipsoids by the C method were larger than those for S′-mode colors. These two ellisoids were similar in form, again except for yellow. (3) The percentages of false discriminations for (F_0, F_0) were higher in the S′-mode colors than in the A-mode with the following regularity. The change from S′ to A of the size of internal threshold for producing false discrimination is approximately proportinal to the change of the size of the ellipsoid from S′ to A. Results (1) to (3) are consistent with the assumption that the amount of noise inherent in the processes underlying the perception of a color χ remains the same for the two modes of appearance but that the internal criterion for binary judgments becomes smaller when the mode changes from A to S′ (Indow, 1993b).

Ellipsoids around the yellow F_0 estimated using the C-method behaved differently from other ellipsoids. This is another example of anomalous behavior of yellow. Others were mentioned earlier in this section and also in previous studies (Indow, 1987; Indow and Stevens, 1966).

We now have color discrimination ellipsoids for S′-mode colors. Then it is an interesting question to ask whether they turn out to be circles of the same size when plotted in the Munsell space. The radius of an ellipsoid along the Y-axis (the luminance JND, ΔY) was converted using Equation 1.3 to the JND in terms of V, ΔV. The conversion of an ellipse in (x, y) to H/C cannot be done through the table of CIE equivalents of the Munsell renotation because JNDs are too small compared with the step size in the table. Hence, two JNDs in the directions of H and C, ΔH and ΔC, were indirectly estimated from ellipses in (x, y) by a procedure described in Indow (1993). These JNDs (ΔH, ΔC, and ΔY or ΔV when available), were calculated for all ellipsoids (means of two subjects) around F_0 by the C method for the A- and S′-mode colors, as well as for ellipsoids of the A-mode colors from the Brown and MacAdam matching experiment, and ellipses of painted samples (S-mode) from Luo and Riggs (1986). From these two studies, ellipsoids or ellipses of F which are close to F_0 were used.

Weber's ratios, $\Delta Y/Y_0$, ranged from 0.007 to 0.067 and ΔV from $0.012V$ to $0.155V$. If heterogeneity of experimental conditions is taken into account, these results are not too surprising. For the ellipsoids around different F_0 in the same experiment, these values were more stable. As stated earlier, Munsell et al. (1933) defined the V scale by the equisection method. They also measured JND

ΔY's at five levels of Y. By integrating these ΔY's, they defined another scale of luminance as a function of Y. This scale was in good agreement with $V(Y)$ by the equisection (Figure 12 in Munsell et al., 1933), which suggests that ΔV is a constant independent of level of Y. I converted their ΔY to ΔV and found that ΔV behaves in that way in each subject, except for the lowest level of Y close to the stimulus threshold. The representative value of ΔV is $0.07V$, i.e., in the same range as stated above. Hanes (1949) reported the same level of agreement between two scales for brightness of achromatic aperture color, one by fractionaion and the other by integration of JNDs (Figure 5 in his article).

On the other hand, both ΔC's and ΔH's vary more according to F_0, which casts doubt on the possibility that the H/C Munsell plane meets the local criterion for UCS. We have no way to convert discrimination ellipses in (x, y) into $\{P_j\}$ in Figure 1.4, because we do not have enough information on the local structure of $\{P_j\}$. However, even if this converstion be made, it will not alter the situation. In other words, as far as H/C are concerned, it is not likely that JNDs are represented by segments of constant size in the Munsell system or in the space into which the Munsell system is embedded by MDS.

3.4. Discussion

The Munsell solid can be regarded as a structure $\{P_j\}$ having Euclidean or elliptic metric representing uni- and multiattribute color differences, with the following reservations. (a) Color differences are larger than JND level but do not exceed the limit in human capability for evaluating color differences (see the second subsection). (b) When color difference d is scaled by matching with gray difference, d and interpoint distance $\hat{d}$ in $\{P_j\}$ are proportional but with the scatter as shown in Figure 1.4b. (c) Because d fails to be additive, the relationship between the scale color difference d and perceptual color difference δ as a latent variable may not be so straightforward as that between the scaled distance d in VS and perceptual distance δ (second section).

Lightness JNDs, no matter whether are measured with achromatic or chromatic colors, are represented as segments of approximately the same size on the scale V, $\Delta V \approx 0.05$ (see the previous subsection). That seems not to be the case, however, for ΔC and ΔH. In other words, JND cannot be regarded as the extreme limit of suprathreshold color difference, and the two criteria for UCS, local (JND) and global (Munsell), cannot be taken as equivalent. Even with the V scale, more extensive experiments are necessary to see whether the equisection and the accumulation of JNDs really give the same scale.

4. UNIFIED SCALES OF FOUR QUALITIES OF TASTE

4.1. Practical Use of Psychophysical Methods

In many areas in industry and technology, the problem of how to obtain quantitative representation of human sensation has puzzled engineers and technicians (e.g., the taste of food products; body of texture; picture quality in TV, copy, and photography; noise in various areas). When I was in Tokyo, I was active in the Research Committee of Sensory Evaluation of the Union of Japanese Scientists and Engineers. That committee published the *Sensory Evaluation Handbook,* the first version in 1962 and a revised and expanded version in 1973. The 920 pages includes various methods of psychophysical scaling, procedures to determine stimulus and differential thresholds, and statistical tests of significance of discriminability between samples of a fixed number (triangle test, duo-trio test, Scheffé's analysis of variance, Bradley's method, etc.). Scientists and engineers desire to have functional relationships between physical variables **s** and perception θ. Some examples of scaling that I performed are the relation between picture quality of prints and Gamma (slope on log scale) of the printing process (1955), subjective hardness of pencil lead from HB to 9H (1959), and quality of TV pictures as a function of noise and bandwidth (1960). (These papers are all in Japanese.) To illustrate this type of work, I describe here the construction of standard taste scales for the food industry (Indow, 1966, 1969).

4.2. Unified Scales of Four Qualities of Taste

Lewis (1948), using the method of fractionation, constructed a scale for each of four qualities of tastes: (A) sweetness, τ_A, (B) bitterness, τ_B, (C) sourness, τ_C, and (D) saltiness, τ_D. These results were tested and named the "gust" scale by Beebe-Center and Waddell (1948) and Beebe-Center (1949). Because of the nature of judgments involved in the scaling, when (gust of a)/(gust of b) = r, we predict that for each quality the subject will say "τ_a is r times stronger than τ_b." No heteroqualitative judgments were included and the relative size of the unit "gust" between different qualities remained unspecified. I tried to obtain a scale having a common unit "tau" for the four qualities from which we can predict judgments on differences between two strengths of taste ($\tau_a \ominus \tau_b$) including homo- and heteroqualitative pairs. In general, we cannot predict both difference and ratio judgments from the same scale x in the form $x_a - x_b$ and x_a/x_b (e.g., Indow and Ida, 1977). Hence, denote the tau scale (based on difference judgments and with a common unit) and the gust scale (based on ratio judgments with quality-specific units) by $u(s_\alpha)$ and $v_\alpha(s_\alpha)$ respectively, where s_α is a level of concentration of substance α in terms of gram solute per 100 cc of distilled water, α = A, B, C, and D.

With the collaboration of a leading Tokyo food company, the tau scale $u(s_\alpha)$ was constructed as follows. First, I ascertained from taste specialists which judgments they wished to predict from a scale. They were unanimous that their main interest was a difference measure characterizing what changes of ingredients do, for example, to the sweetness of the food. The following experimental procedure was used:

(1) Eight solutions of the substances (A) sucrose, (B) quinine sulfate, (C)

tartaric acid, (D) sodium chloride were prepared for the scaling experiment ($s_{\alpha j}$, $\alpha = 1$ to 4, $j = 1$ to 8). Out of the 496 possible pairs of stimuli (s_a, s_b), where each s represents one of $s_{\alpha j}$ and a, $b = 1, 2, \ldots, 4 \times 8$, 140 pairs were presented to each subject i in a panel consisting of five taste-testing specialists from the company and five novices. Pairs include two different concentrations of the same substance α as well as a solution of α and a solution of β.

(2) Two solutions were repeatedly tasted with rinsing each time until the impression of intensity difference ($\tau_a \ominus \tau_b$) was formed. Subject i assigned an integer G_{iab}, -3 to 3 including 0, according to this impression. It was assumed that a judgment $G_{iab} = g$ is given if and only if $l_{ig-1} \leq x_{iab} < l_{ig}$, where x_{iab} represents $\tau_a \ominus \tau_b$ for subject i and l_{ig-1}, l_{ig} are boundaries of the interval corresponding to the category g, both being defined on the personal scale τ_i in the black box. The use of these categories may differ among subjects and so the boundaries have the suffix i.

(3) Using a modified form of the latent model of test scores (Lord, 1952; Lord and Novick, 1968), x_{iab} was shown to be decomposable into two components, X_{ab} and e_{iab}, where X_{ab} denotes the difference ($\tau_a \ominus \tau_b$) on the interpersonal scale as the core of x_{iab}'s over subjects and e_{iab} are the idiosyncratic part that are assumed to be independent between subjects. It was important to the company to learn whether the taste-testers and consumers judge taste on a common basis. Including all 10 panel members, the homogeneity of judgments was well supported (eigenvalues of the 10×10 matrix representing correlations [extended form of tetrachoric] of judgments between subjects were 6.63, 0.44, 0.24, . . .). Scrutinizing the results more carefully, two subjects, one a specialist and one a novice, could be called outliers. All of the following results were obtained on the basis of the remaining eight subjects (whose eigenvalues were 5.52, 0.25, . . .).

(4) Individual sets of six *judgment characteristic curves* $P_{ig}(X)$, the probability that subject i gives G_{iab} equal to or less than a category g (-3 to 2) for pairs (s_a, s_b), are defined where X is the taste difference on the interpersonal scale τ (Figure 1 in Indow, 1966).

(5) Define $z_{ab} = \Sigma G_{iab}$, $i = 1$ to 8. Then, for each value of X_{ab}, the conditional probabilities of z_{ab} when X_{ab} is fixed, $f(z_{ab}|X_{ab})$, were determined through the synthesized curves of $P_{ig}(X)$ over eight subjects. If the marginal distribution of X_{ab}, $f(X_{ab})$, is known, each $f(z_{ab}|X_{ab})$ can be converted to a density of the 2-D distribution (z_{ab}, X_{ab}) from which the following three sets of information are obtained: (1) for each value of z_{ab}, the most likely values $\hat{X}_{ab}$ or differences ($\tau_a \ominus \tau_b$) on the interpersonal scale τ; (2) reliability intervals for these estimations; (3) the marginal distribution $\hat{f}(Z_{ab})$ for Z_{ab}, the theoretical values of z_{ab}. Two assumptions were invoked to complete steps (3) to (5): a bivariate Gaussian relationship between x_{iab} and $x_{i'ab}$, where i and i' represent two subjects, and also that between X_{ab} and x_{iab}. The latter implies that the distribution of $f(X_{ab})$ is Gaussian. According to the definition of X_{ab}, its distribution is symmetric around zero, but being Gaussian is an additional assumption. The acceptability of these

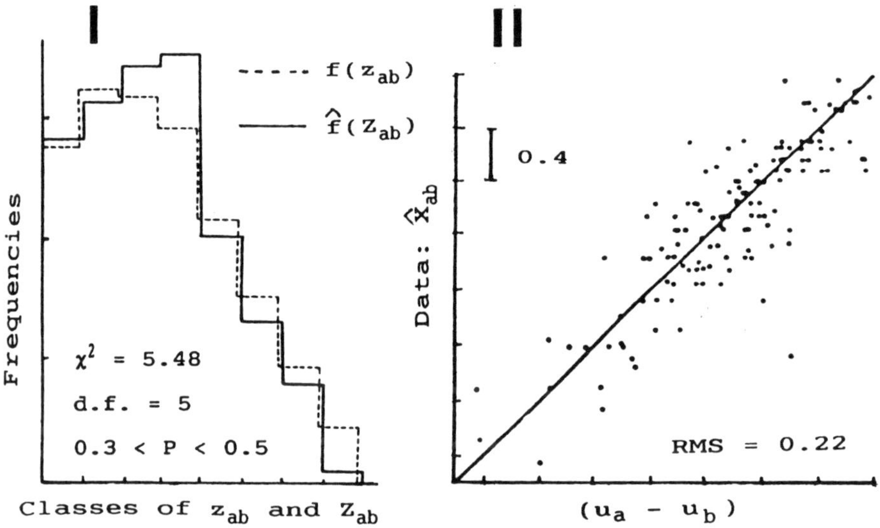

FIG. 1.5. A unified scale τ for four qualities of taste. (I) Coincidence between empirical and theoretical histograms in the process of scaling. (II) Reproducibility of scaled taste intensity difference $\hat{X}_{ab}$ by the difference of final values on τ, $(u_a - u_b)$. (Reprinted by permission from T. Indow [1966], *Japanese Psychological Research, 8,* Figs. 3 and 4 on p. 117.)

assumptions was tested in two ways, and both seem satisfactory. One is shown in the left panel I in Figure 1.5; the goodness of fit between two histograms for z_{ab} and Z_{ab}. This is a comprehensive test for all conditions from (3) to (5).

(6) Values of u_a and u_b were determined for the incomplete matrix $(\hat{X}_{ab})$ so as to minimize $\Sigma\Sigma\{\hat{X}_{ab} - (u_a - u_b)\}^2$ (Gulliksen, 1956). The right panel II in Figure 1.5 shows the plot of $\hat{X}_{ab}$ versus $u_a - u_b$, analogous to Figure 1.4b.

(7) It was found that for each α, $u(s_\alpha) = A_\alpha \log(s_\alpha / s_{\alpha 0})$ holds very well. The estimated value A_α is almost constant (2.27–2.83), whereas the stimulus threshold $s_{\alpha 0}$ varies from 0.00011 for B (quinine sulfate) to 0.48 for A (sucrose). On the basis of these results, four series of $s_{\alpha I}$, $I = 1$ to 10, were proposed to define the perceptually equidistance scale with a unified unit tau for the four qualities. For $s_{\alpha 0}$, tau $= 0$, and the increase of sweetness from the solution of 1.0 g sucrose to 3.0 g sucrose in 100 cc distilled water is defined to be tau $= 1$. When subjects matched the four qualities of taste of various solutions (mixtures of two substances) to the tau scale, very systematic interactions between substances were observed (Indow, 1969). For example, mixing a certain amount of salt with sucrose considerably enhances the sweetness. Adding salt in cooking sweets is a traditional trick of Japanese chefs. Average standard deviations of matches were on the order of 0.3 to 0.5 tau. Since tau ranges from 0 to 5.5, these are much larger than those of V in Figure 1.4b.

(8) Log gust$_\alpha$ was found to be linear with tau, which means that gust$_\alpha$ is a power function of s_α. Beebe-Center (1949) showed that when 5% sugar is added to black coffee, gust$_A$ increased from 2.0 to 3.2, whereas gust$_B$ decreased from 42.3 to 23.8. We can, therefore, expect that people will say "the sweetness increases 1.6 (=3.2/2.0) times and the bitterness decreases 1.8 (=42.8/23.8) times," i.e., a slightly larger change in bitterness. If measured by tau, the same changes are from 0.76 to 2.06 for sweetness and 5.10 to 4.39 for bitterness. Thus, in this measure the increase of sweetness is 1.30 (=2.06–0.76) whereas the decrease of bitterness is 0.71 (=5.10–4.39), which says the change in sweetness is larger than that in bitterness. Which fits your impression? However, I cannot expect that people will say, when asked, that the increase of sweetness is 1.83 (1.30/0.71) times of the decrease of bitterness.

4.3. Discussion

The main role of the scale tau is to predict judgments of general consumers concerning differences in taste intensity when concentration is changed from a level to another, including the four qualities. I did not go into the problem of how these judgments are generated from the latent variable (θ) in the black box (Figure 1.1). This problem was not the main concern to practitioners in this field. When I visited a research laboratory in Japan in 1985, I was surprised to see that the scale was still in use. However, for scientific study of taste sensation, to understand the relationship between stimulus (s) and the latent process (θ) will be of primary interest.

Until 1979, I had been also a member of the motivation research group in Dentsu, the largest advertisement company in Japan, and had been involved in scaling of preference toward various commodities (e.g., Figures 86, 87 in Stevens, 1975). It is not difficult to obtain scales to predict people's judgments or verbally expressed opinions, but I always felt uneasy about assuming a close relation between opinion and actual behavior of consumers. Again it is a problem, for behavioral scientists, to understand interrelationships between preference (latent variable), verbal responses, and actual behavior.

REFERENCES

Beebe-Center, J. G. (1949). Standards for the use of the gust scale. *Journal of Psychology, 28*, 411–419.

Beebe-Center, J. G., & Waddell, D. (1948). A general psychological scale of taste. *Journal of Psychology, 26*, 517–524.

Berger, M. (1987). *Geometry I and II*. New York: Springer-Verlag.

Blank, A. A. (1958). Axiomatics of binocular vision: The foundation of metric geometry in relation to space perception. *Journal of the Optical Society of America, 48*, 328–334.

Blank, A. A. (1959). The Luneburg theory of binocular space perception. In S. Koch (Ed.), *Psychology: A study of a science: Vol. 1. Sensory, perceptual, and physiological Foundations* (pp. 395–426). New York: McGraw-Hill.

Blumenfeld, W. (1913). Untersuchungen über die scheinbare Grösse im Sehraume. *Zeitschrift für Psychologie, 65,* 241–404.

Brown, W. R. J., & MacAdam, D. L. (1949). Visual sensitivities of combined chromaticity and luminance differences. *Journal of the Optical Society of America, 39,* 808–834.

Busemann, H. (1955). *The geometry of geodesics.* New York: Academic Press.

Duncker, K. (1929). Über induzierte Bewegung. *Psychologische Forschung, 12,* 180–259.

Falmagne, J-C. (1985). *Elements of psychophysical theory.* New York: Oxford University Press.

Green, D. M., & Swets, J. A. (1966). *Signal detection theory and psychophysics.* New York: Wiley.

Gulliksen, H. (1956). A least squares solution for paired comparisons with incomplete data. *Psychometrika, 21,* 125–135.

Hanes, R. M. (1949). A scale of subjective brightness. *Journal of Experimental Psychology, 39,* 438–452.

Indow, T. (1966). A general equi-distance scale of the four qualities of taste. *Japanese Psychological Research, 8,* 136–150.

Indow, T. (1968). Multidimensional mapping of visual space with real and simulated stars. *Perception and Psychophysics, 3,* 45–64.

Indow, T. (1969). An application of the τ scale of taste: Interaction among the four qualities of taste. *Perception and Psychophysics,* 5, 347–351.

Indow, T. (1979). Alleys in visual space. *Journal of Mathematical Psychology, 19,* 221–258.

Indow, T. (1974). Application of multidimensional scaling in perception. In E. C. Carterette & M. P. Friedman (Eds.), *Handbook of perception: Vol. II.* (pp. 493–525). New York: Academic Press.

Indow, T. (1980). Global color metrics and color apperance systems. *Color, Research and Application, 5,* 5–12.

Indow, T. (1982). An approach to geometry of visual space with no a apriori mapping functions: Multidimensional mapping according to Riemannian metrics. *Journal of Mathematical Psychology, 26,* 204–236.

Indow, T. (1987). Psychologically unique hues in aperture and surface colors. *Die Farbe, 34,* 253–260.

Indow, T. (1988a). Alleys on apparent frontoparallel plane, *Journal of Mathematical Psychology, 32,* 259–284.

Indow, T. (1988b). Multidimensional studies of Munsell color solid. *Psychological Review, 95,* 456–470.

Indow, T. (1991). A critical review of Luneburg's model with regard to global structure of visual space. *Psychological Review, 98,* 430–453.

Indow, T. (1993a). Indiscriminable regions, color differences, and principal hue vectors in Munsell space. *Colour '93,* The 7th Congress of the International Color Association, Technical University of Budapest.

Indow, T. (1993b). *Parallel shift of psychometric function in cutaneous and color discrimination* (Tech. Rep. No. MBS 93-24). Irvine: University of California Irvine, Institute for Mathematical Behavioral Sciences.

Indow, T., & Ida, M. (1977). Scaling of dot numerosity. *Perception and Psychophysics, 22,* 265–276.

Indow, T., Inoue, E., & Matsushima, K. (1962). An experimental study of the Luneburg theory of binocular space perception: 2. Alley experiment. *Japanese Psychological Research, 4,* 17–24.

Indow, T., Inoue, E., & Matsushima, K. (1963). An experimental study of the Luneburg theory of binocular space perception: 3. The experiments in a spacious field. *Japanese Psychological Research, 5,* 1–27.

Indow, T., & Morrison, M. L. (1991). Construction of discrimination ellipsoids for surface colors by the method of constant stimuli. *Color Research and Application, 16,* 42–56.

Indow, T., Robertson, A. R., von Grunau, M., & Fielder, G. H. (1992). Discrimination ellipsoids of aperture and simulated surface colors by matching and paired comparison. *Color Research and Application, 17,* 6–23.

Indow, T., & Stevens, S. S. (1966). Scaling of saturation and hue. *Perception and Psychophysics, 1,* 253–271.

Indow, T., & Watanabe, T. (1984a). Parallel- and distance-alleys in the horizontal plane. *Perception and Psychophysics, 35,* 144–154.

Indow, T., & Watanabe, T. (1984b). Parallel- and distance-alleys on the horopter plane in the dark. *Perception, 13,* 165–182.

Indow, T., & Watanabe, T. (1988). Alleys on an extensive apparent frontoparallel plane: A second experiment. *Perception, 17,* 647–666.

Jameson, D., & Hurvich, L. M. (1955). Some quantitative aspects of an opponent-color theory: I. Chromatic responses and spectral saturation. *Journal of the Optical Society of America, 45,* 546–552.

Köhler, W. (1929). Ein altes Scheinproblem. *Naturwissenschaften, 17,* 395–401.

Krantz, D. H. (1972). A theory of magnitude estimation and cross-modality matching. *Journal of Mathematical Psychology, 9,* 168–199.

Krantz, D. H. (1975a). Color measurement and color theory: I. Representation theorem for Grassmann structure. *Journal of Mathematical Psychology, 12,* 283–303.

Krantz, D. H. (1975b). Color measurement and color theory: II. Opponent-color theory. *Journal of Mathematical Psychology, 12,* 304–327.

Lewis, D. R. (1948). Psychological scales of taste. *Journal of Psychology, 26,* 437–446.

Lord, F. M. (1952). A theory of test scores. *Psychometric Monograph,* No. 7.

Lord, F. M., & Novick, M. R. (1968). *Statistical theories of mental test scores.* New York: Addison-Wesley.

Luneburg, R. K. (1947). *Mathematical analysis of binocular vision.* Princeton, NJ: Princeton University Press.

Luneburg, R. K. (1950). The metric of binocular visual space. *Journal of the Optical Society of America, 50,* 637–642.

Luo, M. R., & Riggs, B. (1986). Chromaticity-discrimination ellipses for surface colors. *Color Research and Application, 11,* 25–42.

MacAdam, D. L. (1942). Visual sensitivities to color differences in daylight. *Journal of the Optical Society of America, 32,* 247–274.

McCamy, C. S. (1993). The primary hue circle. *Color Research and Application, 18,* 3–10.

Metzger, W. (1930). Optische Untersuchungen am Ganzfeld II. Zur Phänomenologie des homogenen Ganzfeld. *Psychologische Forschung, 13,* 6–29.

Munsell, A. E. O., Sloan, L. L., & Godlove, I. H. (1933). Neutral value scale: I. Munsell neutral value scale. *Journal of the Optical Society of America, 23,* 394–425.

Newhall, S. M., Nickerson, D., & Judd, D. B. (1943). Final report of the O.S.A. subcommittee on the spacing of the Munsell colors. *Journal of the Optical Society of America, 33,* 385–418.

Schiler, P. (1950). *The image and apperance of the human body,* New York: International Universities Press.

Shepard, R. N. (1978). On the status of "direct" psychological measurement. In C. W. Savage (Ed.), *Minnesota Studies in the Philosophy of Science: Vol. IX. Proceedings of Minnesota Philosophical Meeting.* Minneapolis: University of Minnesota Press.

Spivak, M. (1979). *A comprehensive introduction to differential geometry* (Vol. 5). Berkeley, CA: Publish or Perish.

Stevens, S. S. (Edited by Geraldine Stevens) (1975). *Psychophysics: Introduction to its perceptual, neural, and social prospects.* New York: Wiley.

Suppes, P., Krantz, D. H., Luce, R. D., & Tversky, A. (1989). *Foundations of measurement: Vol. II. Geometrical, threshold, and probabilistic representations.* New York: Academic Press.

Thurstone, L. L. (1927). A law of comparative judgements. *Psychological Review, 35,* 273–286.
Torgerson, W. S. (1952). Multidimensional scaling: I. Theory and method. *Psychometrika, 17,* 401–419.
Torgerson, W. S. (1958). *Theory and methods of scaling.* New York: Wiley.
Wyszecki, G., & Fielder, G. H. (1971). New color-matching ellipses. *Journal of the Optical Society of America, 61,* 1135–1152.
Wyszecki, G., & Stiles, W. S. (1982). *Color science: Concepts and methods, quantitative data and formulae* (2nd ed.). New York: Wiley.

APPENDIX:

Tarow Indow: A Brief Biography and a Bibliography of His Papers in English

Tarow Indow was born on August 22, 1923. His entire education, elementary school through university, occurred at Keio-Gijuku. Keio was Japan's first institute of higher education, founded in 1858 by Yukichi Fukuzawa, who had become the most influential scholar in introducing Western civilization to Japan and served as an interpreter of the first Japanese delegation to the United States in 1860. Indow's father and most of his uncles and cousins are also alumni of Keio. After completing junior high school, his formal education was disrupted by the Japanese wars, first with China and then with the United States.

During this era, most Japanese universities had just one professor of psychology. At Keio University, he was Professor M. Yokoyama, who received his Ph.D. from Clark University under the guidance of E. G. Boring. Yokoyama acknowledged himself to be a Titchenerian, and to his students he spoke often, with great respect, about the honorary Society of Experimental Psychologists founded by Titchener. In 1983 Indow was elected a Fellow of that Society, and his only regret was that Yokoyama had already passed away.

Indow was not impressed by the ideas of introspectionism. Rather, he was far more influenced by the Gestalt psychologists, especially W. Köhler. From the beginning, his interest lay in quantitative analysis. He studied psychophysical methods, scaling, psychometrics, factor analysis through articles of L. L. Thurstone, and mathematical modeling through chapters of S. Hecht and W. J. Crozier in the *Handbook of General Experimental Psychology* (Ed. C. Murchison, 1934).

During the occupation period following World War II, General MacArthur's General Headquarters developed a program to send Japanese leaders from various areas for three-month visits to the United States. In 1951, this provided Indow with his first opportunity to visit the United States. His task was to observe the use of psychological tests in both schools and companies. During his stay at the Eduational Testing Service (ETS), he read the Ph.D. theses of Princeton University's ETS Psychometric Fellows, including Warren Torgerson's, which introduced multidimensional scaling, and Frederick Lord's, which formulated the latent-trait model. In the Princeton University bookstore he discovered the thin paperback *Mathematical Analysis of Binocular Vision* by R. K. Luneburg. These several chance encounters determined the direction of Indow's subsequent research career.

In 1953, Indow married Minako Kawamura, whose father was one of the officers in the Japanese Ministry of Foreign Affairs who resigned in protest to his country's alliance with Nazi Germany.

Starting in 1955 Indow began multidimensional investigations of the Munsell color system. These studies attracted the attention of color scientists. At first, he undertook this problem because it was relatively inexpensive. This was prior to the age of the color monitor, and one needed only a standard light source and the Munsell color chips. Soon, however, the psychophysical approach to color perception itself became one of his main interests. His studies about the geometry of visual space began in 1959. Together with F. Samejima, he constructed a reasoning ability test based on Lord's latent-trait model which was published in 1962. The computations were initially done on a Tiger hand-driven calculator, followed by a Monroe electric desk calculator. His first use of an electronic computer was in 1959. Since 1970, he has added concept identification and memory to his repertoire of research.

Indow received his Ph.D. from Keio University in 1959, and he was appointed as a full professor of his alma mater. Subsequently he taught not only in the Department of Psychology, but also in the School of Engineering at Keio, as well as in Tokyo University, Tokyo University of Education, Kyoto University, and Kanazawa University. He was also active in applying quantitative methods to practical problems as a member of the Society of Illuminating Engineering, the Committee of Sensory Evaluation in Engineering, and the Motivation Research Group in Dentsu Co.

Indow's second opportunity to visit the United States resulted from an invitation to participate in a 1966 symposium of the American Psychological Association held in Washington, DC. There, Professor S. S. Stevens invited him to visit Harvard's Laboratory of Psychophysics for two years as a research fellow. He was also afforded an opportunity to stay four months at the Carnegie Institute of Technology with Professors H. A. Simon and A. Newell.

On returning to Tokyo in 1969, he began to serve as a liaison between Japan and the international community in the fields of psychology, human factors, color science, and sometimes statistics. Indow participated in various meetings in the United States and Europe. From 1969 to 1981, he served as a member of the Executive Committee of AIC (Association Internationale de la Couleur) and became its President from 1973 to 1977. In 1989 he was honored with the Deane B. Judd AIC Award.

In 1971, Indow visited for a year at the Institute for Advanced Study in Princeton, NJ, at the invitation of Dr. R. D. Luce, then a visiting professor at the Institute; they had first met in 1965 during Indow's Harvard stay with Stevens.

Four years later he and Dr. J. D. Carroll, then a member of the AT&T Bell Laboratories, ran a U.S.-Japan Seminar on "Theory, Methods, and Applications of Multidimensional Scaling and Related Techniques" at the University of California, San Diego. Immediately after this meeting, he attended the Eighth Annual Mathematical Psychology Meeting at Purdue University, where Luce was present. Although he was about to move to Harvard from the University of California, Irvine, where he had gone following his three-year stay at the Institute of Advanced Study, Luce raised the possibility of Indow's moving to UCI on a permanent basis. Both UCI and Indow found the idea attractive, but because of his commitment in Keio University, he could not resign until 1979.

Since joining UCI, Irvine has become Indow's home ground and that has made direct participation in various research activities in the United States and Europe much easier. During the period 1981–84 he was adjunct professor of the Rensselaer Polytechnic Institute helping with the psychophysical aspects of colorimetric studies being carried out in the Department of Chemistry.

Although Indow retired in 1993 and is now professor emeritus of both UCI and Keio University, he continues to be actively engaged in research and teaching. He has continued, on average, to present five papers and publish two articles every year.

PUBLICATIONS

Books in Japanese

1. (1951). (T. Indow, M. Makita and T. Hidano) *Psychological measurements.* Kanekoshobo.
2. (1957). *Probability and statistics.* Coronasha.
3. (1962). (T. Indow and F. Samejima) *LIS. Measurement of reasoning ability. Non-verbal.* Nippon Bumkasha.
4. (1969). (ed.) *Mathematical psychology.* Tokyo University Press.
5. (1973). (ed.) *Mathematical models in psychology.* Tokyo University Press.
6. (1977). (T. Indow, H. Ikeda and S. Ono) *Measurements in psychology and learning theories.* Morikita.

Articles in Japanese

Over 100 academic articles and book chapters have been published in Japanese.

Articles in English

7. (1956). (T. Indow and T. Shiose) An application of the method of multidimensional scaling to perception of similarity or difference in color. *Japanese Psychological Research, 1*(3), 45–64.
8. (1958). (T. Indow and T. Shiose) A note on an application of the methods of multidimensional scaling to perception of similarity or differences in colors. *Japanese Psychological Research, 1*(5), 21.
9. (1958). The mental growth curve defined on the absolute scale: Comparison of Japanese and foreign-data. *Japanese Psychological Research, 1*(6), 35–48.

10. (1958). (T. Indow, U. Kuno and T. Yoshida) Studies on the induction in visual process taking electrical phosphene as an index (2)—Experiments on the propagation of the induction across the blind spot (1). *Psychologia*, 1, 175–181.
11. (1959). (T. Indow, T. Koyazu and T. Yoshida) Studies on the induction in visual process taking electrical phosphene as an index (5)—Experiments on the propagation of the induction across the blind spot (2). *Japanese Psychological Research, 1*(7), 17–28.
12. (1960). (T. Indow, T. Koyazu and T. Kozaki) Studies on the propogation in visual process taking electrical phosphene as an index (6)—Experiments on the propagation of the induction across the blind spot (3). *Japanese Psychological Research, 1*(9), 38–47.
13. (1960). The logic of factor analysis—Discussions from the methodological point of view. *Annals of the Japan Association for Philosophy of Science, 1*(5), 313–328.
14. (1960). (T. Indow and T. Uchizono) Multidimensional mapping of Munsell colors varying in hue and chroma. *Journal of Experimental Psychology, 59,* 321–329.
15. (1960). (T. Indow and K. Kanazawa) Multidimensional mapping of Munsell colors varying in hue, chroma, and value. *Journal of Experimental Psychology,* 59, 330–336.
16. (1960). (T. Indow and T. Kawai) A construction of uniform lightness scales on various backgrounds in unified terms. *Japanese Psychological Research, 2,* 1–12.
17. (1960). (T. Indow and T. Koyazu) Experiments on induction in the binocular field composed of the independent monocular fields (1). *Japanese Psychological Research, 2,* 142–151.
18. (1961). (T. Indow and T. Koyazu) Experiments on induction in the binocular fields composed of the independent monocular field (2). Binocular summation. *Japanese Psychological Research, 3,* 42–56.
19. (1962). (T. Indow, E. Inoue and K. Matsushima) An experimental study of the Luneburg theory of binocular space perception (1). The 3- and 4-point experiments. *Japanese Psychological Research, 4,* 6–16.
20. (1962). (T. Indow, E. Inoue and K. Matsushima) An experimental study of the Luneburg theory of binocular space perception (2). The alley esperiments. *Japanese Psychological Research, 4,* 17–24.
21. (1962). Multidimensional mapping of Munsell colors varying in hue and chroma and Multidimensional mapping of Munsell colors varying in hue, chroma and value. *Japanese Science Review, 12,* 77–80.
22. (1962). (T. Indow, K. Sano and H. Namiki) A mathematical model for interpretation in projective tests: An application to Seiken SCT. *Japanese Psychological Research, 4,* 163–172.
23. (1963). (T. Indow, E. Inoue and K. Matsushima) An experimental study of the Luneburg theory of binocular space perception (3): The experiment in a spacious field. *Japanese Psychological Research, 5,* 10–27.
24. (1963). Two kinds of multidimensional scaling methods as tools for investigating color space from the macroscopic point of view. *Acta Chromatica, 1,* 60–71.
25. (1964). Man as an object of scientific study. *Annals of the Japan Association for Philosophy of Science, 2,* 218–225.
26. (1964). On distance function in multidimensional scaling. (Multidimensional analysis of similarity data). *Acta Psychologica, 23,* 347–348.
27. (1966). (T. Indow and S. S. Stevens) Scaling of saturation and hue. *Perception & Psychophysics, 1,* 253–271.
28. (1966). Two interpretations of binocular visual space: Euclidean and hyperbolic. *XVIII International Congress of Psychology,* 149.
29. (1966). *A computer program to fit a straight line when both variables are subject to error.* (Tech. Rep. No. PPR-317). Laboratory of Psychophysics, Harvard University.
30. (1966). A general equi-distance scale of the four qualities of taste. *Japanese Psychological Research, 8,* 136–150.
31. (1966). Professor Matsusaburo Yokoyama. *Psychologia, 9,* 185–186.
32. (1966). (T. Indow and F. Samejima) On the results obtained by the absolute scaling model and

the Lord model in the field of intelligence. Hiyoshi Rep. No. 3, The Psychological Laboratory on the Hiyoshi Campus, Keio University, Japan.
33. (1967). Two interpretations of binocular visual space: Hyperbolic and Euclidean. *Annals of the Japan Association for Philosophy of Science, 3,* 51–64.
34. (1967). Saturation scales for red. *Vision Research, 7,* 481–495.
35. (1968). Multidimensional mapping of visual space with real and simulated stars. *Perception and Psychophysics, 3,* 45–53.
36. (1968). Hue-discrimination thresholds and hue-coefficients—Comparisons among data (with C. Takagi). *Japanese Psychological Research, 10,* 179–190.
37. (1969). An application of the τ scale of taste: Interaction among the four qualities of taste. *Perception and Psychophysics, 5,* 347–351.
38. (1969). (T. Indow and S. Sono) The amount of transmitted information. *Japanese Psychological Research, 11,* 1–12.
39. (1969). Judgment characteristic curves and selection of panel members in sensory tests. ICQC '69'—Tokyo (International Conference on Quality Control). *Proceedings,* 447–450.
40. (1969). (T. Indow and S. Suzuki) Application of four stochastic models to learning of numerical sequences. *Japanese Psychological Research, 11,* 183–197.
41. (1969). (T. Indow and K. Matsushima) Local multidimensional mapping of Munsell color space. *Acta Chromatica, 2,* 16–24.
42. (1970). (T. Indow and K. Togano) On retrieving sequence from long-term memory. *Psychological Review, 77,* 317–331.
43. (1970). Models of responses of customers with a constant rate. *Journal of Marketing Research, 7,* 498–502.
44. (1970). The formation of subjective value: A theoretical and empirical analysis: I. *Japanese Psychological Research,* 1970 *12,* 184–191.
45. (1970). *Outline of sensory test research group* (JUSE *ISIT-Rep.*), 1–16.
46. (1971). Models for responses of customers with a varying rate. *Journal of Marketing Research, 8,* 78–84.
47. (1971). The uniformity among eight measures for color differences. *COLOR 69,* Stockholm, 1971, 664–669.
48. (1971). Formation of subjective value: A theoretical and empirical analysis: II. *Japanese Psychological Research,* 1971 *13,* 1–10.
49. (1972). (T. Indow and K. Ohsumi) Multidimensional mapping of sixty Munsell colors by nonmetric procedure. In Vos, J. J., Friele, L. F. C., & Walraven, P. L. (Eds.), *Color metrics* (pp. 124–133). AIC/Holland, c/o Institute for Perception TNO Soesterverg.
50. (1972). (T. Indow and S. Suzuki) Strategies in concept identification: Stochastic model and computer simulation: I. *Japanese Psychological Research, 14,* 168–175.
51. (1973). (T. Indow and S. Suzuki) Strategies in concept identifications: Stochastic model and computer simulation II. *Japanese Psychological Research, 15,* 1–9.
52. (1973). (T. Indow and A. Murase) Experiments on memory and visual scannings. *Japanese Psychological Research, 15,* 136–146.
53. (1974). Colour atlases and colour scaling. In *Association Internationale de la Couleur, Colour' 73* (pp. 137–152). London: Adam Hilger.
54. (1974). Mathematical models of perception: Summary Report. In *Proceedings of XXth International Congress of Psychology* (pp. 210–211). Science Council of Japan.
55. (1974). On geometry of frameless binocular perceptual space. *Psychologia, 17,* 50–63.
56. (1974). Discussion. International Association of Applied Psychology, *XVIIth International Congress Actes Proceedings.*
57. (1974). Applications of multidimensional scaling in perception. In E. C. Carterette & M. P. Friedman (eds.), *Handbook of perception* (Vol. II, pp. 493–525). Academic Press.
58. (1974). Scaling of saturation and hue shift: Summary of results and implications. In H. R. Moskowitz, B. Scharf, & J. C. Stevens (Eds.), *Sensation and measurement—Papers in honor of S. S. Stevens* (pp. 351–362). Dordrecht, Holland: D. Reidel.

59. (1974). (T. Indow and K. Matsushima) Euclidean representation of four color difference formulae. *Die Farbe, 23,* 291–306.
60. (1974). (T. Indow and K. Matasushima) On the trainability of visual assessment color differences. *Die Farbe, 23,* 279–290.
61. (1975). (T. Indow, S. Dewa and M. Tadokoro) Strategies in attaining conjunctive concept: Experiment and simulation. *Japanese Psychological Research, 16*(3), 132–142.
62. (1975). Dr. Guirao from Argentina. *Acta Chromatica, 2*(5), 216.
63. (1975). An application of MDS to study of binocular visual space. In *U.S.-Japan Seminar; Theory. Methods and Applications of Multidimensional Scaling and Related Techniques* (pp. 103–111). University of California, San Diego.
64. (1975). (T. Indow, M. Kobayashi and S. Dewa) Concept identification with natural material. *Acta Psychologica, 39,* 131–139.
65. (1975). (T. Indow and M. Ida) On scaling from incomplete paired comparison matrix. *Japanese Psychological Research, 17*(2), 98–105.
66. (1975). On choice probability. *Behaviormetrika, 23,* 13–31.
67. (1976). Professor F. W. Billmeyer, Jr. from the U.S.A. *Acta Chromatica, 3*(1), 24.
68. (1977). A mathematical analysis of binocular visual space under illumination with no a priori assumption on mapping functions. Collection of Papers Dedicated to Keizo Hayashi (commemorative report 9th) (pp. 21–35). *The Psychological Laboratory on the Hiyoshi Campus,* Keio University.
69. (1977). (T. Indow, H. Kanazawa and R. Sugano) Principal hue components in Munsell colors. Association Internationale de la Couleur, *Colour' 77* (pp. 341–344). Adam Hilger.
70. (1977). (T. Indow and M. Ida) Scaling of dot numerosity. *Perception and Psychophysics, 22*(3), 265–276, 1977.
71. (1978). Scaling of saturation and hue in the nonspectral region. *Perception & Psyhophyics, 24*(1), 11–20.
72. (1979). Alleys in visual space. *Journal of Mathematical Psychology, 19*(3), 221–258.
73. (1980). Errors in simple and monotone tasks: A proposal of testing on site. *Behaviormetrika, No. 7,* 1–12.
74. (1980). Global color metrics and color-appearance systems. *Color, Research and Application, 5*(1), 5–12.
75. (1980). (T. Indow and M. Watanabe) Absolute identification of colors in the Munsell notation: Trainability and systematic shifts. *Color, Research and Application, 5*(2), 81–85.
76. (1980). *Two comments on factor analysis: Why it was born in psychology and how individual information is represented* (Hiyoshi Rep. No. 11). Psychological Laboratory in Hiyoshi Campus, Keio University.
77. (1981). (T. Indow and Hiroko Takada) Recognition memory of retrieved sequence of words. *Acta Psychologica, 407,* 207–228.
78. (1981). Some Characteristics of word sequence retrieved from a given category. In R. S. Nickerson (Ed.), *Attention and performance* (VIII, pp. 621–636). Hillsdale, NJ: Erlbaum.
79. (1981). Book Review: Color science handook: New version. *Color Research and Application 6,* 123–124.
80. (1982). Remarks on filling the gap between studies of two different modes of appearance: aperture color and object color. Proceedings of the Third Taniguchi Foundation symposium on Neurobiological and Psychophysical Aspects of Color Vision, November 25–29, Katata, Japan, *Color Research and Application, 7,* 290–212.
81. (1983). An approach to geometry of visual space with no a priori mapping functions: Multidimensional mapping according to Reimannian metrics. *Journal of Mathematical Psychology, 26,* 204–236.
82. (1983). (T. Indow and N. Aoki) Multidimensional mapping of 178 Munsell colors. *Color Research and Application, 8,* 145–152.
83. (1983). Physiological models and geometry of visual space. *Behavioral and Brain Sciences, 6,* 667–668.

84. (1984). (T. Indow and T. Watanabe) Parallel- and distance-alleys with moving points in the horizontal plane. *Perception and Psychophysics, 35,* 144–154.
85. (1984). (T. Indow and T. Watanabe) Parallel- and distance-alleys on horopter plane in the dark. *Perception, 13,* 164–182.
86. (1985). The second report on principal hue components in object colors. *Mondial Couleur* 85. Tome 1, No. 19. Actes du 5 eme Conqres de l'Association Internationale de la Couleur.
87. (1988). Alleys on an apparent frontoparallel plane. *Journal of Mathematical Psychology, 32,* 259–284.
88. (1988). Multidimensional studies of Munsell color solid. *Psychological Review, 88,* 456–470.
89. (1988). (T. Indow and T. Watanabe) Alleys on extensive apparent frontoparallel plane: a second experiment. *Perception, 17,* 647–666.
90. (1990). On geometrical analysis of global structure of visual space. In H. G. Geissler (Ed.), *Psychological explorations of mental structures* (pp. 172–180). Toronto: Hogrefe & Huber.
91. (1991). (T. Indow and M. L. Morrison) Construction of discrimination ellipsoids for surface colors by the method of constant stimuli. *Color, Research and Application, 16,* 42–56.
92. (1991). Book Review. Color image scale by Kobayashi Shigenobu. *Color, Research and Application, 16,* 130–131.
93. (1991). A critical review of Luneburg's model with regard to global structure of visual space. *Psychological Review, 98,* 430–453.
94. (1991). Different effects due to context for the same local stimulation. *Proceedings of VIIth Meeting of International Society for Psychophysics,* Duke University.
95. (1991). Spherical model of colors and brightness discrimination by Izmaijlov and Sokolov. *Psychological Science, 2,* 260–262.
96. (1991). Discrimination ellipsoids of surface color. *Proceedings of Advances in Color Vision in Honor of* R. M. Boynton.
97. (1992). (T. Indow and A. R. Bobertson, M. von Grunau, and G. H. Fielder) Discrimination ellipsoids of aperture and simulated surface colors by matching and paired comparison. *Color, Research and Application, 17,* 6–12.
98. (1992). JND and supra-threshold difference in Munsell color solid. *Proceedings of VIIIth Meeting of International Society for Psychophysics* (pp. 105–110). Stockholm University.
99. (1993). (T. Indow and W. Batchelder) Book Review. Teaching by axiomatic approach to formal psychology. A review of quantitative psychology by Jan Drösler. *Journal of Mathematical Psychology, 37,* 111–118.
100. (1994). Analysis of events counted on time-dimension: A soft model based on extreme statistics, *Behaviormetrika, 20,* 109–124.
101. (1944). Metrics in color spaces: Im kleinen und in Grossen. In G. H. Fisher and D. Laming (Eds.), *Contribution to mathematical psychology, psychometrics, and methodology* (pp. 3–17). Springer.
102. (1994). Psychophysical scaling of visual distance and color difference. *Fichner's Day 94, Proceedings of 10th Annual Meeting of International Society for Psychophysics,* University of British Columbia.

I SPACE

Donald D. Hoffman

INTRODUCTION

Space and time are central concepts in physics and psychology. And in both disciplines these concepts have had to be revised in ways that, at least at first, shock one's intuitions.

In classical physics we had the comfortable picture of space as three-dimensional and Euclidean, a global framework in which all events occurred, neatly ordered by a global sequence of time. This picture seemed so natural that it was viewed by Kant and others to be essential to our concepts of space and time, to be, as they put it, a priori. But it was replaced in relativistic physics, by a dynamically curving and four-dimensional space-time, after Einstein realized that the concepts of space and time had to be brought "down from the Olympus of the a priori in order to adjust them and put them in a serviceable condition" (Einstein, 1922). Einstein's conception of spacetime as a continuum, in turn, may need revision in light of quantum theory, since this theory implies that a continuum model is untenable (Wheeler, 1990), particularly at scales smaller than the Planck length. Thus the concepts of space and time are still in flux for physics.

The same is true for these concepts in psychology. The concept of perceptual space as Euclidean was generalized, by Luneberg, to a space of constant curvature. Given this concept, the task of psychologists was to measure, by psychophysical experiments, the sign and magnitude of curvature of this (global) perceptual space. This task was done by Indow and his colleagues in a series of careful experiments. They found that their estimates of curvature varied considerably from experiment to experiment, and even from day to day for an individual subject on a single experiment. What this means for the concept of perceptual space is a central issue in current studies of spatial vision and, in consequence, is much discussed in the chapters of this section.

Suppes suggests that the experiments of Indow, Foley, Wagner, and others require us to abandon Luneburg's concept of perceptual space as one having constant curvature. Indeed, he suggests that the concept of perceptual space as a single unified structure must also be abandoned. MacLeod and Willen agree. Experiments they conducted based on the Zöllner and the Müller-Lyer figures are incompatible with the concept of perceptual space as a global and unified structure. Moreover, they note, the relevant visual neurobiology seems not to support a unified conception. Different aspects of visual space (position, motion, orientation) are represented piecemeal by different neurons. So it appears that different, special-purpose, neural mechanisms underlie different spatial judgments.

In this spirit, Lakshminarayanan and Santhanam, building on the work of Carlton (1988), discuss how the brain might use an irreducible representation of the Lorentz group in its modeling of the motions of rigid bodies. Drösler discusses how perceptual invariances are powerful tools in the study of monocular and binocular spatial vision and in the representation of color vision. Albert and Hoffman discuss the power of a "principle of genericity" to explain aspects of the visual perception of line drawings, illusory surfaces, illusory contours, and three-dimensional objects and their parts. And Dzhafarov considers the empirical meaningfulness of sentences involving measurement functions, an issue of relevance to empirical claims about spatial vision and to empirical claims more generally.

As this brief introduction suggests, and as the chapters in this part illustrate in more detail, the concept of perceptual space is now undergoing radical revision. Where this revision will go is not yet clear. But perhaps the resulting psychological concept of space will be no less striking than the physical.

REFERENCES

Carlton, E. H. (1988). Connection between internal representation of rigid transformation and cortical activity pattern. *Biological Cybernetics, 59,* 419–429.

Einstein, A. (1922). *The meaning of relativity.* (p. 2). Princeton, NJ: Princeton University Press.

Wheeler, J. (1990). Information, physics, quantum: The search for links. In W. Zurek (Ed.), *Complexity, entropy and the physics of information* (pp. 3–28). Reading, MA: Addison-Wesley.

2 Some Foundational Problems in the Theory of Visual Space

Patrick Suppes
Stanford University

ABSTRACT

The most important general feature of visual space is that it is context dependent, a characteristic of physical systems rather than classical geometrical ones. Using results from some classical experiments by Foley and by Wagner, arguments are given to show that no reasonably simple unified set of axioms in the spirit of qualitative synthetic geometry can be given for the structure of visual space.

PRELIMINARIES

Perhaps the most important feature of visual space that must be taken account of in a thorough analysis is that visual space is more like a physical system than a geometrical one. What I mean by this is that there are strong context affects which do not exist at all in classical geometry. The motion of two particles in classical mechanics is very much affected by bringing into close proximity a third particle, but the geometric relation at a given time of these two particles is not affected by the presence or absence of this third particle. On the other hand, perceptual judgments of symmetry or of congruence are known from many experiments to be much affected by context. This is a strong warning from the very beginning that we cannot hope to have the theory of visual space placed too firmly and thoroughly within the framework of classical geometry.

The second point that follows from such strong contextual effects is that multiple geometrical models of visual space should be required because different contexts will lead to different models. The geometric relations will not be the same from context to context. Once we recognize this aspect then what we can hope for, insofar as we are attempting to obtain quantitative results, is to find models that are parametrically stable in a given kind of context. But even this hope is too optimistic, as we shall see shortly.

SPACES OF CONSTANT CURVATURE

The Luneburg quantitative approach to visual space is perhaps still the best theoretically developed program, even now after nearly half a century since it was first introduced. In orthogonal sensory coordinates the line element *ds* can be represented in terms of sensory coordinates α, β and γ (these Greek letters are not the ones used by Luneburg) by

$$ds^2 = \frac{d\alpha^2 + d\beta^2 + d\gamma^2}{[(1 + \frac{1}{4}K(\alpha^2 + \beta^2 + \gamma^2)]^2}, \tag{2.1}$$

where $K = 0$ for Euclidean space,
$K < 0$ for hyperbolic space,
$K > 0$ for elliptic space.

The best recent treatment is Indow (1979).

It is important to recognize that the Luneburg approach is strictly a psychophysical approach to visual space. It assumes a transformation of physical Euclidean space to yield a psychological one. In this sense it is not a measurement-theoretic approach or a qualitative approach to the axioms of visual space.

By far the best experimental studies of the Luneburg ideas are those that have been conducted by Indow and his colleagues over many years, although a large number of other investigators have also contributed. A review of this extensive literature is to be found in Volume II of *Foundations of Measurement* (Suppes, Krantz, Luce, & Tversky, 1989). It is also the case that the best analysis of multidimensional scaling of spaces of constant curvature has been given by Indow (1974, 1982).

As I want to make clear from these references, I myself have learned more about the theory of visual space from Indow and his colleagues than from anyone else. I was initially enormously skeptical of the Luneburg ideas, and I came to realize they could be converted into a realistic program just because of the extraordinary, careful experiments performed by Indow and his colleagues. In this case the fact that the program has not turned out to be more satisfactory than it is is not because of the weakness of the experiments but in fact because of their very strength. They have given us confidence to recognize that there are fundamental things wrong with the Luneburg approach to visual space. Above all, as Indow and his colleagues have brought out on several occasions, there is a complete lack of parametric stability once we assume that a space of constant curvature, for example, negative curvature in the hyperbolic case, is a reasonable hypothesis. When we estimate the curvature we find that even for the same experiments the results are not stable from day to day for a given subject, and certainly when we transfer from one kind of experiment, for example, the classical alley experiments, to judgments of a different kind, there is little transfer at all of the parametric values estimated in one situation to the next.

TWO IMPORTANT COUNTEREXAMPLES

I now want to turn to two important experimental results based upon qualitative considerations and that fly very much in the face of the Luneburg ideas, which fundamentally assume that visual space is a space of constant curvature as defined by Equation 2.1. In fact these counterexamples are such that they raise problems that I am not able to answer about how we should think about the geometry of visual space even in a quite simple range of experiments.

The first is a beautiful counterexample due to Foley (1972). The situation is shown in Figure 2.1. The instructions to the subject are to make judgments of perpendicularity ($\perp$) and congruence ($\simeq$) as follows, with the subject given point A as fixed in the depth axis directly in front of his position as observer O:

1. Find point B so that $AB \perp OB$ & $AB \simeq OB$.
2. Find point C so that $OC \perp OB$ & $OC \simeq OB$.
3. Judge the relative lengths of OA & BC.

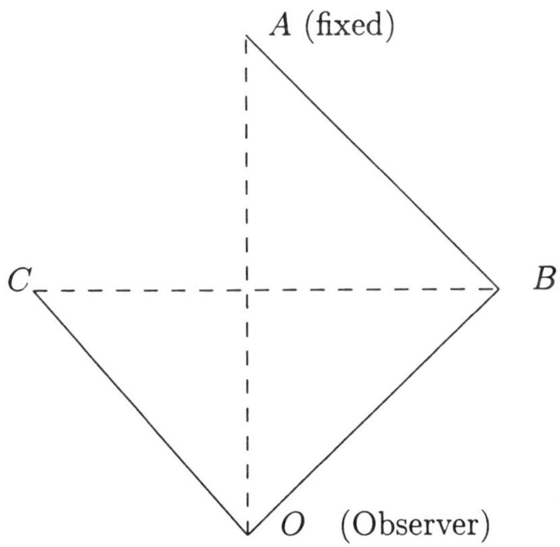

FIG. 2.1. Foley experiment.

The results are that 24 subjects in 40 of 48 trials judged BC significantly longer than OA. This judgment that BC is longer than OA contradicts properties of any space of constant curvature, whether that curvature is positive, negative, or zero. This simple experiment of Foley pushes us at once completely outside the framework of spaces of constant curvature and leads us to abandon for detailed purposes any hope of fully satisfying the Luneburg psychophysical postulates.

The second experiment is Wagner (1985), which dealt with perceptual judgments about distances and other measures among 13 white stakes in an outdoor viewing area. In physical coordinates let x equal measurement along the depth axis and y along the frontal axis, and let perceptual judgments of distance be shown by primes. Then the general result of Wagner is that $x' \simeq 0.5y'$ if physically $x = y$. Notice how drastic the foreshortening is along the depth axis. This result of Wagner's is not anomalous or peculiar but represents a large number of different perception experiments showing dramatic foreshortening along the depth axis.

PARTIAL AXIOMS FOR THE FOLEY AND WAGNER EXPERIMENTS

I am not able to give a fully satisfactory geometric analysis of the strongly supported experimental facts found in these experiments, but I do think there are some things to be said of interest that will help clarify the foundational situation. I have divided the axioms that I propose into various groups.

1. Affine Plane. These I take to be standard axioms. We would of course, reduce them and not require the whole plane but that is not important here. We can take as primitives either betweenness or parallelness and some concept like that of a midpoint algebra. Again the decision is not critical for considerations here. The reader can think just in terms of betweenness as an awkward but thoroughly developed theory of ordered affine planes. Let B be the ternary relation of betweenness. I use the simple and suggestive following notation for this relation. For any three points a, b, c, $B(a, b, c)$ if and only if $a|b|c$; i.e., point b is between points a and c, with weak inequality intended. In other words, if φ is a real-valued function on the set of points, the intended numerical interpretation is

$$a|b|c \quad \text{iff} \quad \varphi(a) \leq \varphi(b) \leq \varphi(c) \quad \text{or} \quad \varphi(c) \leq \varphi(b) \leq \varphi(a).$$

Axioms on betweenness needed to characterize an affine plane are given in the Appendix.

We now add judgments of perceived congruence ($\simeq$).

2. Three Distinguished Points. Intuitively, let o_1 be the center of the left eye, o_2 the center of the right eye, and o the bisector of the segment o_1, o_2. Explicitly, they satisfy the following two axioms.

2a. The three points are collinear and distinct;

2b. $o_1 o \simeq o o_2$.

3. Axioms of Congruence

3a. Opposite sides of any parallelogram are congruent.

3b. If $aa \simeq bc$, then $b = c$.

3c. $ab \simeq ba$.

3d. If $ab \simeq cd$ & $ab \simeq ef$, then $cd \simeq ef$.

3e. If $a|b|c$ & $a'|b'|c'$, $ab \simeq a'b'$ & $bc \simeq b'c'$, then $ac \simeq a'c'$ (this is a familiar and weak additivity axiom).

The *frontal axis* is the line containing o_1, o, and o_2 and the *depth axis* is the half-line through o such that for any point a on the axis $o_1a \simeq o_2a$. (Notice that we cannot characterize the depth axis in terms of a general notion of perpendicularity for that is not available, and in fact will not be available within the framework of these axioms. The depth axis is only a half-line because a subject cannot see directly behind the frontal axis.)

3f. *First special congruence axiom.* If $a \neq c$, a, c on frontal axis and b is on the depth axis, and $ao \simeq oc$, then $ab \simeq bc$ (see Figure 2.2).

3g. *Second special congruence axiom.* If $a \neq c$, a, c on the frontal axis, $ao \simeq oc$, ab and $cd \parallel$ to depth axis, $ab \simeq cd$, then $ob \simeq od$ (see Figure 2.3).

The last two special axioms of congruence extend affine congruence to congruence of segments that are not parallel, but only in the case where the segments are symmetric about the depth axis as is seen from Figures 2.2 and 2.3. This means that we have a weak extension of affine congruence. An extension that is far too weak even to give us the axioms of congruence for absolute spaces (see *Foundations of Measurement II*, Chapter 13). (Absolute geometry can be thought of this way. Drop the Euclidean axiom that through a given point a exterior to a line α there is at most one line through a that is parallel to α and lies in the plane formed by a and α. Adding this axiom to the axioms of absolute geometry gives us Euclidean geometry, as is obvious. What is much more interesting is that adding the negation of this axiom to those of absolute geometry gives us hyperbolic geometry.)

We can prove the following theorem.

Theorem 2.1. Let the half-space consisting of all points on the same side of the frontal axis as the depth axis be designated the frontal half-plane:

(1) The frontal half-plane is isomorphic under a real-valued function φ to a two-dimensional affine half-plane over the field of real numbers, with the x-axis the depth axis and the y-axis the frontal axis. Moreover, congru-

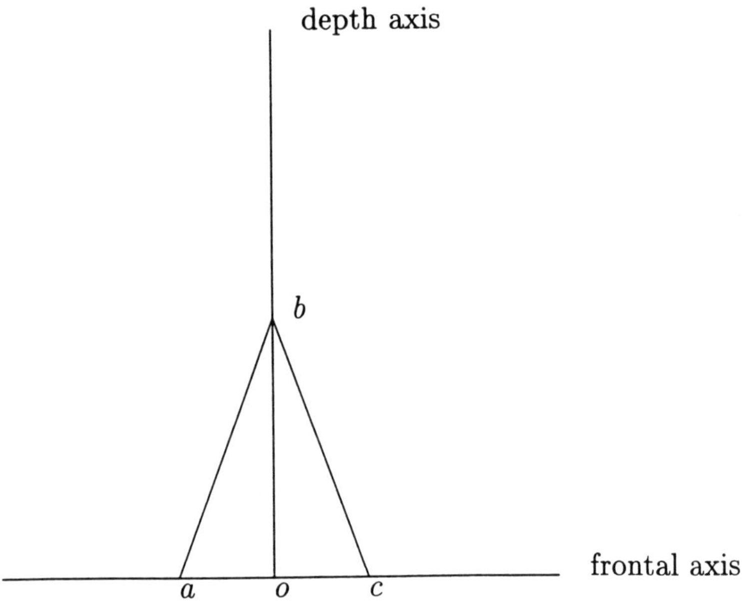

FIG. 2.2. Congruence Axiom 3f.

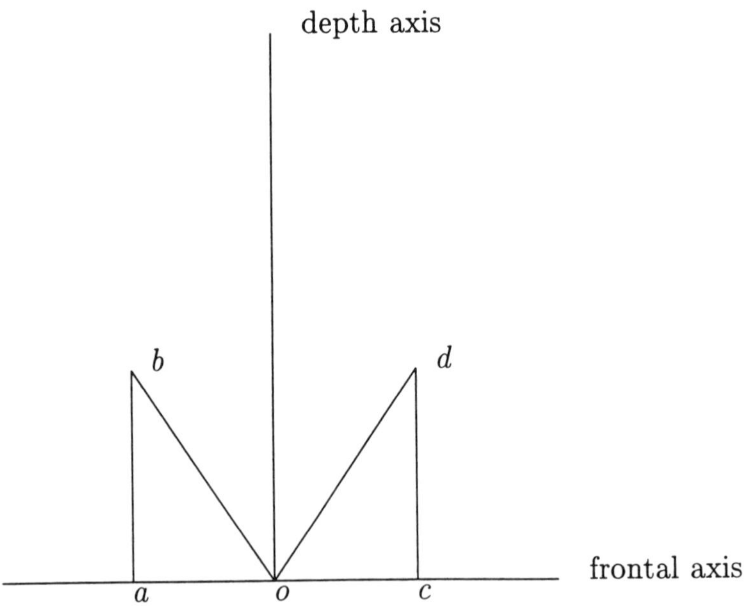

FIG. 2.3. Congruence Axiom 3g.

ence as measured by the Euclidean metric is satisfied when line segments are congruent; i.e., if ab $\simeq$ cd then $\Sigma_{i=1}^{2}(\varphi_o(a) - \varphi_i(b))^2 = \Sigma_{i=1}^{2}(\varphi_i(c) - \varphi_i(d))^2$.

(2) The only affine transformations possible are those consisting of stretches α of the frontal axis and stretches β of the depth axis with α, $\beta > 0$.

The proof of (1) is obvious from familiar results of classical geometry. The proof of (2) follows from observing that the affine transformations described are the only ones that preserve the symmetry around the depth axis required by the two special congruence axioms.

It is clear that the results of Theorem 2.1 are quite weak. The theorem is not contradicted by the Foley and Wagner experiments, but this is not surprising, for the apparatus of betweenness plus symmetric congruence about the depth axis cannot describe the results of either experiment. If we were to add a concept of perpendicularity, as required by the Foley procedure, then we would essentially get a Euclidean half-plane, and the resulting structure would contradict the Foley results.

Correspondingly, we cannot describe the Wagner psychophysical results of extensive perceptual foreshortening along the depth axis without adding some psychophysical assumptions radically different from those of Luneburg. Of course, the Foley results also are best interpreted as perceptual foreshortening along the depth axis. The natural conclusion is that we cannot consistently describe visual geometric relations in any space close to the specificity and simplicity of structure of a space of constant curvature.

Even the weak affine structure of Theorem 2.1 is too strong and probably should be replaced by a standard version of absolute geometry, but with the congruence axioms weakened as given above. Such an absolute geometry can be extended to hyperbolic geometry, but the affine structure cannot. What we seem to end up with is a variety of fragments of geometric structures to describe different experiments. A hyperbolic fragment perhaps for alley experiments, a fragment outside the standard ones for the Foley experiments, etc.

The goal of having a unified structure of visual space adequate to account for all the important experimental results now seems mistaken. A pluralistic and fragmentary approach seems required.

APPENDIX

Let A be a nonempty set, and let B be a ternary relation of betweenness as discussed in the text. Then we have the following definition, where a *line ab* is the set of all points c such that $a|c|b$ or $c|b|a$ or $c|b|a$.

Definition 2.1. A structure (A, B) is an *affine plane* if and only if the following axioms are satisfied for a, b, c, d, e, f, g, a′, b′ and c′ in A:

1. If $a|b|a$, then $a = b$.
2. If $a|b|c$, then $c|b|a$.
3. If $a|b|c$ and $b|d|c$, then $a|b|d$.
4. If $a|b|c$ and $b|c|d$ and $b \neq c$, then $a|b|d$.
5. (Connectivity) If $a|b|c$, $a|b|d$, and $a \neq b$, then $b|c|d$ or $b|d|c$.
6. (Extension) There exists f in A such that $f \neq b$ and $a|b|f$.
7. (Pasch's Axiom) If abc is a triangle, $b|c|d$, and $c|e|a$, then there is on line de a point f such that $a|f|b$.
8. Desargues's axiom. If $d|a|a'$, $d|b|b'$, $d|c|c'$, $a|b|e$, $a'|b'|e$, $a|c|f$, $a'|c'|f$, $b|c|g$, $b'|c'|g$, not $d|a|b$, not $a|b|d$, not $b|d|a$, not $d|b|c$, not $b|c|d$, not $c|d|b$, not $d|c|a$, not $c|a|d$, not $a|d|c$, and $a \neq a'$, then $e|f|g$.
9. (Axiom of Completeness) For every partition of a line into two non-empty sets Y and Z such that
   (i) no point b of Y lies between any a and c of Z and
   (ii) no point b′ of Z lies between any a′ and c′ of Y,
   there is a point b of $Y \cup Z$ such that for every a in Y and c in Z, b lies between a and c.
10. (Dimensionality). There are three noncollinear points a_0, b_0, c_0 in A such that for any point d′ in A there are points e′ and f′ such that $e' \neq f'$, e′ and f′ lie each on one of the three lines a_0b_0, a_0c_0, and b_0c_0, and d′, e′, and f′ are collinear.

The formulations of Axioms 8 and 10 just in terms of betweenness are rather complicated. But Foley (1964), in an excellent experimental study, found that a number of subjects made judgments satisfying the Desarguesian property (Axiom 8).

REFERENCES

Foley, J. M. (1964). Desarguesian property in visual space. *Journal of the Optical Society of America, 54,* 684–692.

Foley, J. M. (1972). The size-distance relation and intrinsic geometry of visual space: Implications for processing. *Vision Research, 12,* 323–332.

Indow, T. (1974). Applications of multidimensional scaling in perception. In E. C. Carterette (Ed.), *Handbook of perception: Vol. 2. Psychophysical judgment and measurement* (pp. 493–531). New York: Academic Press.

Indow, T. (1979). Alleys in visual space. *Journal of Mathematical Psychology, 19,* 221–258.

Indow, T. (1982). An approach to geometry of visual space with no a priori mapping functions:

Multidimensional mapping according to Riemannian metrics. *Journal of Mathematical Psychology, 26,* 204–236.

Suppes, P., Krantz, D. H., Duncan, R., Luce, & Tversky, A. (1989). *Foundations of measurement* (Vol. II). New York: Academic Press.

Wagner, M. (1985). The metric of visual space. *Perceptions and Psychophysics, 38,* 483–495.

3 Is There a Visual Space?

Donald I. A. MacLeod
J. Douglas Willen
University of California, San Diego

ABSTRACT

The compelling phenomenological reality of visual space has rarely been questioned, let alone objectively tested, yet a unitary visual space stands as one of the key assumptions of most characterizations of human spatial vision. Here we evaluate the claim that all of our spatial judgments are determined by perceived locations of things in some personal phenomenal space. We show that if distortions of phenomenal visual space are spatially continuous (hence locally correlated) we can account for Weber's law in length judgments, as well as the fall-off from Weber's law observed at greater lengths. But experiments in the detection of a sinusoidal ripple fail to support the use of locations in a unitary space and suggest instead that features are located through distance or orientation measures relative to the objects to which the features belong. Experiments with the Zöllner and Müller-Lyer illusory figures fail to support the idea that apparent position completely determines apparent orientation (or vice versa). Instead, we suggest that special-purpose hardware underlies different spatial discriminations.

Feature space, color space, knowledge space: these all illustrate the use of spatial representation as an analytical tool in the effort to understand perception. But there is one *real* space in visual perception: the phenomenal space that represents the world as we see it; and no one doubts the psychological reality and spatial character of that. As Indow describes it, visual space is "the most comprehensive percept that includes all individual objects appearing in front of the perceived self" (Indow & Watanabe, 1988). It is the substrate for everything in the rich phenomenal world of visual experience. This space is also the subject of one of the most splendid and well-defined theoretical structures in psychology, the

Luneburg model that Tarow Indow has done so much to expound and refine. But the aim of this chapter is to question the natural assumption that we experience a visual space. Can our visual experience of the world really be characterized as a sequence of apparitions occurring at definite locations in our personal visual space?

Obviously, in some sense the existence of visual space, and its genuinely spatial character, can hardly be doubted. You have merely to look around: surely being in a space is what vision feels like; or you can read the literature and find that mathematical models of visual space take it for granted. Although the spatial character of "visual space" has a compelling basis in experience, and is naturally embodied in geometrically formulated perceptual theories, it is not so obvious how (or whether) it can be objectively demonstrated or put to experimental test.

To proceed toward such a test, we first need to rephrase the point at issue in more definite terms. The claim we wish to evaluate is that our visual experience consists of nothing more than objects and events occurring in some personal phenomenal space. We take this to imply that *all of our spatial judgments, including judgments about such things as distance, movement, and orientation, are determined by perceived locations* of things in that space. In other words, all the spatial information to which we have conscious access is implicit in the changing perceived configuration of things in our personal visual space. We might call this the Tidy Mother model: the one necessary and sufficient principle for the organization of the phenomenal world is "a place for everything, and everything in its place."

The term "implicit in" could mean either that the derived spatial measures such as distance and orientation are directly given by the perceived locations of the features defining them, or that they are completely determined by those perceived locations. The responses of an observer in an experiment on spatial judgments cannot distinguish between these two possibilities; but the idea that distances, orientations, movements, and spatial frequencies are given *directly* by the configuration of things in subjective visual space can, we believe, be rejected as inadequate or incomplete on logical grounds. Suppose you are shown a closed curve (Figure 3.1) with some clearly visible feature inside or outside it. You would have no trouble pressing the correct button to indicate whether you are seeing the inside or the outside case. Now it could plausibly be maintained that you have in your head a representation of space where internal variables of some kind supply perceived coordinates, for each point on the curve and for the enclosed or excluded feature. But notice that with that assumption we still have not fully accounted for your ability to press the correct button. In fact, no one has ever, as far as we know, given a complete, connected, causal account of the ability of humans to perform this task. If you imagine being given a list of the physical coordinates of successive points as an intermediate result to work from, the difficulty of the remaining part of the job can be appreciated. It is just as hard

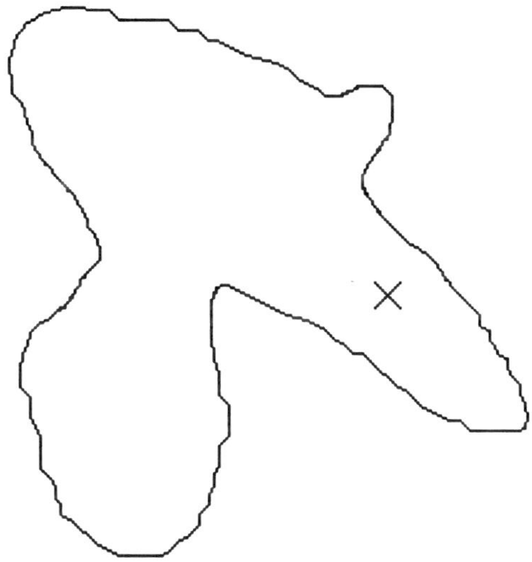

FIG. 3.1. Closed curve. Is the X inside or outside the curve? An easy discrimination. However, are the coordinates of the X and of the elements of the curve represented explicitly in some kind of internal model? What would the advantage of such a representation be over the representation present already at the level of the retina?

as the corresponding task for color space[1] where we usually consider the spatial representation to be only a metaphor. The initial replacement of physical coordinates with subjective coordinates is almost no help at all.

To acknowledge this difficulty, we might say that the "visual space" model under discussion requires a homunculus, who is left with the job of computing distances, orientations, etc., from perceived locations. If he (or she) does this using a consistent metric and without introducing further error in addition to whatever errors affect the perceived locations, we might be justified in taking the operations of the homunculus (or homuncula) for granted, by considering phenomenal visual space as the end product of the perceptual process. We require the homunculus to be infallible, because if that little person (or the equivalent perceptual or postperceptual mechanism) introduces substantial error of its own, you have to give up the key claim that the phenomenally registered positions of things in your visual space are what determine your spatial judgments.

With the "visual space" assumption thus defined, we now consider its experimentally testable consequences. We will deal first with a very simple case: the judgment of length or distance in the frontoparallel plane.

[1]The corresponding task for color is as follows. You are presented with (1) a strip along which color varies continuously, perhaps along a complex curve in color space, as in Figure 3.1, returning at its far end to its initial color; and (2) a homogeneously colored test patch. You are asked, does the range of colors on the strip enclose the test color, or does the test color fall outside them?

A DIFFICULTY AT THE OUTSET: WEBER'S LAW

If, as the visual space model requires, perceived separations are completely determined by the perceived locations of the two objects or features whose separation is judged, then any error in the separation judgment must be traceable to errors in those two perceived locations. A difficulty for this model at the outset is that Weber's law holds (at least roughly: some of the relevant studies are listed by Ogle, 1962) for judgments of length or separation: the average absolute error is a constant fraction of the judged length, increasing in proportion to the length or distance judged. How can this be if the perceived distance is simply a difference between two perceived locations, and the variability in those perceived locations is independent of the distance between them?

We can reconcile Weber's law for distance with the visual space assumption by imagining visual space as analogous to an elastic ruler, subject to fluctuating forces that vary continuously both over time and across space, and that locally stretch or compress the subjective ruler. The perceived locations of identifiable points in visual space are read off from the ruler with some error that depends upon the ruler's current state. If the deformations of the measuring ruler are spatially continuous, the registered positions of two nearby points will tend to fluctuate together, rather than independently. The correlated component of the error in the registered positions of the two points will cancel when the subject (or the homunculus) evaluates the distance between them. The resulting distance estimate is relatively precise because only the uncorrelated component of error contaminates it. Moreover this uncorrelated component of error, and hence the error in the distance estimate, decreases as the physical separation between the points decreases, as Weber's law requires.

To see how we can get the quantitatively proportional relationship implied by Weber's law out of this, let $\mathbf{x}_1$ and $\mathbf{x}_2$ represent the horizontal positions of the two points in the observer's frontoparallel plane whose separation is being judged. The corresponding subjectively registered positions, $\mathbf{x}'_1$ and $\mathbf{x}'_2$, are randomly perturbed from their mean values, $\mathbf{x}_1$ and $\mathbf{x}_2$, by an error term with mean zero and variance σ^2. The homunculus derives the subjectively registered distance between $\mathbf{x}_1$ and $\mathbf{x}_2$ without added error as $\mathbf{x}'_1 - \mathbf{x}'_2$. The variance of this is equal to $2\sigma^2(1 - \rho)$, where ρ is the correlation between the errors affecting $\mathbf{x}'_1$ and $\mathbf{x}'_2$. The root-mean-square error in the distance estimate is the square root of that and is thus proportional both to σ and to $(1 - \rho)^{1/2}$. For spatially continuous deformations of the subjective ruler, the correlation ρ will increase smoothly toward 1 as the distance $\mathbf{s} = \mathbf{x}_1 - \mathbf{x}_2$ between $\mathbf{x}_1$ and $\mathbf{x}_2$ decreases toward zero, and the factor $(1 - \rho)^{1/2}$, and hence the error, will decrease toward zero. This makes quantitative the intuitive expectation that the distribution of subjective separation measurements will become tighter as the judged separation decreases. For Weber's law, all that is needed is that $\rho(\mathbf{s})$, the spatial correlation function defining the correlation between the errors of position for points separated by $\mathbf{s}$, be a bell-shaped function of **s** or a smooth, monotonically decreasing function |**s**| such as the positive valued lobe of a bell-shaped function. Then for sufficiently small **s**, $1 - \rho(\mathbf{s})$ will be proportional to the square of **s**, and its square root (and hence the root-mean-square error in the subjective estimate of **s**) will be proportional to **s**.

In two dimensions, similarly, we might compare the visual field with an elastic sheet undergoing spatially continuous deformations that introduce errors in spatial judgment. Another suggestive physical parallel is heat haze. As evidence that such deformations occur in perception, we note that quite pronounced continuous modifications of perceived shape are reported under the influence of mescaline (Kluver, 1966), and undrugged but fatigued observers sometimes report that the scene appears to "swim" noticeably. The mechanisms of such effects are not known, but there are some plausible candidates. Constancy-scaling mechanisms might impose a fluctuating scaling factor on regions of the visual field. On a slower time scale, topographic maps in the visual system are clearly plastic and are continuously revised and refined by experience (e.g., Kaas, 1987). Error-correcting mechanisms, or processes that reorganize the map to represent the range of spatial stimuli uniformly and efficiently (Kohonen, 1989) could inject time-varying deformations by continuously and locally readjusting the coordinate system.

A few years ago, Levi, Klein, and Yap (1988) suggested that Weber's law for spatial separation might be due to quite different factors. The precision of localization falls off from fovea to periphery, and this could limit the precision with which we judge large distances, because when you judge a large distance, at least one of the defining stimulus features must necessarily fall far from the fovea at any given time. They investigated the influence of this retinal position factor by presenting stimuli in brief flashes during fixation and ensuring that the eccentricity of the test stimuli was just as great for the small as for the large separation stimuli (see Figure 3.2). Their first results suggested that the precision of distance judgments in these isoeccentric presentations might actually be independent of the distance involved, in contravention of Weber's law. They therefore proposed that Weber's law arises (in less carefully contrived conditions) simply because errors of localization increase in proportion to retinal eccentricity. Such a proportionality would be consistent with a visual space that is logarithmically compressed, if the apparent positions of features in the compressed representation have a uniform statistical dispersion; and the complex log transform actually has good anatomical support as a rough idealization of the mapping from retina to cortex (Schwartz, 1980).

But this finding does mean the correlation-based elastic sheet model for length discrimination that we have been developing is in trouble—not for failing to predict Weber's law, but because it also predicts Weber's law under isoeccentric conditions, where, according to Levi et al., it is not found. Fortunately Levi and Klein (1990) have found in further work, prompted by contrary results from other labs (e.g., Morgan & Watt, 1989) and confirmed by our own observations,

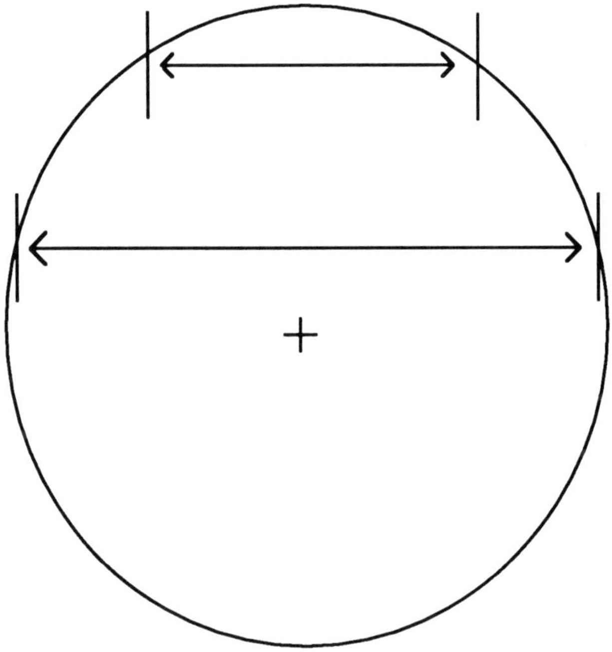

FIG. 3.2. Overview of the stimulus configuration used by Levi and Klein (1990). Each pair of vertical bars represents a given separation that subjects must evaluate. The bars always fall on the isoeccentric arc (made explicit here but not presented to subjects). Separations are judged relative to a remembered, reference separation.

that Weber's law can hold even at constant eccentricity, if the separations or distances judged are not too large. Their newer results, however, continue to show an unusual failure of Weber's law under constant eccentricity conditions, in that the precision of the judgment becomes asymptotically independent of the distance involved when that distance is large. This is consistent with the elastic sheet model. As the judged distance or separation becomes larger, it will eventually progress beyond the range across which positional errors are correlated. At that point, the errors in apparent position become practically statistically independent and the factor $1 - \rho(\mathbf{s})$ becomes independent of the separation $\mathbf{s}$.

In Figure 3.3 we show the constant-eccentricity results of Levi and Klein (1990), fit by the elastic sheet model, assuming a simple bell-shaped form for the spatial correlation function for errors of localization: $\rho(\mathbf{s}) = 1/(1 + (\mathbf{s}/\mathbf{s}_0)^2)$. The just reliably detectable length difference, $\Delta\mathbf{s}$ is obtained by substituting this expression of the spatial correlation function for localization errors into the expression $\sigma(2(1 - \rho(\mathbf{s})))^{1/2}$ for root-mean-square error:

$$\Delta s = (2\sigma^2)^{1/2}\left(\frac{(\mathbf{s}/\mathbf{s}_0)^2}{(1 + (\mathbf{s}/\mathbf{s}_0)^2}\right)^{1/2}.$$

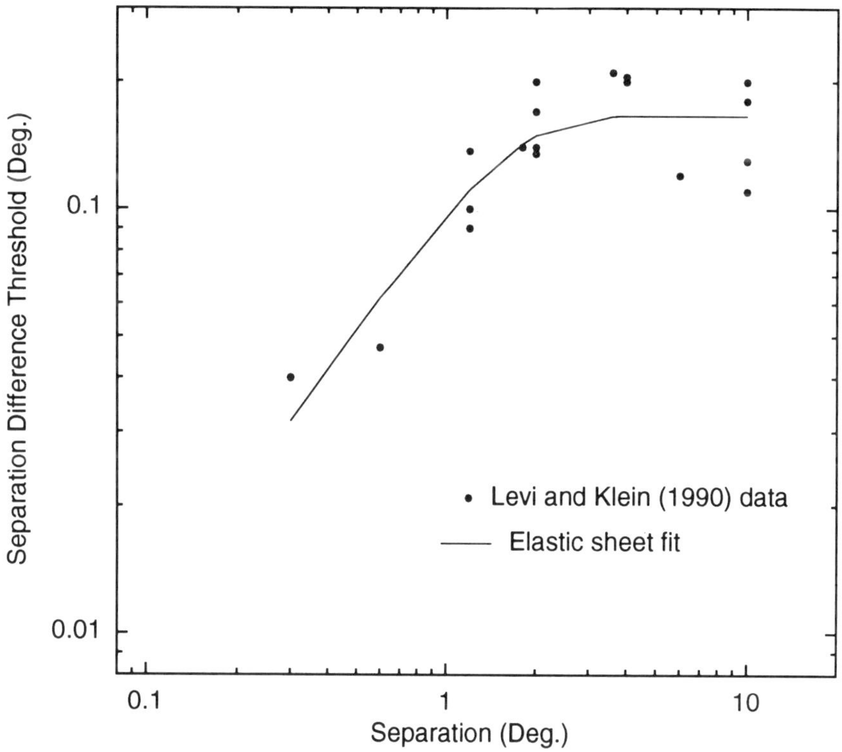

FIG. 3.3. Data of Levi and Klein, fitted assuming that errors of localization are spatially continuous, with the correlation function given in the text.

This correctly generates Weber's law for the isoeccentric condition for relatively small **s,** as well as the constant asymptote for large **s.** The best-fitting value of s_0, the half-width at half-height of the spatial correlation function for errors of localization, was 1.45° of visual angle. An attractive aspect of the elastic sheet conception of visual space perception is that it similarly allows prediction of the precision of any spatial judgment once the spatial correlation function for errors of localization is given; but comprehensive analyses of this sort have yet to be undertaken.

This general approach to the statistical analysis of spatial judgments is far from new. Cattell (1893) started from the assumption that errors of subjective magnification, rather than of subjective position, occur independently at different positions within the field of view. He showed that the error in length discrimination should then increase as the square root of the length judged rather than in direct proportion to it. This introduced the "square root law" into psychology. Following a suggestion of Fullerton, Cattell also showed how Weber's law would arise instead, provided only that the errors of subjective magnification for different small segments of the judged line are positively correlated over trials—they need not be perfectly correlated, but need only be positively correlated on the average over all pairs of segments. Weber's law emerges because in any judgment the same error tends to be repeated for all parts of the line. Fullerton and Cattell's account in terms of errors of magnification is closely related to ours in terms of errors of apparent position, since the correlation functions for magnification and for position are interdependent. The main differences between the two schemes are (1) for compatibility with Weber's law the correlation function for position errors must be bell shaped, whereas the correlation function for magnification need only generate a positive average correlation between segments of the judged distance; and (2) if it is the errors of magnification that become statistically independent at large separations, we would expect a square root law asymptote for length discrimination with long lengths, and not the constant asymptote that is predicted (Figure 3.2) if it is the errors of localization that become asymptotically independent.

EFFECTS OF PROXIMITY BETWEEN TEST AND REFERENCE

Thus far, the visual space assumption has survived confrontation with experiment well, with the proviso that errors of localization (or of magnification) must fluctuate continuously across space in order to generate Weber's law. Next, however, we introduce some findings that create difficulties for the model.

The Weber's Law experiment itself generates a difficulty on closer consideration. In the above analysis, we have implicitly assumed that the just noticeable difference in distance is proportional to the standard deviation of the judged distance. In practice, however, the just noticeable difference is almost always determined by comparing two distances. To be reliably detected, a difference in distance must exceed the standard deviation of the *difference* between the perceived values of the two distances compared. This introduces once again the same statistical considerations that arise in the relation of perceived distance to perceived locations of the defining pair of features. The variability of the distance-difference is dependent on the correlation between the two perceived distances, and on the model under discussion this will depend on the spatial separation between the test and comparison lines or between the two defining pairs of features. In fact, with a bell-shaped correlation function for position, the expected outcome is (as we hope to show elsewhere) the proportionality of the just-noticeable difference in distance both to the judged distance (as implied by Weber's Law) and to the test-to-comparison distance. The latter proportionality is not generally observed, even approximately: proximity of test and comparison stimuli does not improve distance discrimination in the way required by the visual space assumption (Andrews, 1967; Lennie, 1972).

We have looked at this problem in the context of what Tyler (1973) calls "periodic vernier acuity." Tyler found that a horizontal line with minimal vertical sinusoidal ripple could be discriminated from a straight one if its peak-to-peak vertical excursion was (over some range) a constant fraction of its horizontal spatial period. Tyler's interpretation was that the subject requires a minimum orientation variation along the line for detection of the ripple. Alternatively, however, his result can be predicted from the elastic sheet model without invoking any representation of orientation as such and without abandoning the visual space assumption. We need only suppose, by analogy with our treatment of Weber's Law for length, that the correlation of vertical errors of localization is a bell-shaped function of horizontal (as well as vertical) distance. Now if this is the correct explanation for Tyler's result, we might expect that an adjacent, straight "landmark" or reference line would improve performance, particularly under conditions where the separation between the landmark and the sinusoid is much less than the period of the sinusoid. In this landmark condition, subjects should be able to assess the distances of the peak and trough of the test line to the reference line with greater precision when those distances are relatively small, and this cue should undercut the one based on comparisons made along the sine wave. However, in experiments to test this point (Willen & MacLeod, 1991), we were unable to demonstrate any benefit of the landmark or reference line, even when the distance to the reference was much smaller than the spatial period of the test sinusoid.

This failure to benefit from a nearby reference line in the detection of sinusoidal ripple is consistent with a model like Tyler's in which orientation-sensitive mechanisms process the test line (more or less independently of its context, almost as if landmark and test were in separate spaces) for signs of ripple. But it is difficult to reconcile with our alternative, in which ripple detection depends only upon the appropriate use of subjective locations in a unitary space that have been perturbed by spatially continuous error.

More generally, measures of such things as distance and orientation derived from each judged stimulus object (Baylis & Driver, 1993; Lennie, 1972; Watt, 1988) may be contaminated by errors that are not traceable to localization errors, but instead originate in the neural representation of the derived quantity (e.g., distance or orientation). Such errors need not show the proximity-dependent correlation required by the visual space assumption. Physiological observations also support the orientation model, inasmuch as visual cortex is packed with orientation-selective neurons. We next consider some other possible psychophysical consequences of that.

FRAGMENTATION: ZÖLLNER AND MÜLLER-LYER FIGURES

The current picture of visual processing suggested by anatomy and physiology is indeed the antithesis of the one favored by naive realism on the one hand and mathematical psychology on the other. Anatomy and physiology have revealed a plethora of specialized mechanisms for representing such specific spatial attributes as orientation, motion and spatial frequency. It would be surprising if those different representations were not each subject to their own systematic biases and their own sources of random error. But where in this confusingly complex scenario, with its independent and potentially inconsistent representations of different elements, is visual space? Is it possible that we construct a coherent, self-consistent spatial representation from all these potentially inconsistent signals, much in the way that Ullman's (1984) model for structure from motion takes the results of different local analyses and tries to fit them together?

We have examined the mutual consistency of perceived distance and orientation, in an experiment in some ways resembling the alley experiments which in Blumenfeld's (1913) hands originally provided the main basis for the Luneburg model. There a Riemannian geometry was invoked to explain what in Euclidean geometry would have been an inconsistency. But we use frontoparallel presentation with small fields, where Indow (1991) has found that the Euclidean description is valid with great precision. Our question is whether a geometrical illusion figure can create an inconsistency between distance and orientation judgments: first for the Zöllner figure and then the Müller-Lyer figure. The Zöllner figure is usually regarded as an orientation illusion, the Müller-Lyer as one of length. We wished to examine the mutual consistency of judgments that might depend on assessments of orientation or on length when these figures are viewed.

Our Zöllner stimulus (Figure 3.4) was composed of illuminated lines on a CRT screen. Subjects adjusted the vertical separation of the central pair of dots in this figure to match the separation of the upper or lower endpoints of the nearly vertical lines—the apparent orientation of which is normally distorted by the oriented crossing lines of the Zöllner figure. We compared the settings for the two ends of the illusory taper, to yield a measure of the illusion's effect on the apparent position of the ends of the lines. In a separate phase of the experiment, we also asked subjects to adjust the orientations of the vertical lines so that they were subjectively parallel. Thus, we have both a position-based measure of the extent of the illusion and an orientation-based measure of the extent of the illusion. These will be consistent only if lines judged subjectively parallel are also judged equidistant. Our data so far are somewhat equivocal. In the average over subjects, the positional measure and the orientation measure are quite consistent, a result that seems surprising if one assumes, as physiology invites one to do, that orientation is represented more or less independently of position in visual processing. There is, however, significant individual variation among subjects, with some subjects demonstrating statistically significant differences ($F(10,202) = 3.72, p < .0001$) between their parallelism and positional settings, different in direction for different subjects.

For the Müller-Lyer figure, we similarly asked subjects to set the vertical separation of a pair of dots equal to the horizontal length of the upper or lower

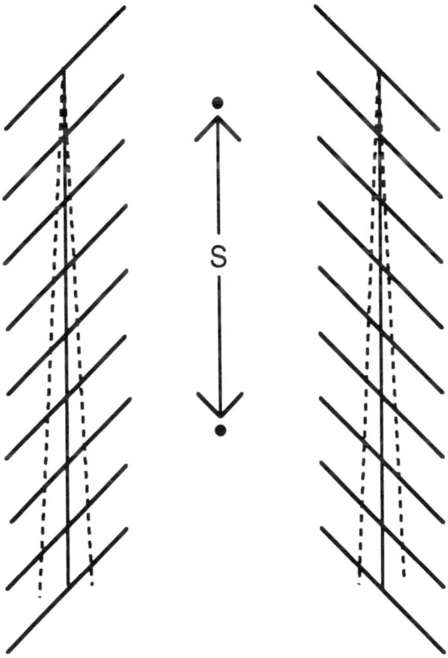

FIG. 3.4. Zöllner figure. Subjects were asked to adjust the separation between the vertically separated dots so that it would match the distance between the specified (either upper or lower) endpoints of the (nearly) vertical lines. The actual orientation of the vertical lines varied randomly within a small range (indicated by the dashed lines), but the relevant endpoints were always fixed to be the same distance apart.

lines in the Müller-Lyer stimulus presented as part of Figure 3.5. Subjects were also asked to adjust the length of the two lines to be subjectively equal to each other while viewing exactly the same stimulus. To make the orientation judgment more explicit, we then added lines connecting the endpoints of the horizontals and asked subjects to again make a comparative separation judgment, and again to set the horizontals to be of equal length.

In the control condition where no Müller-Lyer inducing lines are present, the results of the two types of judgment were fully consistent. Small but statistically significant ($F(1, 9) = 5.63$, $p < .05$) differences emerged, however, when the illusion was in force (Figure 3.5), with a tendency for conditions in which orientation was made salient to show less illusion than those where distance was assessed with a vertical reference dot pair, and/or where no parallels were provided in the stimulus. This supports the notion that the Müller-Lyer illusion is to some degree specific to distance judgment, and is not consistently reflected in perceived orientations within the figure.

In a further experiment with the Müller-Lyer figure (without physically present parallels) we asked subjects to make two different settings: the usual setting for equality of length, and a setting for parallelism of imaginary lines connecting the endpoints. The results uphold the statement made without supporting evidence in Suppes, Krantz, Luce, and Tversky (1989, p. 135) that the illusion is

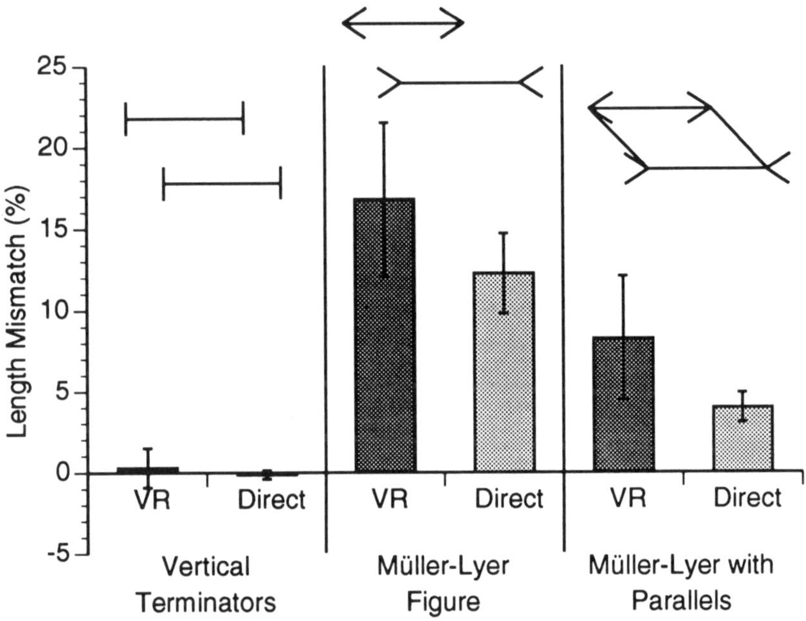

FIG. 3.5. Length mismatch in subjects' settings with the Müller-Lyer figure (displayed above the data for each condition) as measured by comparison of the indicated length to a vertical reference (VR) pair of dots, or as measured by directly adjusting (Direct) the length of the horizontals in the figure. The horizontal and vertical positions of the two lines was varied randomly from trial to trial within a limited range.

less apparent in the parallelism judgment. The length mismatch was 15% in the settings of the vertically separated reference dot pair, but only 11% in the parallelism settings. Thus, there was a difference of approximately 3–5 percentage points in length mismatch between the parallelism and equal length settings, averaged over subjects. Again as in the Zöllner case, individual subjects made parallelism and equal length settings that differed systematically in one direction or another ($F(9,760) = 6.46$, $p < .0001$).

Thus, in both the Zöllner and the Müller-Lyer cases, the results for individual subjects cannot be said to support the idea that apparent orientation is completely determined by apparent position (or vice versa). However, the overall approximate consistency of the orientation and position judgments could support an alternative scenario in which each subject's visual system tries (without complete success) to generate a unitary spatial representation that strikes a good single compromise between inconsistent reports about orientation and position.

So this evidence suggests either that there is no reconstruction of a unitary visual space, or else that an attempt to reconstruct one from independent data about orientation and position is made but without complete success. But there are cogent logical objections to any reconstruction-of-visual-space scenario. Why would we want to produce such a reconstruction? Useful processing has been done in the visual pathway to explicitly encode spatial frequency, motion, orientation, and who knows what other behaviorally useful information. To merely use this to reconstruct a spatial representation would be to cancel (doubtless with great difficulty) all the gains of all that preconscious perceptual processing and go back to the retinal image. We would then need the homunculus as much as if we did not have a visual cortex. Conversely, that nonexistent homunculus is the only guy who needs a reconstructed unitary visual space. It seems more likely that the fragmentation physiologically revealed is also psychologically real. Doubtless the results of different local computations have to interact, but their interaction need not, should not, and probably does not take the form of constructing a unified spatial representation.

CONCLUSIONS

We are left with a view in which special-purpose hardware underlies different spatial judgments. Motion detection by direction-selective cells need not be based on an internalized representation of position any more than a car's speedometer reading or a Doppler shift speedometer requires such a representation. Similarly, orientation mechanisms need not internalize positional measurements. A complex cell in primary visual cortex (or more strictly a collection of them) is probably as good an example as any of a device that represents orientation without retaining information about position.

Visual experience provides its own support for this counterintuitively fragmented conception of spatial processing. A few examples to add to the experimental results reviewed above: first, it is a familiar observation that the motion after-effect is subjective motion often unaccompanied by subjective translation, presumably because it originates in activity specific to a motion-signaling system (Gregory, 1990). Second, the spatial-frequency after-effect of Blakemore and Sutton (1969) and the similar simultaneous induction effect of MacKay (1973) are situations in which particular types of feature (e.g., grating stripes) undergo perceptual expansion or contraction without a geometrically consistent perceived expansion or contraction of the window in which they appear. Third, the impossible objects of Penrose and Penrose (1958), and the prints of Escher, in which first one and then another representation is constructed from local and fragmentary information, or the Frazer spiral, which is really a circle but can never look like one, indicate that the phenomenal integrity of visual space is itself an illusion. The sense of paradox in these cases arises because we think, at any moment, that we have a complete and consistent spatial representation of what is out there, when in fact we have no such thing, but are incorrectly inferring the whole from fragmentary information.

The mother of all illusions is the illusion of objectivity. A part of that may be the illusion of spatiality: the notion of a perceived visual space, natural though it is, may not capture important realities of visual space perception. Visual space perception starts with a space, but it probably does not end with a space.

REFERENCES

Andrews, D. P. (1967). Perception of contour orientation in the central fovea part: 1. Short lines. *Vision Research, 7,* 975–997.

Baylis, G. C., & Driver, J. (1993). Visual attention and objects: Evidence for hierarchical coding of location. *Journal of Experimental Psychology: Human Perception and Performance, 19*(3), 451–470.

Blakemore, C., & Sutton, P. (1969). Size adaptation: A new aftereffect. *Science, 166,* 3902.

Blumenfeld, W. (1913). Untersuchungen über die scheinbare Grösse in Sehräume. *Zeitschrift für Psychologie, 65,* 241–404.

Cattell, J. M. (1893). On errors of observation. *The American Journal of Psychology, 5*(3), 285–293.

Gregory, R. L. (1990). *Eye and brain* (4th ed.). Princeton, NJ: Princeton University Press.

Indow, T. (1991). A critical review of Luneburg's model with regard to global structure of visual space. *Psychological Review, 98*(3).

Indow, T., & Watanabe, T. (1988). Alleys on an extensive apparent frontoparallel plane: A second experiment. *Perception, 17,* 647–666.

Kaas, J. H. (1987). *The organization and evolution of neocortex.* New York: Wiley.

Kluver, H. (1966). *Mescal, and mechanisms of hallucinations.* Chicago: University of Chicago Press.

Kohonen, T. (1989). *Self-organization and associative memory* (3rd ed.). New York: Springer-Verlag.

Lennie, P. (1972). *Mechanisms underlying the perception or orientation.* Unpublished doctoral dissertation, University of Cambridge, Cambridge.

Levi, D. M., & Klein, S. A. (1990). The role of separation and eccentricity in encoding position. *Vision Research, 30*(4), 557–585.

Levi, D. M., Klein, S. A., & Yap, Y. L. (1988). "Weber's law" for position: Unconfounding the role of separation and eccentricity. *Vision Research, 28*(5), 597–603.

MacKay, D. M. (1973). Lateral interaction between neural channels sensitive to texture density? *Nature, 245,* 5421.

Morgan, M. J., & Watt, R. J. (1989). The Weber relation for position is not an artefact of eccentricity. *Vision Research, 29*(10), 1457–1462.

Ogle, K. N. (1962). The Optical Space Sense. In H. Davson (Ed.), *The eye: Visual optics and the optical space sense.* New York: Academic Press.

Penrose, L. S., & Penrose, R. (1958). Impossible objects: A special type of visual illusion. *British Journal of Psychology, 49,* 31–33.

Schwartz, E. L. (1980). Computational anatomy and functional architecture of striate cortex: A spatial mapping approach to perceptual coding. *Vision Research, 20,* 645–669.

Suppes, P., Krantz, D. H., Luce, R. D., & Tversky, A. (1989). *Foundations of Measurement (Vol. II).* San Diego, CA: Academic Press.

Tyler, C. W. (1973). Periodic vernier acuity. *Journal of Physiology (London), 228*(3), 637–647.

Ullman, S. (1984). Maximizing rigidity: The incremental recovery of 3-D structure from rigid motion. *Perception, 13*(3).

Watt, R. J. (1988). *Visual processing: Computational, psychophysical, and cognitive research.* Hillsdale, NJ: Lawrence Erlbaum.

Willen, J. D., & MacLeod, D. I. A. (1991). *Sinusoid discrimination with a landmark.* Paper presented at the annual meeting of the Association for Research in Vision and Ophthalmology, Sarasota, FL [Abstract] *Invest. Ophthalmology & Visual Science Supplement, 32*(4), 1024.

4 Representation of Rigid Stimulus Transformations by Cortical Activity Patterns

V. Lakshminarayanan
School of Optometry and Department of Physics and Astronomy, University of Missouri—St. Louis

T. S. Santhanam
Department of Science and Mathematics, Parks College of St. Louis University

ABSTRACT

It has been shown that the group representation linking rotation of an object in space and the pattern of cortical, neural activity as proposed by Carlton (1988) is biologically restrictive. We have used the method of group extension to extend the Euclidean group of rotations and translations to the Lorentz group, and the resultant representation answers the major criticism of Carlton's hypothesis.

INTRODUCTION

The mathematical construct of a group acting on a function space has been long exploited in various aspects of modern physics. Lie algebras, in particular, have proven to be a very useful mathematical formalism (for optical applications, see Sanchez-Mondragon & Wolf, 1986). There have been a few applications of such group-theoretical methods in perception/visual science. Consider the fact that, although our optical retinal image of the world is in continuous flux and with nonuniform projections, our perception of the real world is one that is stable and contiguous. The reason is that we can extract properties of the image that are invariant under these transformations. These invariant properties such as angle, size, color, distance, etc., are called constancies in the perception literature. This implies that, in perceiving, the visual system preserves certain affine properties of the retinal image (e.g., DeValois, Lakshminarayanan, Nygaard, Schlussel, & Sladky, 1990; Lakshminarayanan, Parthasarathy, & DeValois, 1991). Some researchers have studied the invariant properties of some aspects of spatial vision using certain well-defined transformation. Hoffman (1966, 1978) in particular, has formulated the Lie transformation group to provide a metalanguage for perceptual processes (see also Dodwell, 1970). There is some experimental evidence for the application of this formalism to pattern or contour extraction (Caelli 1976, 1977). Foster (1973, 1978) has shown how the affine transformation groups are representative of permissible apparent paths in motion perception. The Lie group approach can be used in the invariance coding problem for both the amplitude and phase encoding of information (Ferrao and Caelli, 1988).

The study of brain processing of mental images is a very active area of research in both cognitive psychology and neuroscience (for a good review, see Kosslyn, 1988). In this chapter we will confine our attention to a particular hypothesis dealing with the linking of perceived mental rotations of a three-dimensional object (as exemplified in the pioneering work of Shepard & Metzler, 1971; Shepard & Cooper, 1982) with the sequence of internal activities in the visual cortex (brain state-space). This hypothesis was first put forth by Carlton (1988). Carlton's formalism links a path in geometrical space to a path in brain state-space. This mathematical construct bridges complex mental percepts with neurophysiological states which presumably can be measured using SQUID (superconducting quantum interference devices; Williamson, Kaufman, & Brenner, 1979), neuromagnetic imaging (Aine, George, Supek, & Maclin, 1991; Marg, 1991) or other electrophysiological (e.g., neurocognitive pattern analysis; Gevins et al., 1987) technologies.

This link between real and brain state-space is basically a linear representation of the Euclidean group (three dimensional) E^+ on to an appropriate functional space of square-integrable functions $L^2(R^4)$ in the visual cortex. Soto-Andrade and Varela (1991) have pointed out certain valid criticisms of this hypothesis and discuss the fact that the construction is biologically restrictive and cannot be completed in the desired way (vide infra). However, the hypothesis is elegant and mathematically sophisticated, linking group actions to a harmonic analysis of a functional space in the spirit of Mackey (1980). In this chapter we extend the group representation and analysis, which answers the major criticism of the model. In the rest of this chapter we assume the reader's familiarity with the work of Carlton (1988).

CARLTON'S HYPOTHESIS

Shepard and co-workers have extensively studied the phenomena of apparent motion and mental rotation of rigid bodies in three-dimensional Euclidean space. In particular, a hypothesis has been advanced (Cooper, 1975; Cooper and Shepard, 1985; Robins and Shepard, 1977; Shepard, 1984; Shepard and Cooper, 1982) that mental rotation has "analog" properties in that "intermediate stages of the internal process represent intermediate positions of the transformed object"

(Shepard, 1984). In other words, ongoing mental transformation has a one-to-one correspondence with external rotation of an object. It is appropriate to quote Shepard "although the experienced motion (in the case of apparent motion) is perceptual and involuntary, the paths of experienced transformation evidently are the same as those traversed in voluntary imagined transformations suggesting that the same representational system is operative in both cases." As an aside it should be pointed out that both propositional processing (Pylyshyn, 1981) and nonpropositional cognition (Weiskrantz, 1988) have been advanced as forming the basis for mental transformations. However, recent studies argue for the existence of analog transformations (e.g., Farah, 1989; Georgopoulos, Lurito, Petrides, Schwartz, & Massey, 1989; Lurito, Georgakopoulos, & Georgopoulos, 1991).

The functional space R^4 is based on descriptions of visual cortical receptive fields (Daugman, 1980, 1985; DeValois, Albrecht, & Thorell, 1978; Jones and Palmer, 1987; Marcelja, 1980). In particular, it has been shown that Gabor functions (Gabor, 1946; Helstrom, 1966) can fit receptive field data measured in both spatial (retinal position) and spatial frequency domains (e.g., Kulikowski, Marcelja, & Bishop, 1982; Marcelja, 1980; Watson, 1983). This convenient mathematical choice allows for representation as square-integrable functions of four real variables corresponding to two spatial positions ($p1$, $p2$) and two preferred spatial frequencies ($u1$, $u2$) of the receptive field (Pribram and Carlton, 1986).

Using this as a basis, Carlton (1988) has characterized mental rotation as geodesic paths in the Euclidean motion group of translations and rotations of a rigid body. The reader is referred to the excellent discussion of groups and representations given in the Appendix of Carlton (1988; see also Boerner, 1963; Caelli, 1981). The Euclidean group E^+ is written as a semidirect product of its translation R^3 and rotation subgroups SO(3). The Euclidean group has the property of leaving the space invariant under translations and rotations and is the relevant group for rigid-body motion. That is,

$$E^+ = R^3 \text{ s } \mathrm{SO}(3). \tag{4.1}$$

The rotation subgroup SO(3) has a canonical representation via the group of unit quarternions S^3 which is isomorphic to the group of 2×2 complex unitary matrices SU(2) as a group of unitary operators on the Hilbert space $L^2(R^4)$, with R^4 being a quarternion (Hamilton, 1853) representation in a spinor space C^2. The action of E^+ on $L^2(R^4)$ is the connection between perceived motion and cortical activity. s represents the semi-direct product operation.

The cortical activity at any time t is an element of $L^2(R^4)$. The Euclidean group E^+ acts on $L^2(R^4)$, the cortical activity representation function. The motion percept (an entire path in E^+) is nothing more perceptually than the action of successive E^+ elements on the cortical activity function representing the initial activity pattern and would correspond to the flow of cortical activity.

The action of E^+ on the four-dimensional space R^4 is simply

$$(a, \mathbf{b})(\mathbf{x}) = axa^{-1} + \mathbf{b}, \tag{4.2}$$

with $a \in s^3$, $\mathbf{b} \in R^3$, and $\mathbf{x} = x_0e_0 + \mathbf{x}\cdot\mathbf{e}.$ Using the well-known relation between the group SU(2) (complex unitary unimodular) and SO(3) (real rotations in three dimensions), the above construction yields a unitary representation of the Euclidean group onto $L^2(R^4)$, the "phase space" of the visual cortex.

CRITICISM OF THE MODEL

In an elegant short paper, Soto-Andrade and Varela (1991) discuss Carlton's model in detail and show that the representation of Carlton (an extension of Bargmann's representation [1962] for the rotation subgroup), although correct, does so in a biologically questionable way and is inadequate. They point out that a linear reducible representation is still wanted. A representation is said to be reducible if the transformations leave a subspace invariant (or unchanged). For instance, in this case, the four-dimensional space has a one-dimensional subspace which is unchanged as shown above. An irreducible representation, on the other hand, by definition does not leave any subspace invariant and will connect all components of the space. In the present context, an irreducible representation is more relevant, though a reducible representation is still needed.

The construction as given by Carlton can only yield a reducible representation, leaving the x_0 component unchanged as shown above. In other words, it is only the $\mathbf{x}\cdot\mathbf{e}$ of $\mathbf{x}$ that is acted upon by E^+, which leaves x_0e_0 totally unaffected. This again clear from the fact that the group E^+ is noncompact (due to the translation group R^3) and any finite-dimensional representation of E^+ (like the natural representation of dimension 4 acting on R^4) has to be necessarily reducible (in this case $4 = 1 + 3$) on simple trivial lifts from three dimensions.

AN ALTERNATE REPRESENTATION

The notion of group extensions can be used to imbed a given space in a "larger" space that contains the former. The larger space is constructed by the method of extension of the representation of the smaller space (Boerner, 1963). The Lorentz group $O(3, 1)$, which gained importance in the theory of special relativity, is a group which leaves a four-dimensional "length," $l^2 = x_1^2 + x_2^2 + x_3^2 - x_4^2$ unchanged under a rotation in a four-dimensional space (Naimark, 1964). In visual perception, Hoffman (1966), for example, has used the idea of Lorentz group to discuss motion invariance. It is possible to use the ideas of group extension to extend $E^+ = \mathrm{SO}(3)$ s R^3 to the Lorentz group $O(3, 1)$. The idea is as follows (Sankaranarayanan, 1968; Wigner, 1939): Let the three generators, whose exponents give the elements of the group of SO(3), be denoted by **J** and the three generators of the translation group by **P.** Their Lie algebra commutation relationship is given by

$$[J_i, J_j] = i\epsilon_{i,j,k}J_k,$$
$$[J_i, p_j] = i\epsilon_{i,j,k}p_k,$$
$$[p_i, p_j] = 0. \tag{4.3}$$

Here, $\epsilon_{i,j,k} = 0$ for any two (or more) of the indices being equal and has a value of ± 1, depending on the permutations of i, j, k. Then, as is obvious from the commutation relationships, the p_i form an Abelian invariant subgroup (since p_i and p_j commute) and the J_i form the rotation subgroup. Since the J_i close by themselves, they form a subgroup.

We can construct, using the extension theorem, the new generators (in vector notation):

$$\mathbf{K} = (1/2\mu)(\mathbf{P} \times \mathbf{J} - \mathbf{J} \times \mathbf{P}), \tag{4.4}$$

where μ^2 is the eigenvalue of what is known as the Casimir operator $\mathbf{P}\cdot\mathbf{P}$ of E^+. A Casimir operator is a function of the generators that commutes with all the generators. In the space on which the generators act, the Casimir operator has to be a multiple of identity. For instance, in the case of the rotation group SO(3), the Casimir operator is $J_1^2 + J_2^2 + J_3^2$, whose eigenvalue (J) labels a representation and the generators do not change this eigenvalue. In the case of the Lorentz group there are two Casimir operators whose eigenvalues label a representation.

The crucial point to note is that the **J** and **K** so constructed to form the Lie algebra of the Lorentz group $O(3, 1)$ satisfy the commutation relations

$$\mathbf{J} \times \mathbf{J} = i\mathbf{J},$$
$$\mathbf{K} \times \mathbf{K} = -i\mathbf{J},$$
$$\mathbf{J} \times \mathbf{K} + \mathbf{K} \times \mathbf{J} = 2i\mathbf{K}. \tag{4.5}$$

J and **K** thus generate an algebra isomorphic to the Lie algebra of $O(3, 1)$. The construction of **K** is representation dependent (since it involves the Casimir operator specifying a definite basis). Indeed, it yields the infinite-dimensional Majorana representation, which is well known in particle physics. A representation of the Lorentz group is usually specified by J, M, which are the eigenvalues of $\mathbf{J}^2$ and J_3 [expressed in terms of its subgroup SO(3)]. The generators **K** in general, will take a representation labeled by J to $(J + 1, J, J - 1)$. In the Majorana representation it takes it only to the eigenvalues $(J + 1, J - 1)$. This was studied by Majorana in connection with a relativistic wave equation which bears his name (Majorana, 1932). For these, the second Casimir operator of $O(3, 1)$, which is $\mathbf{J}\cdot\mathbf{K}$, is identically zero.

In this chapter we have shown that the Euclidean group of rotations and translations can be extended to the Lorentz group. This idea can be carried out by the same procedure as used by Carlton (1988), that is, by utilizing the Lie group germ. The result maintains the homomorphism properties that have been postulated. We wish to remark that, although the representation obtained using this procedure is infinite dimensional, it is still unitary and irreducible, which means that no subspace remains unchanged by the action. Thus, this powerful result answers the major criticism of Soto-Andrade and Varela. In future work, we expect to build on this foundation to derive a finite-dimensional representation [for example, a $O(2, 2)$ type] of this perceptual model. It would also be necessary to extend the analysis to the study of E^+ in function spaces of higher functional dimensions, as well as other mathematical formulations of the receptive fields (e.g., Bessel functions, difference of Gaussians, Gaussian derivatives, etc.). The role of the Lorentz group in the context of theoretical physics is well known. In the present context, it remains open to discussion which restrictions are introduced by the group for the presentation of visual stimuli or for the neurophysiological measurements in the visual cortex. Theoretical formulations such as this may provide a bridge connecting different areas of experimental research. These group-theoretical analyses may lead to prediction of new measurable effects. These aspects will be explored in future research.

REFERENCES

Aine, C. J., George, J. S., Supek, S., & Maclin, E. L. (1991). Noninvasive studies of human visual cortex using neuromagnetic techniques. In *Technical Digest on Noninvasive Assessment of the Visual System, 1991* (Vol. 1, pp. 162–165). Washington, DC: Optical Society of America.

Bargman, V. (1962). On the representations of the rotation group. *Reviews in Modern Physics, 34,* 829–845.

Boerner, H. (1963). *Representations of Groups.* Amsterdam: North-Holland.

Caelli, T. (1976). The prediction of interactions between visual forms by products of Lie operations. *Mathematical Biosciences, 30,* 191–205.

Caelli, T. (1977). Psychophysical interpretations and experimental evidence for the Hoffman LTG/NP theory of perception. *Cashiers de Psychologie, 20,* 107–134.

Caelli, T. (1981). *Visual perception theory and practice.* Oxford, UK: Pergamon Press.

Carlton, E. H. (1988). Connection between internal representation of rigid transformation and cortical activity pattern. *Biological Cybernetics, 59,* 419–429.

Cooper, L. A. (1975). Mental rotation of random two-dimensional shapes. *Cognitive Psychology, 7,* 20–43.

Cooper, L. A., & Shepard, R. N. (1984). Turning something over in the mind. *Scientific American, 251,* 106–114.

Daugman, J. G. (1980). Two-dimensional spectral analysis of vertical receptive field profiles. *Vision Research, 20,* 847–856.

Daugman, J. G. (1985). Uncertainty relation for resolution in space, spatial frequency and orientation optimized by two-dimensional visual vertical filters. *Journal of the Optical Society of America A, 2,* 1160–1169.

De Valois, R. L., Albrecht, D. G., & Thorell, L. G. (1978). Cortical cells: Bar and edge detectors or spatial frequency filters? In S. J. Cool & E. L. Smith (Eds.), *Frontiers in visual science* (pp. 544–556). Berlin: Springer-Verlag.

De Valois, R. L., & De Valois, K. K. (1985). *Spatial vision.* New York: Oxford University Press.

De Valois, K. K., Lakshminarayanan, V., Nygaard, R. W., Schlussel, S., & Sladky, J. (1990). Discrimination of relative spatial position. *Vision Research, 30,* 1649–1660.

Dodwell, P. (1970). *Visual pattern recognition.* New York: Holt, Rinehart & Winston.

Farah, M. J. (1989). The neural basis of mental imagery. *Trends in Neuroscience, 12,* 395–399.

Ferrao, M., & Caelli, T. M. (1988). Relationship between integral transform invariances and Lie group theory. *Journal of the Optical Society of America A, 5,* 738–742.

Foster, D. H. (1973). An experimental examination of a hypothesis connecting visual pattern recognition and apparent motion. *Kybernetick, 14,* 63–70.

Foster, D. H. (1978). Visual apparent motion and the calculus of variations. In E. Leeuwenberg & H. Buffart (Eds.), *Formal theories of visual perception* (pp. 231–246). New York: Wiley.

Gabor, D. (1946). Theory of communicaton. *Journal of the IEE (London), 93,* 429–457.

Georgopoulos, A. P., Lurito, J. T., Petrides, M., Schwartz, A. B., & Massey, J. T. (1989). Mental rotation of the neuronal population vector. *Science, 243,* 234–236.

Gevins, A. S., Morgan, N. H., Bressler, S. L., Cutillo, B. A., White, R. M., Illes, J., Greer, D. S., Doyle, J. C., & Zeitlin, G. M. (1987). Human neuroelectric patterns predict performance accuracy. *Science, 235,* 580–585.

Hamilton, W. R. (1853). *Lectures on quarternions.* Dublin: Hodges and Smith.

Helstrom, C. W. (1966). An expansion of a signal in Gaussian elementary signals. *IEEE Transactions, IT-12,* 81–82.

Hoffman, W. C. (1966). The Lie algebra of visual perception. *Journal of Mathematical Psychology, 3,* 65–98.

Hoffman, W. C. (1978). The Lie transformation group approach to visual neuropsychology. In E. Leeuwenberg and H. Buffart (Eds.), *Formal theories of visual perception* (pp. 27–66). New York: Wiley.

Jones, J. P., & Palmer, L. A. (1987). An evaluation of the two-dimensional Gabor filter model of simple receptive fields in the cat striate cortex. *Journal of Neurophysiology, 58,* 1233–1258.

Kosslyn, S. M. (1988). Aspects of a cognitive neuroscience of mental imagery. *Science, 240,* 1621–1626.

Kulikowski, J. J., Marcelja, S., & Bishop, P. O. (1982). Theory of spatial position and spatial frequency relations in the receptive fields of simple cells in the visual cortex. *Biological Cybernetics, 43,* 187–198.

Lakshminarayanan, V., Parthasarathy, R., & De Valois, K. K. (1991). A generalized perceptual space. *Proceedings of the National Academy of Sciences (U.S.A.)* (submitted).

Lurito, J. T., Georgakopoulos, T., & Georgopoulos, A. P. (1991). Cognitive spatial motor processes: 7. The making of movements at an angle from a stimulus direction: Studies of motor cortical activity at the single cell and population levels. *Experimental Brain Research, 87,* 562–580.

Mackey, G. W. (1980). Harmonic analysis as the exploitation of symmetry: A historical survey. *Bulletin of the American Mathematical Society, 3,* 543–698.

Majorana, E. (1932). Relativistic theory of particles with arbitrary intrinsic momentum. *Nuovo Cimento, 9,* 335–344.

Marcelja, S. (1980). Mathematical descrition of the responses of simple cortical cells. *Journal of the Optical Society of America, 70,* 1297–1300.

Marg, E. (1991). Magnetostimulation of vision: Direct noninvasive stimulation of the retina and the visual brain. *Optometry and Vision Science, 68,* 427–440.

Naimark, M. (1964). *Linear representations of the Lorentz group.* London: Pergamon Press.

Pribram, K. H., & Carlton, E. H. (1986). Holonomic brain theory in imaging and object perception. *Acta Psychology, 63,* 175–210.

Pylyshyn, Z. W. (1981). The imagery debate: analogue media versus tacit knowledge. *Psychological Review, 88,* 16–45.

Robins, C., & Shepard, R. N. (1977). Spatio-temporal probing of apparent rotational movement. *Perception & Psychophysics, 22,* 12–18.

Sanchez-Mondragon, J., & Wolf, K. B. (1986). *Lie methods in optics.* Berlin: Springer-Verlag.

Sankaranarayanan, A. (1968). Extension theorem and representations. *Journal of Mathematical Physics, 9,* 611–620.

Shepard, R. N. (1984). Ecological constraints on internal representation: resonant kinematics of perceiving, imagining, thinking and dreaming. *Psychological Review, 91,* 417–447.

Shepard, R. N., & Cooper, L. A. (1982). *Mental images and their transformations.* Cambridge, MA: MIT Press/Bradford Books.

Shepard, R. N., & Metzler, J. (1971). Mental rotation of three dimensional objects. *Science, 171,* 701–703.

Soto-Andrade, J., & Varela, F. J. (1991). On mental rotations and cortical activity patterns. *Biological Cybernetics, 64,* 221–223.

Watson, A. B. (1983). *Detection and recognition of simple spatial forms* (NASA Tech Memo #84353). Moffet Field, CA: Ames Research Center.

Weiskrantz, L. (Ed.) (1988). *Thought without language.* Oxford: Oxford University Press.

Wigner, E. P. (1939). On unitary representations of the inhomogeneous Lorentz group. *Annals of Mathematics, 40,* 149–204.

Williamson, S. J., Kaufman, L., & Brenner, D. (1979). Evoked neuromagnetic fields of the human brain. *Journal of Applied Physics, 50,* 2418–2421.

5 The Invariances of Weber's and Other Laws as Determinants of Psychophysical Structures

Jan Drösler
Universität Regensburg

ABSTRACT

The study shows that Weber's ratio $\Delta S/S$ = const. for the just-noticeable difference (i) constitutes a projective geometric invariant and (ii) can be interpreted as a representation of an empirical bisection operation. These conditions permit the extension of the theory of projective ordinal scaling (Suppes, Krantz, Luce, Tversky, 1989) to a projective interval scaling. The invariance of Weber's ratio (iii) is traced in its consequences for perceptual spaces of higher dimensions. There, the invariants of projective transformations are a line, a plane or, respectively, a hyperplane as compared with a point in the unidimensional case. Besides projective invariance as a consequence of Weber's law, additional projective invariants of perceptual spaces are generated by various natural laws. With monocular vision in perspective, such invariants are the phenomenal line of the horizon and constancy of area. They imply a projective unimodular structure of the visual horizontal plane. In the two-dimensional case of the binocular visual horizontal plane, invariance of the greatest circle around the observer—corresponding to threshold disparity—generates a projective metric hyperbolic structure. This result generalizes Luneburg's (1947) theory of a hyperbolic structure. In the three-dimensional perceptual space of colors, invariance of the extremal of the convex hull of colors—the surface of the color cone—with respect to color adaptation, results in a new formula for color difference. It is related to the "center of gravity rule." The three-dimensional generalization of the invariance of Weber's ratio, different from Helmholtz (1891), leads to a projective metric hyperbolic structure of color space. Analysis of the set of qualitative empirical assumptions sufficient for the derivation of this result yields a representation of color vision which is independent of the classical Grassmann approach. The present study tries to detect the causes for the occurrence of markedly different structures in perceptual spaces. The question whether for instance binocular visual space is of Euclidean, of hyperbolic, or of parabolic structure has been vigorously discussed in the past. Possible causes, however, were not stated. The present study shows that certain well-known psychophysical invariants endow perceptual spaces with structure, each in its own specific way.

WEBER'S LAW

Weber's Ratio

An early psychophysical invariant is *Weber's ratio* as described by Weber (1834). His paper reports, in Latin, a large number of psychophysical experiments from different sense modalities. The results are subsumed by the expression

$$\Delta S/S = \text{const.}$$

for the just-noticeable difference.

Here S is the stimulus intensity which is presented to the subject, and ΔS is that difference in stimulus intensity which is "just noticeably different" for the subject. This difference was defined statistically by Weber. "Pure chance" judgments are given if the subject is unable to discriminate and responds correctly with probabilities of one over the number of alternatives, i.e. .5 with two choices. A "just-noticeable difference" was said to be prevalent as soon as a subject reports the direction of stimulus change correctly in 75% of the trials. This setting of any particular discrimination probability later was characterized as arbitrary. Luce and Galanter (1963), therefore, set $\Delta S = \text{jnd}(\pi)$ and generalized Weber's ratio as *Weber's function*

$$\Delta S/S = c(\pi) \tag{5.1}$$

for the just-noticeable difference of argument π. Here π stands for the probability used for the current definition of the difference threshold. It can take any value $(0 < \pi < 1)$. The generalized Weber ratio has several advantages. It can be applied, for instance, to the sense modality of pitch measured in half-tone steps, even if it is not the case that half-tones can be discriminated only "just noticeably" in the traditional sense. This example is distinguished by historical precedence because the tempered scale was propagated by J. S. Bach (1685–1750) (Figure 5.1).

One consequence of Weber's invariant consists in a necessary *geometric* progression of stimulus intensity in order to elicit a *linear* progression in sensation.

A stepwise development of Equation 5.1 is given in Table 5.1. It starts with a stimulus intensity S_0 and a corresponding sensation R_0. For the kth "just-noticeable difference", $k \in \mathbb{N}$, starting from an absolute threshold R_0 which corresponds to $k = 0$, renders the geometric dependency

$$S_k = [1 + c(\pi)]^k\, S_0.$$

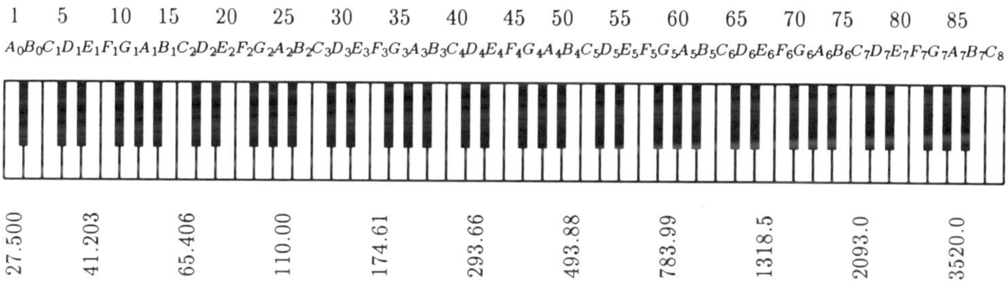

FIG. 5.1. The tempered scale of pitch in its dependence on frequency fulfills the invariance of Weber's law because a geometric increase in the stimulus variable leads to an arithmetic increase of the response.

Because of the dependencies of $c(\pi)$ and letting $k \in \mathbb{R}$, the inverse function

$$k = \frac{1}{\log[1 + c(\pi)]} \log \frac{S_k}{S_0}$$

gives Fechner's law of 1860.

Problem

Studying Weber's invariant today creates a feeling of dissonance regarding psychophysical theory. On one hand, Weber's ratio is the earliest psychophysical invariant in history. On the other hand, contemporary foundations of measurement in psychophysics do not refer to it at all. The present study tries to alleviate this remarkable discrepancy by demonstrating the foundational importance of Weber's contribution to psychophysics.

The unsatisfactory integration of Weber's results into contemporary psychophysical theory becomes apparent, particularly, if one proceeds from the unidimensional case of sensation to multidimensional perceptual spaces. No case is known in which a successful generalization of Weber's ratio to higher dimensions

TABLE 5.1
Dependence of Response Intensity on Stimulus Intensity According to Weber's Law

Stimulus		*Response*
S_0	$\mapsto$	R_0
$S_1 = (1 + c)\, S_0$	$\mapsto$	$R_1 = R_0 + 1$
$S_2 = (1 + c)^2\, S_0$	$\mapsto$	$R_2 = R_0 + 2$
$\vdots$	$\vdots$	$\vdots$
$S_k = (1 + c)^k\, S_0$	$\mapsto$	$R_k = R_0 + k$

has been accomplished. This is so, even though Helmholtz (1891) tried to do so for color space. His formula for the just-noticeable difference in color,

$$\Delta S = \sqrt{\left(\frac{\Delta R}{R}\right)^2 + \left(\frac{\Delta G}{G}\right)^2 + \left(\frac{\Delta B}{B}\right)^2}, \tag{5.2}$$

appears to be a three-dimensional generalization of Weber's ratio for the dimensions of three primary colors R, G, B (red, green, blue). This particular generalization of Weber's invariant to a higher dimensionality in the case of color space led to empirical consequences which could not be validated. For this reason, its theoretical analysis never was carried out very thoroughly. Even superficial study of Equation 5.2 shows that the line element keeps some Euclidean properties. They are manifest in employing summation of quadratic component elements. The elements themselves, however, are non-Euclidean. Thus, Helmholtz postulated a singular geometry for a color space. Its properties are unknown because geometry up to the present never saw empirical reasons for studying that particular combination of Euclidean and non-Euclidean ingredients.

Synthetic and Analytic Geometry

The present study sets out from the particular relation of *arithmetic* progression in sensation to *geometric* progression in stimulus intensity which Weber's ratio demands. It can be shown that the attribute of "geometric" in the present context is indicative of a certain geometry, which is *projective geometry.*

To characterize the essence of a projective invariant, some clarification of projective geometry is called for. Since Weber's ratio concerns unidimensional sensation, it is sufficient, at present, to refer to unidimensional projective geometry. Geometry can be studied in a "synthetic" way, i.e., with reference to points, lines, planes, etc., which is a qualitative way of study. "Analytic" geometry, however, maps the geometric entities into numbers, in the higher-dimensional case into tuples of numbers. Geometric concepts, then, are represented by relations between numbers or tuples of numbers respectively. The transitions from synthetic to analytic geometry is called coordinatization. The opposition of synthetic "elementary geometry," on one hand, and analytical, numerically represented geometry on the other hand, to a certain degree, corresponds to the duality of qualitative (empirical) statements of observers in the laboratory, on one hand, and the quantitative scaling of sensory intensity on the other. In a coordinatization of synthetic geometry, the restrictions of the inverse image are introduced by conventions such as using only compasses and ruler. In psychological scaling, the restrictions enter by the limited faculty of judgment of the observer. Subjects are able, generally, to produce only binary qualitative judgments in a reproducible manner.

Earlier Work

The coordinatization of unidimensional geometry is well known (cf. Redei, 1968). In the monograph of Suppes and others (1989) it is described in a particularly useful way for the present purposes. Since the development of the present study is based on that account, it will be presented here somewhat in detail. The order of real numbers introduces necessary assumptions for coordinatization. Therefore, Suppes et al. (1989) with reference to Crampe (1958) first introduce a qualitative ordering on the line which is invariant with respect to projection. It is based on the quaternary relation of the mutual separation of pairs of points.

If four points $P_0P_1P_2P_3$ are incident with a straight line like

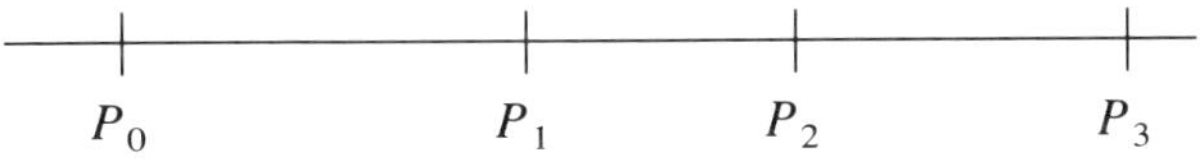

then the pair P_0P_2 and the pair P_1P_3 *separate* each other. The definition of a projective order is based on this quaternary relation, a binary relation on pairs of points (Crampe, 1958; Priess-Crampe, 1983; Suppes et al., 1989).

For points P_0, P_1, P_2, P_3 on a line their cross-ratio is defined as

$$\frac{\overline{P_0P_1}}{\overline{P_0P_3}} \div \frac{\overline{P_2P_1}}{\overline{P_2P_3}} \in \mathbb{R}$$

If $\overline{P_iP_j}$ are the differences in scale values of points P_i and P_j.

Definition 5.1. Let A be a nonempty set and S a quarternary relation on A. Then $\mathfrak{U} = \langle A, S \rangle$ is a *one-dimensional separation structure* if and only if the following six axioms are satisfied for every a, b, c, d, and e in A:

1. It is not the case: $ab\ S\ ac$ or $ab\ S\ cc$.
2. If $ab\ S\ cd$, then $cd\ S\ ab$.
3. If $ab\ S\ cd$, then $ab\ S\ dc$.
4. If $ab\ S\ cd$, then it is not the case: $ac\ S\ bd$.
5. If $ab\ S\ cd$ and $ae\ S\ db$, then $ab\ S\ ce$.
6. If a, b, c and d are distinct, then $ab\ S\ cd$ or $bc\ S\ ad$ or $ca\ S\ bd$.

$\mathfrak{U}$ is *separable* iff there exists a finite or countable order-dense subset of A, i.e., there is a finite or countable subset B or A such that for all a and b in A with $a \neq b$ there exist c and d in B with $ab\ S\ cd$.

Representation Theorem 5.1. (Suppes et al., 1989). Let $\mathfrak{U} = \langle A, S \rangle$ be a structure such that A is a nonempty set and S a quaternary relation on A.

Then $\mathfrak{U}$ is a separable one-dimensional separation structure iff there exists a function φ, $\varphi: A \rightarrow \mathbb{R} \cup \{\infty\}$, such that for all a, b, c, and d in A,
ab S cd iff the cross-ratio of a, b, c, d with respect to φ is negative.

Definition 5.2. Let $R \subseteq \mathbb{R} \cup \{\infty\}$. A function $h : R \rightarrow \mathbb{R} \cup \{\infty\}$ is called *projectively monotone* iff h is strictly monotone or there exists a partition $(\underline{R}, \bar{R})$ of R such that every element of $\underline{R}$ is less than every element of $\bar{R}$ and either

(i) h is strictly increasing on $\underline{R}$ and on $\bar{R}$, and for every x in $\underline{R}$ and every x' in $\bar{R}$, $h(x') < h(x)$, or
(ii) h is strictly decreasing on $\underline{R}$ and on $\bar{R}$, and for every x in $\underline{R}$ and every x' in $\bar{R}$, $h(x) < h(x')$.

Uniqueness Theorem 5.2 (Suppes et al., 1989). Let $\mathfrak{U} = \langle A, S\rangle$ be a separable one-dimensional separation structure, and let φ and φ' be two representing functions satisfying Theorem 5.1. Let R and R' be the ranges of φ and φ' respectively. Then there exists a projectively monotone function $h : R \rightarrow R'$ such that for all $a \in A$, $h[\varphi(a)] = \varphi'(a)$. Moreover, if h is any projectively monotone function defined on R, then $h \circ \varphi$ is another numerical representation of $\mathfrak{U}$.

The coordinatization given by Suppes and others (1989) amounts to the following: If observers can judge the relation of separation for pairs of points on a straight line, then a psychological scaling (e.g., seen position of points on the line) can be achieved. This scaling is invariant with respect to projective transformations (of the stimulus material). The resulting scale is an ordinal scale.

Weber's Ratio as a Representation of a Concatenation Operation

Ordinal scales represent little structure of a subject matter because a concatenation operation is missing. This section will show that the invariance of Weber's ratio imposes restrictions upon the empirical domain which correspond to those of a specific concatenation operation.

The present study for this purpose specializes the Suppes et al. (1989) assumptions for this type of scaling. This is done by introducing more specific relations of points than just the relation of mutual separation of pairs of points. In an experiment subjects would be asked to distinguish different harmonic positions of quadruples of points.

Definition 5.3. Four points P_0, P_1, P_2, P_3 on a straight line are in *harmonic position* if they correspond to the intersections of sides and diagonals of a complete quadrilateral according to Figure 5.2.

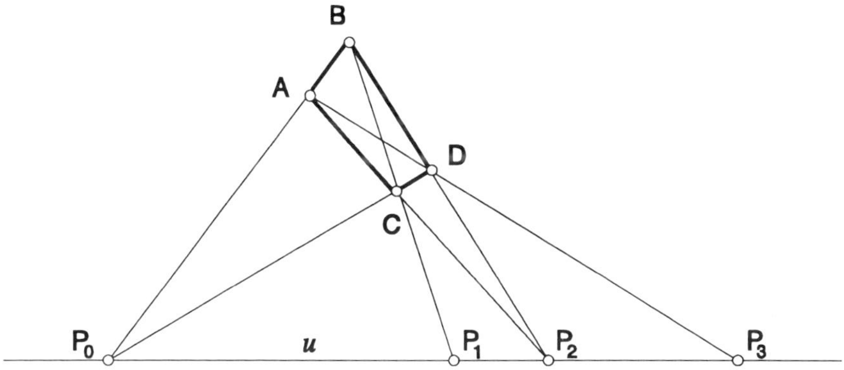

FIG. 5.2. If points P_0 and P_2 are given by the intersections of pairs of sides of a quadrilateral A, C, D, B, and points P_1, P_3 are the intersections of the quadrilateral's diagonals with the line u given by P_0 and P_2, then P_0, P_1, P_2, P_3 are located in harmonic position.

This assumption may, on first sight, look like it were not realizable by observers in the laboratory. It will be shown, however, that the numerical invariance of a certain Weber ratio with respect to projective transformations has as an empirical correspondence to the qualitative properties of a bisection operation.

An essential property of projective geometry on the straight line is the invariance of a certain point on this line with respect to all projective transformations. The point is called P_∞. The following theorem is well known (cf. Redei, 1968): All triples of points P_1, P_2, P_3 with P_0, P_2 S P_1, P_∞ are in harmonic position if P_1, in a specific manner, is "in the middle" of P_0 and P_2.

The following theorem may be formulated as a starting point for the subsequent developments.

Theorem 5.3. Weber's function is invariant with respect to projective monotone transformations.

Its proof is obvious from the development in Table 5.1 and the definition of projective monotone transformations.

For points P_0, P_1, P_2, P_∞ on a line by definition, the calculation of the cross-ratio K, in the case of $P_3 = P_\infty$ reduces to the determination of

$$\frac{\overline{P_0 P_1}}{\overline{P_2 P_1}} = K.$$

The cross ratio in case of harmonic position of the four points assumes the value of $K = -1$. If, furthermore, $P_3 = P_\infty$, then with harmonic position of the four points the point P_1 lies halfway between points P_0 and P_2. In this manner the harmonic position gains an eminent significance for coordinatization. By fixing points $P_3 = P_\infty$ it permits the definition of bisection as a concatenation operation for sections on a straight line.

Harmonic position of P_0, P_1, P_2, P_∞ demands from the intended representation $\varphi : A \rightarrow \mathbb{R} \cup \{\infty\}$, therefore, that

$$\frac{\varphi(P_0) - \varphi(P_1)}{\varphi(P_2) - \varphi(P_1)} = -1 \tag{5.3}$$

holds so that $\varphi(P_1)$ the arithmetic mean $\varphi(P_0)$ and $\varphi(P_2)$.

On the other hand, Weber's law asks directly for a given physical representation of stimulus points $\psi : A \rightarrow \mathbb{R}$, that

$$\frac{\psi(P_1)}{\psi(P_0)} = \frac{\psi(P_2)}{\psi(P_1)} = 1 + c(\pi).$$

From this

$$\psi(P_1) = \sqrt{\psi(P_0)\psi(P_2)}.$$

The first three stimulus intensities ψ of a harmonic quadruple of points with $P_3 = P_\infty$ form a geometric sequence if Weber's law holds. The corresponding sensation intensities φ form an arithmetic sequence.

Both kinds of bisections fulfill the axiom of bisymmetry (Figure 5.3).

The common *arithmetic mean* is an example of a bisymmetric (intensive) operation

$$\begin{aligned}\frac{1}{2}\left[\frac{\phi(a) + \phi(b)}{2} + \frac{\phi(c) + \phi(d)}{2}\right] &= \frac{1}{4}\left[\phi(a) + \phi(b) + \phi(c) + \phi(d)\right] \\ &= \frac{1}{2}\left[\frac{\phi(a) + \phi(c)}{2} + \frac{\phi(b) + \phi(d)}{2}\right].\end{aligned}$$

Also, a *geometric mean* fulfills the axiom of bisymmetry:

$$\begin{aligned}\sqrt{\sqrt{\phi(a)\phi(b)}\ \sqrt{\phi(c)\phi(d)}} &= \sqrt{\sqrt{\phi(a)\phi(b)\phi(c)\phi(d)}} \\ &= \sqrt{\sqrt{\phi(a)\phi(c)}\ \sqrt{\phi(b)\phi(d)}}.\end{aligned}$$

Under the convention that the fourth point P_∞ of the quadruple of points shall assume the scale value of ∞, then the subject's task amounts to a judgment of whether one of the points bisects the other two. Bisection has been analyzed by Pfanzagl (1959, 1968). It is empirically applicable as soon as the judgments fulfill the axiom of bisymmetry.

The bisection operation $\circ$ can be defined by means of the harmonic position.

> **Definition 5.4.** Let $\langle A, S\rangle$ be a one-dimensional separable separation structure with P_∞ fixed under projective monotone transformations. Then for all $P_1, P_2, P_3 \in A$

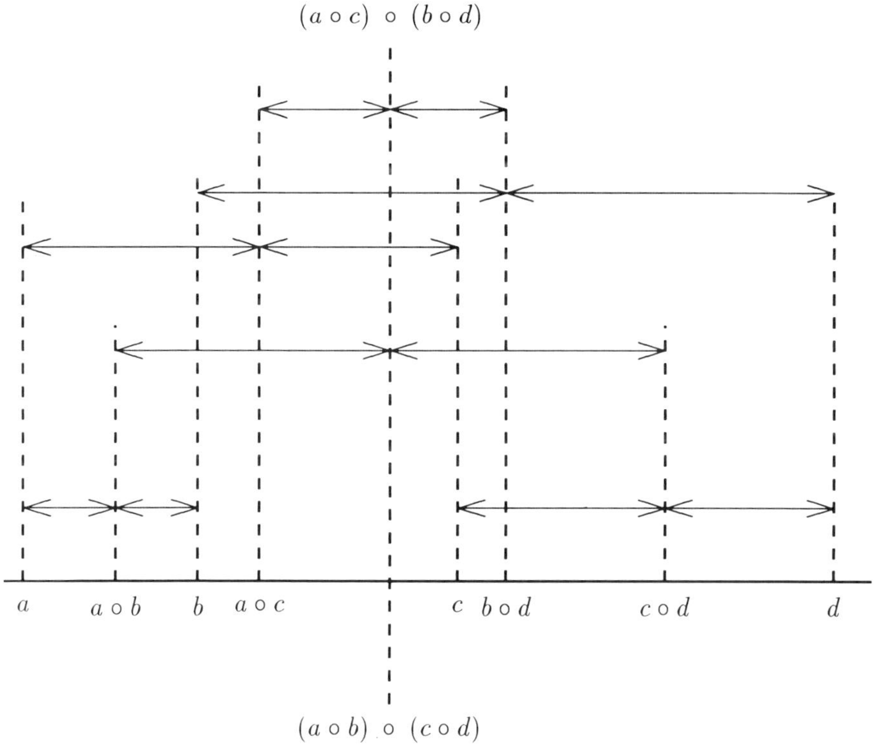

FIG. 5.3. If for given points a, b, c, d on a line the midpoints of different midpoints of pairs of points lead to the identical overall midpoint as indicated in the figure, then the bisymmetry condition holds.

$$\circ : A \times A \rightarrow A, \qquad (P_1, P_2) \mapsto P_3$$

is given iff $P_1P_3 \; S \; P_2P_\infty$ and $P_1P_2P_3P_\infty$ are in harmonic position.

Furthermore, the relation of separation permits the definition of a strict simple order with the negation of $<_{a\infty}$ connected and transitive (cf. Suppes et al., 1989).

Definition 5.5. Let (A, S) be a separable one-dimensional separation structure. Then for all $P_1, P_2, P_3 \in A$

$$P_2 <_{P_1P_\infty} P_3 \quad \text{iff} \quad P_1P_3 \; S \; P_2P_\infty.$$

Since only P_∞ is fixed in this manner, different orders are possible. They are, however, compatible with each other (Priess-Crampe, 1983).

Through these definitions a unidimensional projective scaling is reduced to bisymmetric scaling if Weber's law holds. Under additional assumptions of monotonicity, solvability, and the Archimedean axioms, as they are explained, e.g., in Krantz, Luce, Suppes, & Tversky (1971), the following representation theorem can be proven.

The customary coordinatization of the projective line, which had been adapted to the restrictions of the psychological laboratory by Suppes et al. (1989) in terms of ordinal scaling, can now be strengthened to an interval scaling by means of Theorems 5.4 and 5.5.

Representation Theorem 5.4. Let $\langle A, S, \circ\rangle$ be a separable one-dimensional separation structure and $\langle A, \circ\rangle$ a bisymmetry structure. Then there exists a map $\varphi : A \rightarrow \mathbb{R} \cup \{\infty\}$ such that for $P_1, P_2, P_3, P_\infty \in A$,

$$P_1P_3\ S\ P_2P_\infty \quad \textit{iff} \quad \varphi(P_1) > \varphi(P_2) > \varphi(P_3),$$

and

$$P_1P_2P_3P_\infty \text{ in harmonic position} \quad \text{iff} \quad \varphi(P_1 \circ P_3) = \varphi(P_2).$$

Uniqueness Theorem 5.5. The map φ is invariant up to projective monotone transformations. Furthermore φ is determined by the choice of three points $P_1P_2P_\infty \in A$. If different points are chosen, another representation φ' will result with different zero element a and a different scale factor $[\varphi'(P_2) - \varphi'(P_1)]$, respectively.

After reduction to a bisymmetry structure by means of the definitions given above the proof of the theorems reduces to those for bisymmetry scaling (cf. Krantz et al., 1971).

A constructive proof for the representability of points on a projective line is well known in geometry (cf. Efimov, 1960). It employs the bisection property of the three points which, together with point ∞, are situated in a harmonic position for the successive construction of projective equally spaced coordinates on both coordinate lines. The points can be shown to fulfill the desired criteria.

That subjects in the visual laboratory are able to generate bisection judgments in a reproducible manner has often been established. A more recent study is Burbeck and Yap (1990).

OTHER VISUAL LAWS

Monocular Vision in Perspective

The present section describes the structure of monocular depth perception. In the present context the projections of the physical stimuli onto the retina provide a suitable set of proximal stimuli. After confinement of the stimuli to a horizontal plane at floor level, the effects of eye and head movements of the subject are represented by projective mappings of the retinal image onto itself. The description of these maps meets with minor obstacles. They can be overcome in a way which is standard in projective geometry as the next section will demonstrate. For simplicity, the procedure is discussed in a context of analytic geometry.

Projective Maps

For the description of the process a point z in $\mathbb{R}^3$ is chosen as a *center of projection* (the center of the lens). A plane Y which does not contain z is chosen as the *image plane*. For almost all points $p \in \mathbb{R}^3$ an image point $f(p) \in Y$ can be found: A straight line is drawn, which connects p and z (it corresponds to a ray of light), and the intersection $f(p)$ with Y is found. This will exist for all points p which are not contained in the plane Y' parallel to Y through z. Thus, a map $f : \mathbb{R}^3 \setminus Y' \rightarrow Y$ is given.

Let X be an affine space over an arbitrary field. A *direction* of lines (or an *infinitely distant point*) p_∞ is defined as an equivalence class of parallel straight lines in X. Series of points diverging along parallel lines are defined to converge toward the identical point p_∞.

The point p_∞ is not an element of X. Let X_∞ be the set of all p_∞. Then

$$\bar{X} = X \cup X_\infty$$

is called the *projective closure* of X.

Let r be a point in Y' different from z. A sequence of points (r_i) on a straight line not in Y' which converges toward r, gives rise to diverging images $f(r_i)$ in Y in the direction which is determined by r. Therefore, r is called a *point of alignment* and Y' a *plane of alignment*.

A simple case is given if in a *plane* a straight line Y_1 is projected from a center z on a nonparallel line Y_2. In this case there exists a point $p_1 \in Y_1$ which does not have an image, and there exists an image $p_2 \in Y_2$ which does not possess an inverse image.

This leads to a bijective map $f : Y_1 \setminus \{p_1\} \rightarrow Y_2 \setminus \{p_2\}$. It usually is extended in the following way: The projective closure of Y_1 is generated by adjoining Y_1 with an *infinitely distant point* ∞_1:

$$\bar{Y}_1 := Y_1 \cup \{\infty_1\}.$$

In an analogous manner $\bar{Y}_2 := y_2 \cup \{\infty_2\}$, and the projective closure of Y_2 is generated.

Finally, $\bar{f} : \bar{Y}_1 \rightarrow \bar{Y}_2$ is defined by

$$\begin{aligned} \bar{f}(p) &:= f(p) \qquad \text{for } p \in Y_1 \setminus \{p_1\}, \\ \bar{f}(p_1) &:= \infty_2, \\ \bar{f}(\infty_1) &:= p_2, \end{aligned}$$

and a bijective map of the extended lines results.

Definition 5.6. Let $\mathcal{P}$ and L be nonempty sets and $I \subset \mathcal{P} \times L$ a binary relation. $\langle \mathcal{P}, L, I \rangle$ is an *incidence structure* iff the following conditions are fulfilled:

(i) For all $P, Q \in \mathcal{P}$ there exists a unique $l \in L$ with $P\ I\ l$ and $Q\ I\ l$. (For each pair of points there exists a straight line with which they are incident.)

(ii) For all $l, m \in L$ there exists $P \in \mathcal{P}$ with $P\ I\ l$ and $P\ I\ m$. (For each pair of straight lines there exists a point with which they are incident.)

(iii) There exist $P_1, P_2, P_3, P_4 \in \mathcal{P}$ such that, if $P_i\ I\ l$ and $P_j\ I\ l$ for some $l \in L$, then not $p_k\ I\ l$, with $i, j, k \in \{1, 2, 3, 4\}$ pairwise distinct. (There exist four points any three of which cannot all be incident with a straight line.)

Definition 5.7. A bijective mapping of a pencil of points on a pencil of lines, or vice versa, is called a *perspectivity,* if the corresponding points and lines are incident. Any composition of two such maps which connect two pencils of points or two pencils of lines also forms a perspectivity.

A set of points and a set of lines which are the domain-range pair of a perspectivity are said to be *in perspective*.

Definition 5.8. A composition of perspectivities is called a *projective* map.

Definition 5.9. (Projective form of Desargues' theorem). If two triangles are in perspective with respect to a point, then they are in perspective with respect to a line (cf. Redei, 1968).

If in an incidence structure Desargues' theorem holds, then a Desarguesan incidence structure is given.

The notation $|\varphi(P)|$ signifies the usual absolute value of φ.

The present development has now led up to a point which permits formulation of a representation that obeys a two-dimensional generalization of Weber's law in terms of Theorems 5.6 and 5.7.

Representation Theorem 5.6. Let $\mathcal{P} = \langle P, L, I \rangle$ be a Desarguesan incidence structure for which on each line $l \in L$ there exists a one-dimensional bisymmetric separation structure. Then there exists a map $\varphi : \mathcal{P} \rightarrow [\mathbb{R} \cup \{\infty\}]^2$ such that, for any $l \in L$ with $P_1\ I\ l$ and $P_2\ I\ l$ and $P_3\ I\ l$ and $P_\infty\ I\ l$,

(i) $P_1P_3\ S\ P_2P_\infty$ iff $|\varphi(P_3) - \varphi(P_1)| > |\varphi(P_2) - \varphi(P_1)|$.

(ii) P_1, P_2, P_3, P_∞ in harmonic position iff $\varphi(P_1 \circ P_3) = \varphi(P_2)$.

Uniqueness Theorem 5.7. The map φ is invariant up to projective transformations and is determined by choice of a point of origin P_0 and two coordinate lines $x, y \in L$ incident with the origin, as well as two unit points $X_1\ I\ x$ and $Y_1\ I\ y$. If a different triplet of points is chosen, a map φ' will result with $\varphi' = \alpha\,\varphi + \beta$, α a 2×2 diagonal matrix and $\beta \in \mathbb{R}^2$.

Visual Invariants in Monocular Space Perception

A journalist recently tried to capture particularly realistic film reports from an American football game by mounting a miniature TV camera inside the helmet of one of the players of the team. Even though, technically, the recording of images could be accomplished, viewing of the film amounted to a nauseating experience. The observer's impression was that of teams of dwarfs, playing football inside a running washing machine: neither size constancy nor the level orientation of the playing field could be maintained by the observer.

The present approach represents visual automorphisms as projective maps. They are each characterized by a special visual invariance. The automorphisms form a group which comprises the affine, the affine unimodular and the orthogonal groups (in this order) as subgroups.

The second and third columns of Table 5.2 show the geometric constraints for

TABLE 5.2
The Projective Group of Plane Automorphisms and Its Principal Subgroups

Group	*Automorphism*	*Invariant*	*Numeric Representation*	*Free Par.*
Projective	Projective	Cross-ratio	$\varrho' x'_1 = c_{11}x_1 + c_{12}x_2 + c_{13}x_3$ $\varrho' x'_2 = c_{21}x_1 + c_{22}x_2 + c_{23}x_3$ $\varrho' x'_3 = c_{31}x_1 + c_{32}x_2 + c_{33}x_3$ $\varrho' \neq 0 \in \mathbb{R}$ $\Delta = \begin{vmatrix} c_{11} & c_{12} & c_{13} \\ c_{21} & c_{22} & c_{23} \\ c_{31} & c_{32} & c_{33} \end{vmatrix} \neq 0$	8
Affine	Projective with fixed straight line	Part ratio	$\varrho' x'_1 = c_{11}x_1 + c_{12}x_2 + c_{13}x_3$ $\varrho' x'_2 = c_{21}x_1 + c_{22}x_2 + c_{23}x_3$ $\varrho' x'_3 = c_{33}x_3$ $\Delta \neq 0 \quad c_{33} \neq 0$	6
Affine unimodular	Rotation of axes, translation	Area	$x' = a_1x + b_1y + c_1$ $y' = a_2x + b_2y + c_2$ $D = \begin{vmatrix} a_1 & b_1 \\ a_2 & b_2 \end{vmatrix} = \pm 1$	5
Orthogonal	Rigid rotation and translation	Length	$x' = a_1x + b_1y + c_1$ $y' = a_2x + b_2y + c_2$ $A = \begin{pmatrix} a_1 & b_1 \\ a_2 & b_2 \end{pmatrix}$ $AA' = I$	3

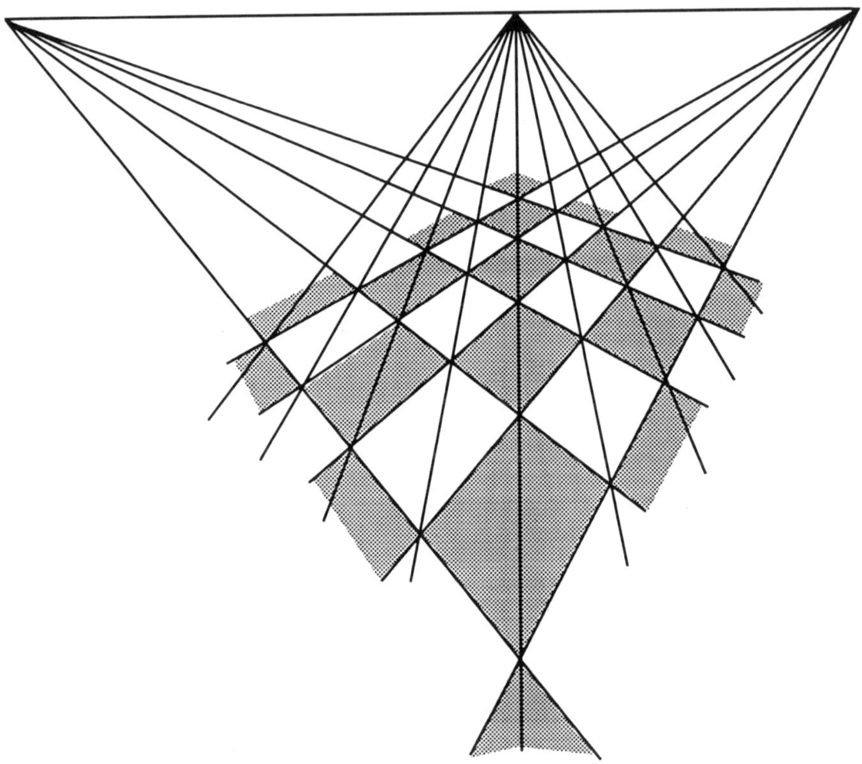

FIG. 5.4. Monocular projection of the horizontal plane at floor level. Under projective automorphisms the "horizon" line remains invariant.

the various projective automorphisms. From the standpoint of a study of vision the condition of a fixed straight line is of particular interest. The visually straight line is interpreted as the horizon (Figure 5.4).

Its fixation in vision becomes particularly apparent under the rare circumstances when it does not remain "in place" under motions like in the filmed game mentioned above. High gravity forces acting on the observer in a curving airplane or altered states of consciousness like dizziness from vertigo or spells of faint are other examples (cf. Drösler, 1992).

Experimental studies of the different invariants of these subgroups as well as the consideration of other empirical observations show that a phenomenally level horizon as well as constancy of area, a weak form of size constancy, are maintained by the observer under head and eye movements. This leads to the conclusion that the structure of monocular space perception is characterized by the affine unimodular subgroup of the projective group. There are indications that the phenomena of size constancy can be represented by affine unimodular maps instead of orthogonal (metric) maps as sometimes has been conjectured.

The present approach of representing visual invariances by specific group structures possesses an advantage over previous studies along similar lines. Hoffman's (1966) work is based on axioms about infinitesimal group germs, the qualitative empirical interpretation of which remains vague. Thus, his assumptions are more difficult to validate than the present assumptions, which directly refer to the visual invariances.

Binocular Vision

Binocular visual space also is mediated through the two eyes by means of projection (Figure 5.5). This geometrical aspect is obscured sometimes by the use of biopolar coordinates with respect to both eyes of the observer (Figure 5.6). Useful as the bipolar coordinates may be for connecting visual phenomena with the *difference in bipolar parallax,* they do not facilitate geometrical analysis. If difference in bipolar parallax (*lateral disparity*) exceeds a certain amount, signals from the two projections cannot be identified with each other or "fused."

In the study of monocular vision the visual horizontal plane at ground level (Figure 5.4) is surveyed by the observer from above. An example of this is given by the view over a large lake. In the binocular case, the horizontal plane consid-

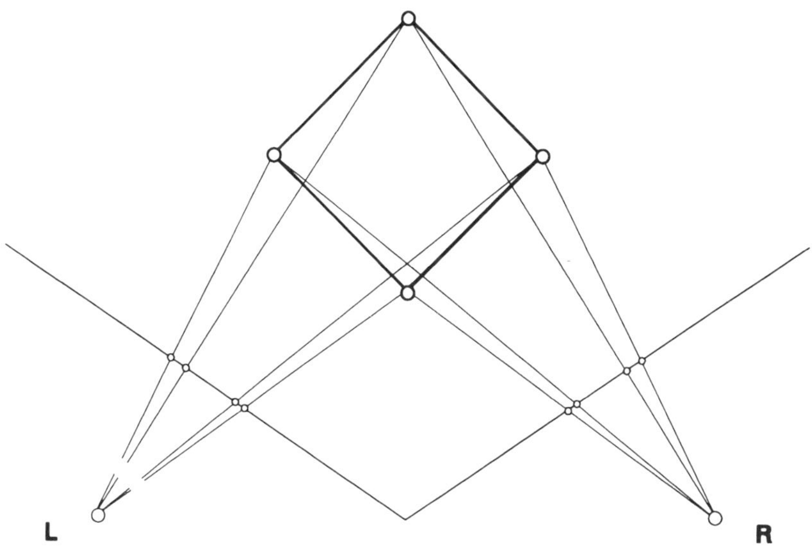

FIG. 5.5. A square configuration of four points in the horizontal plane at eye level is projected by the two eyes L, R, as indicated in the figure.

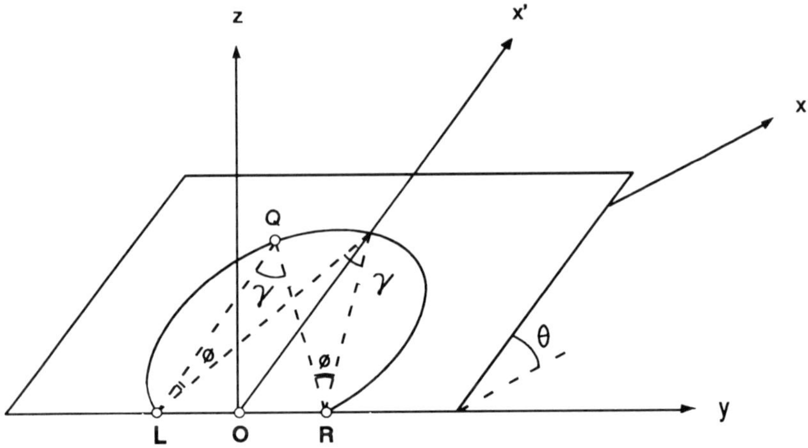

FIG. 5.6. Bipolar coordinates ϕ, γ, θ with respect to the eyes L, R of a point Q in the visual laboratory. The angle γ gives the bipolar parallax of point Q with respect to L, R.

ered is located at eye level in order to exclude all monocular distance cues for the observer. In consequence, there does not exist any phenomenal "horizon." The present analysis, though, identifies a different invariant which generates an alternative geometrical structure.

The Metric of Binocular Visual Space

Binocular vision is characterized by a specific peculiarity. Far away objects which produce a binocular disparity below absolute disparity threshold visually do not vanish in the distance like they would in monocular vision (Figure 5.4). As soon as all monocular visual cues are excluded, these objects are seen at a constant visual distance from the observer. Because this semicircle is invariant under eye movements, the binocular invariant in question is thus identified as a quadratic curve. It is well known that a plane projective geometry, the automorphisms of which leave invariant a quadric, possesses a projective metric hyperbolic structure. This result for the binocular plane (Drösler, 1979) is slightly more general than the corresponding one from Luneburg (1950) who postulated a hyperbolic metric.

The projective transformations which leave invariant a circle constitute an empirical restriction. Thus, the following theorem amounts to an empirical law. It concerns the nonlinear effect of binocular visual motion. (The *signature* of a square matrix is given by the sequence of algebraic signs of its eigenvalues.)

If the coordinates for points in a plane are given with respect to an origin outside the plane, three coordinates describe the position of any point in the plane. If, moreover, these coordinates are determined only up to a multiplicative positive constant, they are called *projective coordinates*. Corresponding conventions are used in higher dimensions. Homogeneous coordinates for points in three-space consist of four numbers, determined up to a common arbitrary constant.

The present approach which employs perceptual invariants leads to a representation of binocular visual space which bears a projective hyperbolic metric structure as is formulated in Theorem 5.8.

Theorem 5.8. Let $C \subset \mathbb{R}^3$ be the unit disk in the horizontal plane of a projective binocular coordinatization in homogeneous coordinates. There exist operators $\mathbf{A} : C \to C$ which form a group under composition such that for all $\boldsymbol{\varphi} \in C$

$$(\mathbf{A}\boldsymbol{\varphi})^T\mathbf{F}\,\mathbf{A}\boldsymbol{\varphi} = 0 \quad \text{iff} \quad \boldsymbol{\varphi}^T\mathbf{F}\,\boldsymbol{\varphi} = 0,$$

with $\mathbf{F}$ a nonsingular 3×3 symmetrical matrix with signature $(+, +, -)$.

It is the choice of homogeneous coordinates which represents a nonlinear projective transformation $\mathbf{A}$ by means of bilinear algebra.

Color Space

Earlier Work

The scaling of colors has been accomplished on the basis of Grassman's laws (1853). Let $\mathcal{A}$ be a set of stimuli, e.g., sectors of color wheel,

$\oplus : \mathcal{A} \times \mathcal{A} \to \mathcal{A}$	temporal superposition (e.g., rotation of color wheel)
$* : \mathbb{R} \times \mathcal{A} \to \mathcal{A}$	control of stimulation intensity (e.g., size of sector on color wheel)
$\sim \subseteq \mathcal{A} \times \mathcal{A}$	a visual equivalence relation

Representation Theorem 5.9. (Grassmann, 1853). Let $C \subset \mathbb{R}^n$ be a convex cone in $\mathbb{R}^n$. There exists an injective linear map $\varphi : \mathcal{A} \to C$ unique up to nonsingular linear transformations if

(i) $\langle \mathcal{A}, \oplus \rangle$ is a commutative cancellation semigroup.
(ii) $*$ is a scalar multiplication on $\langle \mathcal{A}, \oplus \rangle$.
(iii) Grassmann's visual cancellation laws hold: for all $a, b, c \in \mathcal{A}$ and $t \in \mathbb{R}$:
(a) $a \sim b$ iff $a \oplus c \sim b \oplus c$.
(b) $a \sim b$ iff $t * a \sim t * b$.

Krantz (1975) has provided the formal treatment of the original material which contains little mathematical notation. The scaling results in a coordinatization of color space so that visually indiscriminable stimuli receive identical color coordinates. The scaling is unique up to nonsingular linear transformations.

These theoretical results are empirically so well established that they can serve as a referent for the validation of other developments.

The uniqueness relation, unfortunately, has been construed by some researchers as a statement of structure determining automorphisms. An example is v. Kries (1905), who worked under such an implicit assumption. This lead him and many of his followers, up to the present time, to confine research on motions in color space, such as effects of color adaptation, to *linear* transformations in three-space.

An Invariant in Color Space

It was Yilmaz (1962a, b) who pointed out that, for empirical reasons, the convex cone of Theorem 5.9 is an invariant in color space. He could not solve the problem of finding an empirically well founded mathematical expression for the extremal surface of the convex cone C. So he conjectured it to be a quadratic cone. Under the rest of his assumptions, this led him to believe in a pseudo-Euclidean structure of color space, comparable to that of special relativity.

It can be shown (Drösler, 1988, 1994) from experimentally testable assumptions that the extremal surface of the convex cone is indeed that of a quadratic cone. The result follows from assuming that a variant of Bloch's law holds for brightness detection. Visual intensity of a light stimulus is determined by the product of duration and bandwidth. Application of the well-known minimal uncertainty principle from Fourier theory to Bloch's law leads to a Gaussian spectrum for the luminous efficiency curve V_λ. The function describes visual brightness V as a function of frequency ω, $\omega = k\lambda^{-1}$, $k \in \mathbb{R}^+$.

$$V_\lambda(\omega) = \exp\left[\frac{-(\omega - \mu)^2}{2}\right],$$

with μ the center of the visual spectrum at $\lambda = 555$ nm or $\omega = 18{,}000$ lines/cm.

The additional assumption introduced asserts that besides brightness appraisal color vision performs a chromatic evaluation of the wavelength-dependent energy $V_2(\omega)$ of the stimulus according to Planck's law $e = h\,\omega$, where e represents energy and h is Planck's constant, simply by weighting with frequency ω in brightness system response relative to the band center μ:

$$V_2 = (\omega - \mu)(V_\lambda).$$

Furthermore, the second chromatic evaluation arises by weighting the first chromatic response in the time domain with duration t:

$$\mathscr{F}^{-1}\{V_3(\omega)\} = tf_2(t).$$

Here $\mathscr{F}^{-1}$ is the inverse Fourier transform.

The spectra V_λ, V_2, or V_3 generated in this manner constsitute the first three members of an orthogonal system of functions, called Hermite's functions. This reason is that V_2 and V_3 are related to V_λ by successive differentiation with respect to frequency ω. This is so because weighting with duration in the time domain can be represented by differentiation of the spectrum. If $F(\omega)$ and $f(t)$ are Fourier transforms of each other, then

$$\mathscr{F}^{-1}\left\{\frac{d^n F(\omega)}{d\omega}\right\} = (it)^n f(t).$$

Occurrence of the imaginary unit i does not hamper the empirical analysis because what is measured always amounts to spectral power, i.e., squared ordinate values, so that the imaginary unit is eliminated.

The variable ω or, respectively λ can be eliminated by expressing V_λ as a function of V_2 and V_3. A quadratic cone in three dimensions results (Drösler, 1994).

Color Vision, a Projection Operator

Even if Yilmaz's conjecture thus can be supplied with an empirical basis, his conclusions cannot. This is because Yilmaz overlooks the fact that color vision amounts to a *projection* of an infinite-dimensional space of stimuli into a three-dimensional space of colors. This becomes obvious immediately upon regarding the well-established formulas for color coordinates in the novel context of a space of functions. If $P(\lambda)$ is the electromagnetic stimulus spectrum, $\bar{r}(\lambda)$, $\bar{g}(\lambda)$, $\bar{b}(\lambda)$ are *color-matching functions*, then the *color coordinates* $\boldsymbol{\varphi} = [R, B, G]^T \in \mathbb{R}^3$ are given as integrals of the form

$$\varphi_i(P(\lambda)) = \int_{\lambda_1}^{\lambda_2} \bar{r}(\lambda)P(\lambda)\, d\lambda, \qquad i = 1, 2, 3, \tag{5.4}$$

setting $\bar{r}_1 = \bar{r}$, $\bar{r}_2 = \bar{g}$, $\bar{r}_3 = \bar{b}$.

Equations 5.4 can be used to define an operator A,

$$A(P(\lambda)) = \varphi_1(P(\lambda))\bar{r}_1(\lambda) + \varphi_2(P(\lambda))\bar{r}_2(\lambda) + \varphi_3(P(\lambda))\bar{r}_3(\lambda),$$

or, after substitution from Equation 5.4,

$$AP(\lambda) = \sum_{k=1}^{3} \int_{\lambda_1}^{\lambda_2} \bar{r}_k(\lambda)P(\lambda)\, d\lambda\; \bar{r}_k(\lambda). \tag{5.5}$$

The effect of color vision is described by the linear operator A which generates the color coordinates. Equation 5.5 amounts to a function-space *representation* of color-matching data if the axioms of a special *Hilbert space* are empirically true.

Thus, any textbook on color vision containing Equation 5.4 should state the axioms of a Hilbert space of functions implied by Equation 5.4. Since there appears to exist no such reference, the axioms are stated below.

Definition 5.10. Let $\mathbb{C}$ be the set of complex numbers and V a nonempty set. The relational system $\langle V, \boxplus, \star\rangle$ with maps

$$\boxplus : V \times V \rightarrow V, \qquad \star : \mathbb{C} \times V \rightarrow V,$$

is a *vector space* over the field of complex numbers if the following hold:

(i) $\langle V, \boxplus\rangle$ is a commutative group with neutral element $\mathbf{O} \in V$. For all v, $w \in V$, α, $\beta \in \mathbb{C}$, (ii)–(v) hold:

(ii) $(\alpha + \beta) \star v = \alpha \star v \boxplus \beta \star v.$

(iii) $\alpha \star (v \boxplus w) = \alpha \star v \boxplus \alpha \star w.$

(iv) $(\alpha\beta) \star v = \alpha \star (\beta \star v).$

(v) $\mathbf{1} \star v = v.$

The dimensionality of a linear space is given by the maximum number of its elements, which are always linearly independent. Reference to Fourier decompositions would imply an infinite dimensionality of the space.

Let $\boxplus$ signify pointwise addition of function and $\star$ the multiplication of a function by a complex number.

Definition 5.11. Let $I \subseteq \mathbb{R}$ be an open interval. A set $\mathcal{H}$ of continuous functions $F : I \rightarrow \mathbb{C}$ forms a *Hilbert space* if $\langle \mathcal{H}, \boxplus, \star\rangle$ is a vector space and, for all v, $w \in \mathcal{H}$, there exists a *scalar product*

$$(\ ,\ ) : \mathcal{H} \times \mathcal{H} \rightarrow \mathbb{C}, \qquad (v, w) \mapsto \int_{x \in I} v\bar{w}\, dx.$$

In particular, for all $v \in \mathcal{H}$ there exists a *norm*

$$|\ | : \mathcal{H} \rightarrow \mathbb{R}, \qquad v \mapsto (v, \bar{v})^{1/2}.$$

Since the elements of the Hilbert space are used to represent physical electromagnetic spectra in the "visible" interval of wavelengths $I = [\lambda_1, \lambda_2]$, the empirical validity of these axioms does not present problems. The operation $\boxplus$ represents the superposition of spectra, $\star$—for real-valued coefficients—the physical control of the stimulus' radiant power. The norm of a stimulus represents a measure of its electromagnetic power.

Color vision as represented by the linear operator A amounts to a projection of the stimulus spectrum into a three-dimensional subspace. The linearity of A guarantees the linearity of the subspace. Even though the present development is independent of Grassmann's (1853) representation of visual color matching, a connection can be established easily. Grassmann's metameric equivalence classes $A/\sim$ are definable by that partition $P(\lambda)/_{=}$ which is formed by stimuli yielding equal color coordinates according to Equation 5.5. This places the burden of validation upon the empirical status of the analytic basis functions $\bar{r}_i$, $i \in \{1, 2, 3\}$, which were derived above from qualitative empirical assumptions. They can be regarded as determined in their analytical form as a Gaussian and its first two derivatives. Their empirical fit is closer than any fit of two-parameter analytic expressions to color-matching data in the field (Drösler, 1994).

What concerns the perceptual invariant in the representation of the process of color adaptation, the development in the preceding section can be formulated concisely as

> **Theorem 5.10.** There exists a set of basis functions $r(\lambda)$, $g(\lambda)$, $b(\lambda)$ such that a quadratic surface results if one of them is expressed as a function of the two others and the wavelength λ, respectively frequency ω, is eliminated.

The Metric Structure of Color Space

Taking into account that the geometry of color space is projective, the invariant cone no longer generates a pseudo-Euclidean structure as Yilmaz conjectured. Rather, the automorphisms are projective metric hyperbolic. Their empirical interpretation is given by the results of experiments in color adaptation (Burnham, Evans, & Newhall, 1957).

Motions in color space are defined by certain differences in color coordinates. They arise if the scaling of a stimulus yields different color coordinates when presented first to one eye under neutral conditions and then to the other eye which has been preexposed to an adapting stimulus for a certain time. Usually the scaling is conducted by plotting the color coordinates (previously determined under neutral conditions) of that different stimulus which "appears" as metameric to the first stimulus when presented simultaneously to the preadapted eye (Figure 5.7).

These motions can empirically be shown to form a group. They furthermore will not lead to predictions of adaptation effects which would transgress the extremal of the convex cone C, the range of color scaling. This result is remarkable insofar as all linear adaptation theories (e.g., v. Kries, 1905, and his followers) are empirically false on these grounds. The discrepancy between linear theories and observations, apparently, has never been reported, because researchers in their experiments tend to employ colors situated well away from the boundaries of the convex cone.

Theorem 5.11 describes the process of color adaptation. In divergence with v. Kries (1905) and his followers, color adaptation is represented by a projective,

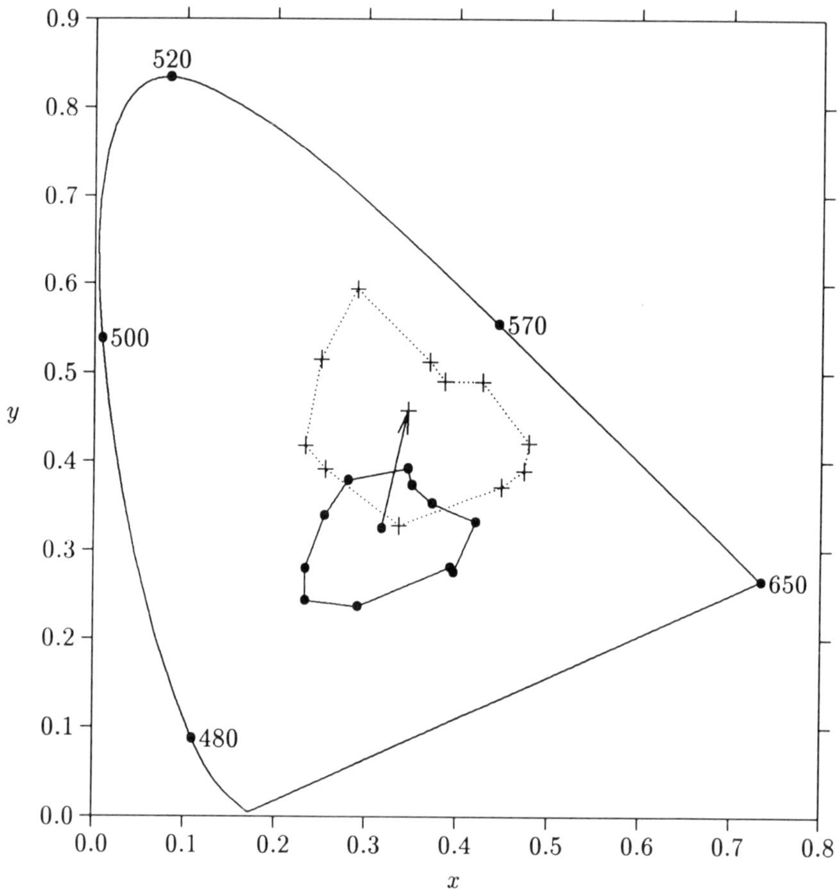

FIG. 5.7. After preadaptation of one eye, two sets of colors "appear" as pairwisely metameric, which normally would be visually distinguished as entered in the chromaticity diagram.

i.e., a nonlinear transformation of color space. It is the use of homogeneous coordinates which still permits a linear algebraic notation at the cost of one extra dimension.

Theorem 5.11. Let $\mathcal{S}$ be an infinite-dimensional Hilbert space of functions and A a linear operator projecting the elements of $\mathcal{S}$ into a convex cone C of a three-dimensional subspace of $\mathcal{S}$. There exists a map $B : C \rightarrow C$ such that all coefficient vectors $\boldsymbol{\varphi}$ of $AP(\lambda)$, according to Equation 5.5 expressed in homogeneous four-dimensional coordinates, fulfill

$$(B\boldsymbol{\varphi})^T\mathbf{F}\,B\boldsymbol{\varphi} = 0 \quad \text{iff} \quad \boldsymbol{\varphi}^T\mathbf{F}\boldsymbol{\varphi} = 0,$$

with $\mathbf{F}$ a constant 4×4 symmetrical matrix with signature $(+, +, +, -)$.

Again it is the choice of homogeneous coordinates which represents a non-linear projective transformation B by means of bilinear algebra.

The representation is independent of the classical Grassmann (1853) representation because it does not require postulation of Grassmann's visual cancellation laws. Their validity in the present development is derivable from corresponding cancellation laws of the physical stimuli and the representation of color vision as a process of projection in Hilbert space.

The metric which characterizes this geometry is well known. It is the negative logarithmic *cross-ratio* referring to the pair of colors whose dissimilarity is under study and the intersections of its connecting line with the limiting cone.

The cross-ratio of four collinear points in three-space is defined analogous to the one-dimensional case.

Definition 5.12. Let y_i, y_j be the coordinates of two colors and s_i, s_j those of the intersections of the connecting line of y_i and y_j with the fundamental or a purple line, then

$$\delta(y_i, y_j) = \frac{|s_i - y_i|}{|s_i - y_j|} \div \frac{|s_j - y_i|}{|s_j - y_j|}$$

is the *cross-ratio* associated with the colors y_i, y_j.

The metric introduced here is

$$d(y_i, y_j) = -c \log(\delta(y_i, y_j)). \tag{5.6}$$

The argument in Equation 5.6 corresponds to the well-known "center of gravity rule," with c an arbitrary imaginary scale factor.

Since the metric 5.6 is invariant with respect to projective transformations, its empirical test can be conducted in any projective plane of the color space, e.g., in the corresponding chromaticity diagram. Since in color science the empirical discussion is for practical reasons often limited to the chromaticity diagram, the automorphisms will be tested in this *plane* first. Wright's (1941) "dashes" supply experimental data for validation.

Their lengths are predicted by Equation 5.6 with high precision (Pearson's coefficient of correlation $\hat{\rho} = 0.66$ globally, and for most subsets $\hat{\rho} > 0.95$).

CONCLUSIONS

The present development calls attention to the fact that psychological scaling and the formulation of empirical laws are related, but clearly distinguishable, steps in the research process. In the language of geometry, psychological scaling amounts to a (quantitative) coordinatization of qualitative subject matter. Measurement theory stresses the point that coordinatization in principle suffers from a lack of uniqueness which can be described by a group of numerical transformations of

TABLE 5.3
Summary of Results

Modality	# of Dim	Perceptual Invariant	Group of Automorphisms
Sensory intensity	1	Weber's fraction	Projective group
Monocular visual plane	2	Fixed horizon and area	Plane affine unimodular group
Binocular visual plane	2	Circle of threshold disparity	Plane projective metric hyperbolic group
Color vision	3	Extremal of the convex cone of colors	Projective metric hyperbolic group

coordinates. This group has sometimes been mistaken for the set of restrictions which impose structure upon the empirical domain in question. An example is v. Kries' (1905) theory of color adaptation. Its linearity is modeled after the admissible linear transformations of Grassmann's (1853) coordinatization. What apparently has been overlooked by v. Kries and many others is that after proper choice of units and scale factors in all coordinates, the set of admissible transformations of any scaling has been reduced to identity. Thus, for the representations of natural laws and their generation of structure, any set of automorphism is available, be they of the form of the original admissible transformations or not.

The present development gives three examples, which are summarized in Table 5.3.

With *monocular vision in perspective,* the structure generating group of automorphisms is the projective unimodular group which preserves a level horizon as well as constancy of visual area. Motions in the *binocular horizontal plane* are representable by the projective metric hyperbolic group. It preserves the visual location of all objects which give rise to subthreshold lateral disparity on a visual circle around the observer.

The process of *color adaptation* is representable by exactly the same group of motions, this time in color space. The details of its development touch upon well known representations in color science. The projection of Equation 5.4 or the center of gravity rule in the present development are consequences of a new representation of color vision which comprises Grassmann's (1853) assumptions as theorems.

The present study shows that the famous last paper of Helmholtz (1891) is based on the fruitful idea of generalizing Weber's (1834) result to higher dimensions. What Helmholtz appears to have overlooked, is that Weber's ratio constitutes a projective, geometric invariant, a restriction which permits generalizations to higher dimensions in a standard geometric way.

REFERENCES

Burbeck, C. A., & Yap, Y. L. (1990). Spatiotemporal limitations in bisection and separation discrimination. Special Issue: Optics, physiology and vision. *Vision Research, 30,* 1573–1586.

Burnham, R. W., Evans, R. M., & Newhall, S. M. (1957). Prediction of color appearance with different adaptation illuminations. *Journal of the Optical Society of America, 47,* 35–42.

Crampe, S. (1958). Angeordnete projektive Ebenen. *Mathematische Zeitschrift, 69,* 435–462.

Drösler, J. (1979). Foundations of multi-dimensional metric scaling in Cayley-Klein Geometries. *British Journal of Mathematical and Statistical Psychology, 32,* 185–211.

Drösler, J. (1988). *Farbensehen als Wirkung eines hermiteschen Operators im Hilbert-Raum.* Paper read at the 36th bi-annual meeting of the Deutsche Gesellschaft für Psychologie at Berlin.

Drösler, J. (1992). Eine Untersuchung des perspektivischen Sehens. *Zeitschrift für experimentelle und angewandte Psychologie, 34,* 515–532.

Drösler, J. (1994). Color similarity represented as a metric of color space. In G. Fischer & D. Laming (Eds.), *Mathematical psychology, psychometrics, and methodology* (pp. 19–37). Berlin: Springer.

Efimov, N. W. (1960). *Höhere Geometrie.* Berlin: Deutscher Verlag der Wissenschaften.

Fechner, G. T. (1860). *Elemente der Psychophysik.* Leipzig: Breitkopf & Härtel.

Grassmann, H. (1853). Zur Theorie der Farbenmischung. *Poggendorffs Annalen der Physik und Chemie, 89,* 69–84.

Helmholtz, H. v. (1891). Versuch einer erweiterten Anwendung des Fechnerschen Gesetzes im Farbensystem. *Zeitschrift für Psychologie und Physiologie der Sinnesorgane, 2,* 1–30.

Hoffman, W. C. (1966). The Lie algebra of visual perception. *Journal of Mathematical Psychology, 3,* 55–98.

Krantz, D. H. (1975). Color measurement and color theory: I. Representation theorem for Grassmann structures. *Journal of Mathematical Psychology, 12,* 283–303.

Krantz, D. H., Luce, R. D., Suppes, P., & Tversky, A. (1971). *Foundations of measurement* (Vol. 1). New York: Academic Press.

Kries, J. v. (1905). Die Gesichtsempfindungen. In W. Nagel (Ed.), *Handbuch der Physiologie des Menschen* (Vol. 3, pp. 109–279). Braunschweig: Vieweg.

Luce, R. D., & Galanter, E. (1963). Discrimination. In R. D. Luce, R. R. Bush, & E. Galanter (Eds.), *Handbook of mathematical psychology* (Vol. 1, pp. 191–244). New York: Wiley.

Luneburg, R. K. (1947). *Mathematical analysis of binocular vision.* Princeton, NJ: University Press.

Luneburg, R. K. (1950). The metric of binocular visual space. *Journal of the Optical Society of America, 40,* 627–642.

Pfanzagl, J. (1959). *Die axiomatischen Grundlagen einer allgemeinen Theorie des Messens.* Schriften des Statistischen Instituts der Universität Wien, Neue Folge Nr. 1. Würzburg: Physica-Verlag.

Pfanzagl, J. (1968). *Theory of measurement.* New York: Wiley.

Priess-Crampe, S. (1983). *Angeordnete Strukturen: Gruppen, Körper, projektive Ebenen.* Berlin: Springer-Verlag.

Redei, L. (1968). *Foundation of Euclidean and non-Euclidean geometries according to F. Klein.* New York: Pergamon Press.

Suppes, P., Krantz, D. H., Luce, R. D., & Tversky, A. (1989). *Foundations of measurement (Vol. II).* San Diego: Academic Press.

Weber, E. H. (1834). De pulsu, resorptione, auditu et tactu. Annotationes anatomicae et physiologicae. Leipzig: Koehler.

Wright, W. D. (1941). The sensitivity of the eye to small colour differences. *Proc. Phys l. Sci.* (London), *53,* 93.

Yilmaz, H. (1962a). On color perception. *Bulletin of Mathematical Biophysics, 24,* 5–29.

Yilmaz, H. (1962b). Color vision and new approach to general perception. In E. E. Bernard & M. R. Kare (Eds.), *Biological prototypes and synthetic systems* (Vol. 1, pp. 126–141. New York: Plenum Press.

6 Genericity in Spatial Vision

Marc K. Albert
Department of Information and Computer Science, University of California

Donald D. Hoffman
Department of Cognitive Science, University of California

ABSTRACT

One principle the brain uses to construct spatial interpretations of retinal images is *genericity*. We describe this principle and illustrate its operation in our perceptions of line drawings, object parts, and subjective surfaces.

INTRODUCTION

Your visual system reports on your environment: its objects, shapes, colors, motions, and spatial layout. You might expect from this report the same objectivity you expect from the local *Times*. A good reporter, you know, never creates news, but just reports it. Opinions and speculations get labeled as such and quarantined to their own section. The front page reports objective facts, free of reporter biases.

You might expect this objectivity from vision, but you will not get it. What you get instead resembles an opening statement from the local DA: a carefully constructed story, part fact, part supposition, clearly biased, sometimes downplaying or ignoring evidence to the contrary.

Why? Your visual system needs to tell a three-dimensional (3D) story about objects, shapes, colors, and motions. The only evidence it has to construct this story are photon catches at receptors laid out in a two-dimensional (2D) mosaic on the retina. The gap between the evidence given and the story to be constructed is enormous, as anyone will testify who has tried to build a working machine vision system: the evidence at the retina is logically compatible with innumerable different stories. Careful detective work is required to bridge the gap.

The problem is so difficult your brain devotes roughly 10 billion neurons to it. Wired into these neurons are many procedures and biases for story construction.

We discuss one of them. Visual psychologists and researchers in computer vision sometimes call it the principle of "genericity" or "nonaccidentalness" (Biederman, 1985; Binford, 1981; Hoffman & Richards, 1984; Koenderink, 1990; Lowe, 1985; Lowe & Binford, 1981; Ullman, 1979; Witkin & Tenenbaum, 1983). We discuss how this principle shapes our perceptions of line drawings, object parts, and subjective surfaces.

GENERICITY AND LINE DRAWINGS

The principle of genericity, in its simplest form, says to reject any 3D interpretation of the retinal image that would place the eye in an "unstable" viewing position. One way to define an unstable viewing position is as follows: it is a viewing position which, if perturbed slightly, would lead to a change in the topological or first order differential structure of the image. Two examples will help.

First a topological case. Suppose the image contains an L junction, i.e., two line segments which meet at a vertex as in the letter *L*. Consider a 3D interpretation consisting of two disconnected line segments at different depths in space. Under this 3D interpretation the reason the image contains an L junction, instead of two separated line segments, must be that your eye is viewing the line segments from a special vantage which makes their endpoints look connected. If you were to move your eye slightly, the image of the L junction would become an image of two separate line segments. (This separation actually occurs in the familiar "Ames chair illusion," in which a set of disconnected sticks in space look like a chair from only one special viewpoint, and otherwise appear to be disconnected. See Kilpatrick, 1952.) This introduction of a gap is a topological change in the structure of the image. Therefore the principle of genericity says to reject this 3D interpretation.

Now a first-order differential example. Suppose the image contains a single line segment. Consider a 3D interpretation consisting of two line segments in space meeting to form a right angle. Under this 3D interpretation the reason the image contains a single line segment, instead of two line segments meeting to form an L junction, must be that your eye is viewing the right angle from a special vantage which hides the vertex. If you were to move your eye slightly, the image of the line would become an image of two lines meeting at an L junction. This introduction of a tangent discontinuity is a change in the first-order differential structure of the image. Therefore the principle of genericity says to reject this 3D interpretation.

The genericity principle has been used by various researchers to justify rules for interpreting images (see, e.g., Lowe, 1985). Some examples are the following:

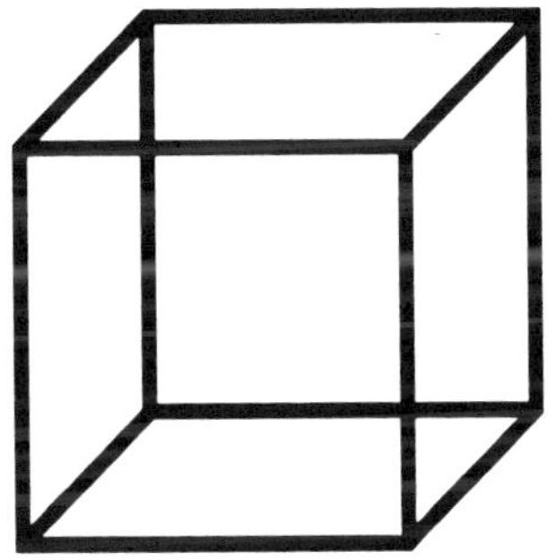

FIG. 6.1. The Necker cube first described by Louis Albert Necker in 1832.

Rule 1. Points collinear in an image are collinear in the world.

Rule 2. Points smoothly connected in an image are smoothly connected in the world.

Rule 3. Points symmetric in an image are symmetric in the world.

Rule 4. Curves terminating at a common point in an image terminate at a common point in the world.

Rule 5. Three or more curves intersecting at a common point in an image intersect in a common point in the world.

These rules greatly constrain the possible interpretations of line drawings. Consider, for instance, the Necker cube shown in Figure 6.1. The straight line segments in this figure must be interpreted as straight lines in space based on Rules 1 and 2. The reasoning is as follows. According to Rule 2, since the points in a line segment are smoothly connected in the image they must be smoothly connected in space. And according to Rule 1, since the points in a line segment are collinear in the image they must be collinear in space. Therefore they must form a line in space.

One can cast this reasoning in a Bayesian format, assuming no noise. Consider the conditional probability that a given straight-line segment S in an image actually arose from the projection of a wiggly curve in space. We can write this as $P(\text{wiggly in world}|S \text{ in image})$ or more simply $P(W \mid S)$. We wish to show that $P(W \mid S)$ is zero. By Bayes' rule we can write

$$P(W \mid S) = \frac{P(S \mid W)P(W)}{P(S)}. \tag{6.1}$$

Here $P(W)$ and $P(S)$ are the prior probabilities, respectively, of wiggly curves in space and of the straight line segment S. For our purposes, we do not need to estimate $P(S)$. We can estimate the numerator, viz., $P(S \mid W)P(W)$, then set $P(S)$ to a value which normalizes the numerator to a probability. Consider, then, the first factor in the numerator, the so-called likelihood $P(S \mid W)$. $P(S \mid W)$ is the

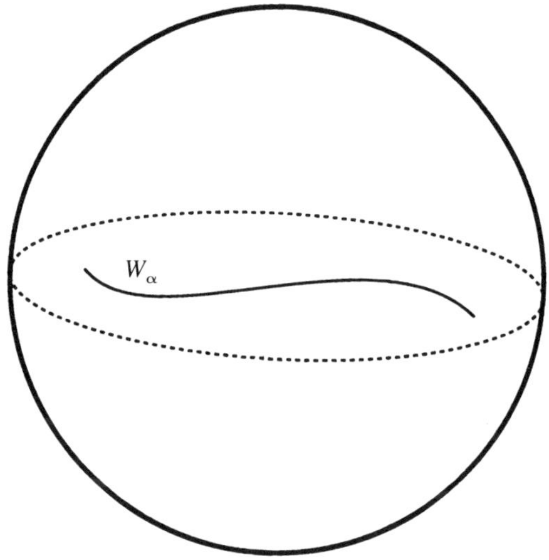

FIG. 6.2. The sphere of viewing directions around a wiggly line, with nongeneric views indicated by dashed lines.

conditional probability that a wiggly curve in space will project to the straight-line segment S. To determine $P(S \mid W)$, genericity says to assume that all viewpoints are equally likely. The set of viewing directions for a wiggly curve W_α in space is illustrated in Figure 6.2 by a sphere surrounding W_α. Since the sphere has finite area, it is possible to place a finite uniform measure on it; in the language of groups we can say that since SO(3) is a compact group it admits finite Haar measures. Given this assumption, the measure of a set of viewing directions is proportional to the area of the set. The set of viewing directions for which W_α projects to S or a scaled version of S is indicated by the dashed great circle on the sphere. This set is one dimensional; therefore, it has no area and, in consequence, zero probability. Moreover *every* wiggly curve W_α in space that can project to the line segment S, or a scaled version of S, can do so only from viewpoints on this *same* great circle. One can see this by noting that S and W_α must be entirely coplanar for W_α to project to S. Thus, not only is $P(S \mid W_\alpha) = 0$ for each α, but so also is $\int P(S \mid W_\alpha)P(W_\alpha)\,d\alpha$. Consequently, $P(S \mid W)$ is zero, which by (1) implies $P(W \mid S)$ is also zero. We conclude that genericity entails the rule: if the prior probability of 3D straight lines is nonzero, interpret any straight line in an image as straight in 3D. A wiggly 3D interpretation is nongeneric; a slight movement of the eye would reveal the wiggles.

Now consider Figure 6.3. Your initial impression is probably that this depicts some sort of pinwheel. You initially see it as flat. With difficulty you might also be able to see it as another view of the Necker cube, seen from a nongeneric view in which two vertices of the cube are precisely aligned. Why is it hard to see the Necker cube? As we have just shown, all line segments in the image must be interpreted as straight lines in space. Moreover, according to Rule 5, each of the

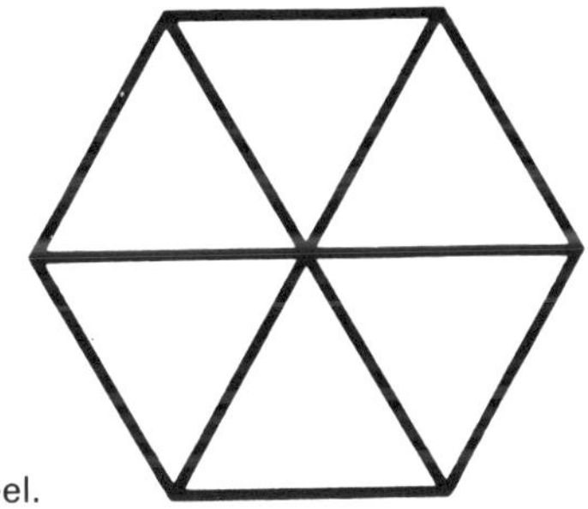

FIG. 6.3. A pinwheel.

three lines that intersect in the center of the figure must be interpreted as intersecting at a common point in space. This precludes a Necker cube interpretation.

You might argue that symmetry, not genericity, is the main principle underwriting a pinwheel interpretation in this figure. This would be the explanation given by the Gestalt principle of Prägnanz: That interpretation is to be made which is simplest. The 2D pinwheel is already highly symmetrical in 2D, so there is no need to go to a 3D interpretation.

But Figure 6.4 suggests that this is not right. Here we see a 3D shape from one generic and one nongeneric view. The generic view leads to a 3D interpretation. The nongeneric view usually does not, even though the 2D interpretation that is seen is much less simple or symmetric. We can explain the differing perceptions based on the rules derived from genericity, just as we did with the pinwheel. We cannot explain the difference by appeal to symmetry.

Genericity is not the only principle used by human vision to interpret images, and in many cases it cannot, by itself, force a unique interpretation. But it is a powerful principle. We will shortly mention other principles that interact with genericity in the generation of 3D interpretations. But Figure 6.5 indicates just

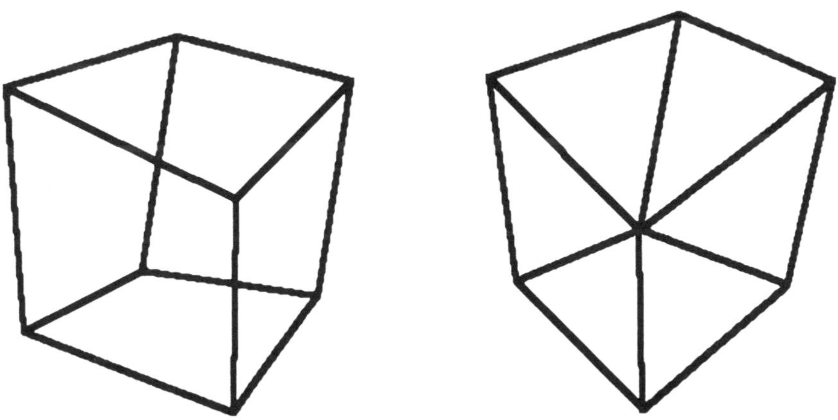

FIG. 6.4. One generic and one nongeneric view of a 3D shape (adapted from Kanizsa, 1975).

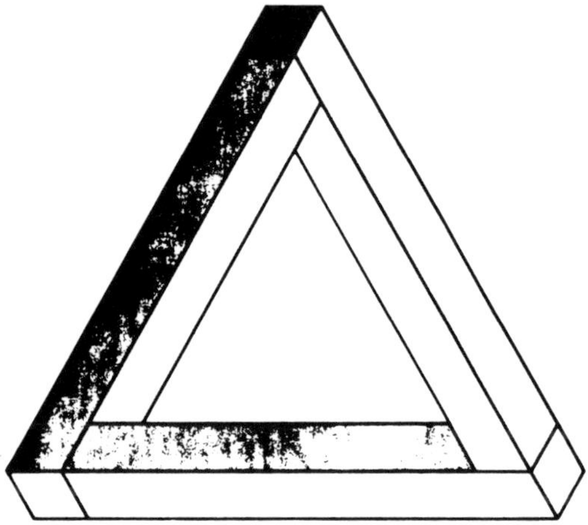

FIG. 6.5. The Penrose triangle.

how committed to genericity we can be. This figure shows the well-known Penrose triangle. At first glance this looks like a normal 3D model of a triangle. A closer look reveals that what you perceive is physically impossible. No real triangle could be built which would project to this image. However, there is a different 3D object which could be built and which would project to this image. It is a triangle broken at one corner with the two edges twisted away from each other. This has been constructed by Gregory (1970). However, to get the image shown in Figure 6.5, one must photograph this 3D model from exactly one viewpoint. Move the camera slightly and the image of the Penrose triangle is ruined. Since this physically possible 3D interpretation requires a special viewpoint, human vision rejects it. We prefer, in this case, to see a 3D interpretation which is physically impossible but satisfies genericity, rather than to see one which is physically possible and violates genericity.

The preceding analyses did not take into account the fact that real-world visual systems have only finite resolution and must tolerate noise. These limitations imply that nongeneric interpretations of images by human vision will have "small" but not zero probability. For this reason the "rules" of image interpretation based on genericity are really like cues: they can be overruled, even by a visual system that is ideal in the sense that it always infers the "most probable" interpretation of the images presented to it. Image interpretation using cues is based on a comparision of the collective weight of the cues (evidence) favoring each interpretation.

Jepson and Richards (1992) have presented counterexamples to the hypothesis that human vision always interprets images in accordance with the generic viewpoint assumption. For example, in Figure 6.6a (due to Jepson and Richards) the bottom edge of the small block on the left appears to be collinear with the bottom edge of the large block on the right. However, if the figure is rotated clockwise

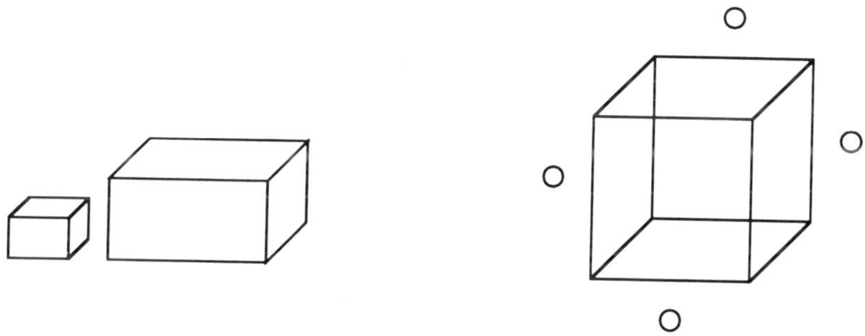

FIG. 6.6. The interaction of genericity with other cues to depth.

by 90° then the interpretation changes. Now those edges are not interpreted as collinear in space. Instead the small block appears to be closer to the observer than the large block. The effect is even stronger when the original figure is rotated counterclockwise by 90°.

A proximity rule of depth assignment seems to be affecting our perception of this display. According to this rule features that are near each other in an image should be interpreted as being near each other in space. The proximity rule is another instance of the principle of genericity (in the sense of "small" probabilities): If two features are widely separated in space, then only a small range of viewpoints would place them near each other in an image. In Figure 6.6b this appears to explain the apparent depth of the small circles relative to the edges of the block.

In Figure 6.6a the bottom edges of the blocks (in the original orientation) are collinear in the image. However, the features on the blocks that are nearest to each other in the image are the right rear edge of the small block and the left front edge of the large block. In the original orientation the bottom edges of the two blocks are at the same height in the visual field. This supports the interpretation that they are at the same depth in space using the "height-in-the-field" rule of depth assignment. This interpretation is also supported by collinearity. The proximity rule, which would predict that the right rear edge of the small block would be at approximately same depth as the left front edge of the large block, appears to have been overruled by the combination of the "height in the field" and collinearity rules in this case.

However, when the display is rotated clockwise by 90° the height-in-the-field rule would be expected to be inoperative since the observer is looking up at the blocks rather than down at them from above; the bottom surfaces rather than the top surfaces of the blocks are now visible. In this case the proximity rule appears to overrule the collinearity rule. When the original display is rotated counterclockwise by 90°, the height-in-the-field rule predicts that the large block should be seen behind the small block, in agreement with the proximity rule. The combination of these rules strongly overrules collinearity. Thus there appears to be a good deal of both cooperation and competition among the various genericity-based rules and other rules of depth assignment.

GENERICITY AND PARTS

Genericity, as we have seen, helps to guide the assignment of 3D structures to 2D images. To recognize these 3D structures as objects, further processing is required. One aspect of this further processing is the decomposition of 3D structures into simpler subunits, or "parts."

Part decompositions aid the recognition process by allowing recognition despite occlusions and despite nonrigid motions of parts, such as legs or arms (Biederman, 1987; Hoffman & Richards, 1984; Marr & Nishihara, 1978). Ideally a part decomposition should be (1) easily computed from images, (2) applicable to all classes of objects, and (3) independent of viewing geometry.

Genericity motivates an approach to part decompositions that is close to ideal. Figure 6.7 illustrates the basic idea. On the left of the figure are two objects. On the right the two have been generically intersected to form a single composite object. Since the intersection is generic, the tangent planes to the surfaces of the two objects are almost never parallel at the points where the two surfaces meet. The two surfaces almost everywhere meet in a concave discontinuity. This is illustrated by the dashed circular contour in Figure 6.7. That surfaces generically intersect in concave discontinuities follows from a transversality theorem of the field of differential topology (Guillemin & Pollack, 1974).

This motivates a simple rule for decomposing 3D shapes into parts: Divide shapes into parts at contours of concave discontinuity (Hoffman & Richards, 1984).

An application of this rule is illustrated in Figure 6.8. On the left is the well-known Schröder staircase. At first this appears to be an ascending staircase, and the two dots appear to lie on the rise and tread of a single step. Note that all the steps are bounded by lines of concave discontinuity, as dictated by genericity.

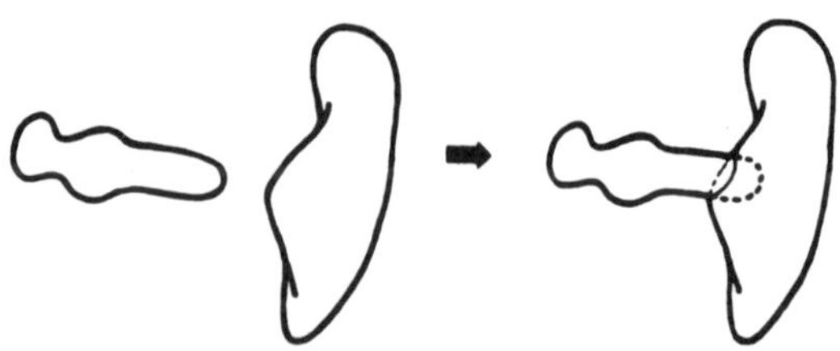

FIG. 6.7. Generic intersections of the surfaces of objects leads to concave discontinuities.

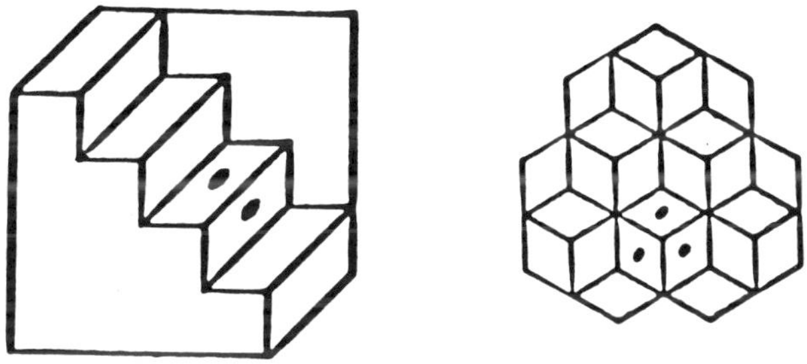

FIG. 6.8. The Schroder staircase and stacked cubes.

Contours of convex discontinuity separate the rise and tread of a single step, but do not serve to carve the staircase into steps. Upon further inspection, this staircase will appear to reverse figure and ground, so that concave discontinuities become convex, and vice versa. Our rule for part decomposition therefore predicts a new organization into parts, with part boundaries along the new lines of concave discontinuity. You can check for yourself that this prediction is fulfilled. The two dots which appeared to be on the rise and tread of a single step now appear to be on two distinct steps. Similar comments hold for the stacked cubes on the right. The cubes are all separated along contours of concave discontinuity, with the three dots at first appearing to lie on a single cube. Upon further inspection, figure and ground reverse and one gets new part boundaries, and new cubes, as predicted by genericity.

Smoothing contours of concave discontinuity leads to extrema of surface curvature, specifically negative extrema in one of the principal curvatures. (An extremum of curvature is a negative extremum if it is in a concave region of the surface.) This suggests that for smooth objects we use negative extrema of the principal curvatures to delineate parts (Hoffman & Richards, 1984). An example of the parts given by this rule is shown for the "cosine surface" illustrated in Figure 6.9. The dashed circular contours indicate the negative extrema of surface curvature and, therefore, the part boundaries. These boundaries organize the surface into a succession of ring-shaped hills. If you turn this illustration upside down, you will notice that the dashed circular contours no longer work as part boundaries. Now they appear to lie in the middle of the hills, instead of between the hills. Your organization of the cosine surface into parts has changed. The reason is that turning the illustration upside down causes your visual system to reverse the choice of figure and ground on the cosine surface. (We reverse figure and ground because, apparently, we prefer to see the surface lying below us rather than floating above us.) This reversal of figure and ground turns concavities into convexities, and vice versa. Thus, negative extrema of the principal curvatures become positive extrema, and vice versa. And, since negative extrema determine the part boundaries, these boundaries must move to the new negative extrema. Consequently we see new parts.

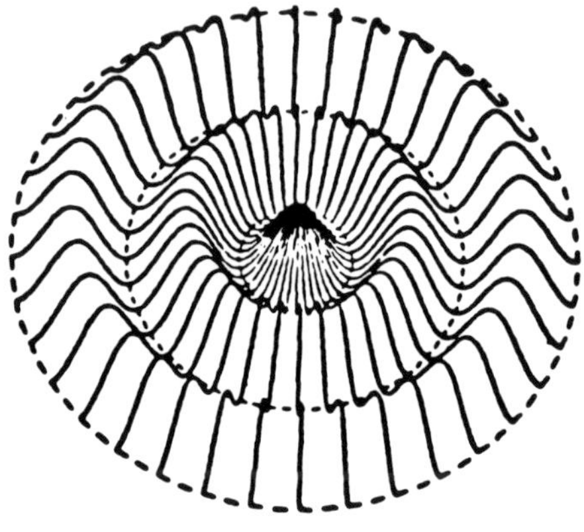

FIG. 6.9. The cosine surface.

Koenderink (1990) has proposed a theory of object recognition based on the idea of generic versus accidental views. In this theory the ambient space of possible viewpoints on a scene is divided into "cells." The cell that contains a particular viewpoint is the largest connected region of the ambient space within which all viewpoints give rise to "qualitatively" equivalent images. The "cell walls" in this theory define surfaces in space. When an observer crosses a cell wall, the qualitative structure of the image changes. If we are considering the case of orthographic projection, then the cells become just patches on the sphere of viewing directions, and the cell walls are the curves that bound those patches. Koenderink's claim is that much of the quantitative, metric information in images is not used by the visual system and that, for most purposes, object recognition proceeds using only qualitative information.

GENERICITY AND ILLUSORY CONTOURS

Genericity turns out to have important implications for the perception of surfaces, including illusory surfaces. Nakayama and Shimojo (1990, 1992) used it to explain phenomena in the area of stereoscopic perception of untextured surfaces, and Kellman and Shipley (1991) used it in their "discontinuity" theory of perceptual unit formation.

When Figure 6.10 (Nakayama & Shimojo, 1990, 1992) is cross-fused (by crossing one's eyes so that the left and right figures superimpose in the middle) most people perceive a horizontal bar overlaying a vertical bar, along with illusory contours in the central region that complete the boundary of the horizontal bar. In other words, the black region appears to split into two distinct surfaces, giving a 3D segmentation. Since the cross in this display is untextured, no disparity information is available in its interior. Also, the horizontal edges carry no disparity information since any point along such an edge in one eye could

FIG. 6.10. Stereo cross (adapted from Nakayama and Shimojo, 1990).

match any point along the corresponding edge in the other eye. (These different correspondences reflect the fact that a physical edge that projects to a straight horizontal edge in both eyes could still oscillate in depth in an arbitrary way, as discussed earlier.) Only the vertical edges carry unambiguous horizontal disparity information.

Nakayama and Shimojo's explanation for this perception is based on the generic viewpoint assumption. This assumption predicts the interpretation perceived by most people and eliminates others that would be predicted by other algorithms. Their argument is essentially the following: According to the generic viewpoint assumption edges that are straight in an image are straight in space, and edges that are collinear in an image are collinear in space. Since the disparity information in the image tells the observer that the outer endpoints of the horizontal edges of the horizontal rectangle are at the same depth (because the vertical edges *do* carry horizontal disparity information), these rules imply that the horizontal edges must be frontoplanar.

Nakayama and Shimojo point out that no simple disparity-spreading scheme is consistent with this perceptual interpretation. The idea here is that if an interpolation algorithm were used to assign depth to the interior of the cross using the disparity information available at its boundaries, then the horizontal arms should be smoothly interpolated in depth between the depths the vertical edges of the horizontal rectangle and the vertical edges of the vertical rectangle (see Figure 6.10). (This assumes that disparity signals generated by the boundaries of a homogeneous connected region can spread freely within that region.)

Kellman and Shipley (1991) have proposed a theory of "perceptual unit formation" using the principle of transversality. The term "perceptual unit formation" refers to illusory surface and contour formation, as well as occluded (amodal) surface and contour formation. According to their theory a necessary condition for perceptual unit formation is the presence of tangent discontinuities in the boundaries of regions in the image. If a pair of contours leading into distinct tangent discontinuities are "relatable," meaning that their extensions beyond the tangent discontinuitites intersect at an angle greater than or equal to 90°, then interpolation occurs between the contours. In this way perceptual units are formed.

The justification for the role given to tangent discontinuities in Kellman and Shipley's theory is based on the concept of generic occlusion: The transversality principle implies that tangent discontinuities almost always (in the technical sense) occur when occlusion is present. Essentially, the concept of generic occlusion in terms of transversality clarifies and justifies in a formal way the use of T junctions to infer interposition.

Later we will present a different proposal about the role of tangent discontinuities in the perception of illusory surfaces. Our proposal will allow for the fact that some examples of illusory surfaces do not give the impression of being interposed in front of their inducers. This will help to explain some data obtained by Shipley and Kellman (1990) that was inconsistent with their theory.

GENERICITY AND ICOs

In this section we will investigate some implications of genericity for the perception of illusory contours (ICs). In particular, we will propose necessary conditions for the perception of ICs in which the illusory surface appears to partially occlude its inducers. We will refer to these as "ICOs," which is short for "illusory contours that occlude." It will be shown that the generic viewpoint assumption places restrictions on the topological and first order differentiable structure of displays in which ICOs are perceived. However, as mentioned earlier, our analyses will assume infinite resolution and no noise. Human vision, with its finite resolution and inevitable noise can be expected to treat our "necessary conditions" as biases rather than strict rules.

For the case of ICOs induced by "blobs" we will use the principle of transversality. Consider Figure 6.11a. In this display most people see a white square that stands out from the surrounding white area. The white square appears to be in front of and partially occluding black disks. Notice that the principle of transversality, as applied to occlusion, is obeyed here: the tangents of the circles differ from the tangents of illusory square at the points where the contours meet. The occlusion is generic. However, in Figure 6.11b we have smoothed out the sharp convex corners of the "pacmen" in Figure 6.11a. If an ICO were seen in this display, then the occlusion would not be generic. In fact, most people do see a weak IC in Figure 6.11b, but they do not describe it as an ICO. Most people perceive the blobs to be pushed up against the side of the illusory square, as though the blobs were made of a soft, flexible material that has been deformed to fit the square's shape.

Thus, we propose that tangent discontinuities are a necessary condition for ICOs. But we do not claim that they are necessary for all ICs. Shipley and Kellman (1990) found in their experiments that subjects do perceive ICs in displays in which the tangent discontinuities have been removed, although the ICs were usually rated by subjects as weaker than when discontinuities were present. However, the IC in Figure 6.12 is rated by most subjects as relatively strong, although it contains no tangent discontinuities and the short line segments by themselves do not produce a signficant IC. There is not much brightness

FIG. 6.11. (a) Tangent discontinuities at the "lips" of the pacmen are necessary for the perception of an occluding illusory surface. (b) Smooth the discontinuities at the "lips" of the pacmen and no occluding illusory surface is seen.

enhancement in this figure. However we, as well as many other researchers (e.g., Kennedy, 1988) distinguish ratings of IC strength from ratings of brightness enhancement, since it has been shown that strong illusions of contour are not always accompanied by enhanced brightness. Other researchers (Bonaiuto, Giannini, & Bonaiuto, 1991; Kennedy, 1978; Kennedy, 1988; Purghe, 1991; Purghe & Katsaras, 1991) have also emphasized the theoretical significance of the distinction between ICs and ICOs. And from a theoretical point of view, the arguments for the necessity of tangent discontinuities based on transversality and generic occlusion only apply to ICOs.

The concept of generic occlusion can also be applied to the case of ICs induced by the ends of lines. The rules described earlier for the interpretation of line drawings are useful in this connection. Consider Figure 6.13a. This is a typical example of an IC induced by line endings. In this display most people

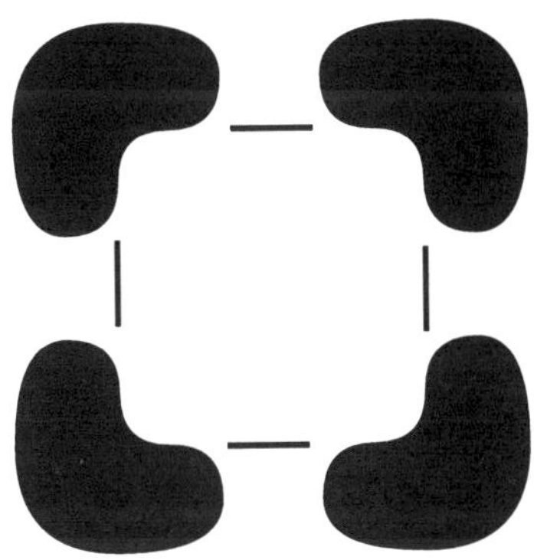

FIG. 6.12. Relatively strong illusory contours can be seen in figures without tangent discontinuities, but these illusory contours do not appear to occlude.

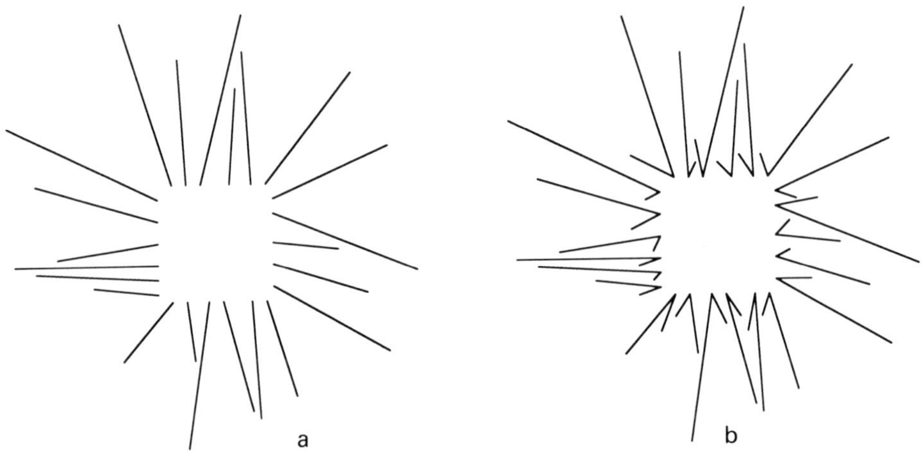

FIG. 6.13. (a) An illusory surface induced by line segments. (b) Adding more line segments to form vertices reduces the strength of the illusory surface and makes it appear to be nonoccluding.

perceive an illusory white square that is interposed in front of the lines. The lines appear to continue underneath the illusory square. Now, in Figure 6.13b we added short line segments to Figure 6.13a to make L junctions at the points where the inducing line endings occurred in Figure 6.13a. This change has weakened the IC, but it has also produced a *qualitative* change in the IC. Its apparent depth has moved back to approximately the same depth as the L junctions of the inducers. The inducing lines no longer appear to continue underneath the illusory surface. Instead, the L junctions of the inducers appear to be stuck into the side of the illusory surface. The IC is not an ICO.

We can account for this qualitative change in appearance using the concept of generic occlusion. The IC in Figure 6.13b passes right through the L junctions of the inducers. So, including the IC, there is in fact a K junction at each point where the IC meets an inducer. From Rule 5 for line-drawing interpretation, we know that if our viewpoint is generic then at a K junction all contours must be at the same depth. Therefore, the IC cannot be an ICO if genericity is obeyed.

Similarly, comparing Figure 6.14a with Figure 6.14b, most observers perceive an ICO in Figure 6.14a, and no IC or a nonoccluding IC in Figure 6.14b. Again, this can be explained by the K junctions formed at the intersection points of the inducers and the potential IC. What is interesting about this example is that the tangents of the inducing lines agree at the vertices. When attention is restricted to a small area around a vertex, it appears more like a termination of an isolated line than an L junction. Yet, the IC is strongly affected. This appears to be inconsistent with the predictions of the line-end-contrast theory of Frisby and Clatworthy (1975) as well as the neural network theory of Grossberg and Mingolla (1985).

FIG. 6.14. (a) An illusory surface induced by semicircles. (b) Adding more curves to form vertices reduces the strength of the illusory surface and makes it appear to be nonoccluding.

WHY OUTLINES OF BLOBS DO NOT INDUCE ICOs

Genericity also provides an interesting new perspective on the question of why "blobs," when drawn only in outline, do not produce significant ICs. This question has been widely discussed among researchers in IC perception (e.g., Kanizsa, 1974; Rock, 1987). For example, consider Figure 6.15a. Most observers report seeing only a very weak if any IC in this display. However, if the short line segments in Figure 6.15a are removed, as in Figure 6.15b, then a strong IC is seen.

We can understand, at least in part, the difference in the way these two displays are perceived by using genericity: Assume that an illusory surface is seen in Figure 6.15a. Then the short line segments cannot be viewed as partially occluded blob-shaped elements since it would be highly improbable that just a very thin edge of those blobs would be visible (also see Kellman & Shipley, 1991). On the other hand, if they are viewed as unoccluded line segments, then the fact that they are lying on (or directly next to) the IC means that they must be at the same depth as the IC. Otherwise it would imply an improbable coincidence of viewpoint. Now the short line segments coterminate with the circular arcs. So by Rule 4, the short line segments must also be seen at the same depth as the circular arcs at the junction points. Therefore, the potential IC must be seen at the same depth as the circular arcs at the junction points, so it cannot be an ICO. A similar argument can be used in the case of outlines of pacman inducers.

Intuitively the idea is that if the circular arcs in Figure 6.15a were perceived as being occluded by an illusory surface, as they are in Figure 6.15b, then the visual system would have to "wonder" why the short line segments terminate exactly where the circular arcs pass underneath the illusory surface in the image.

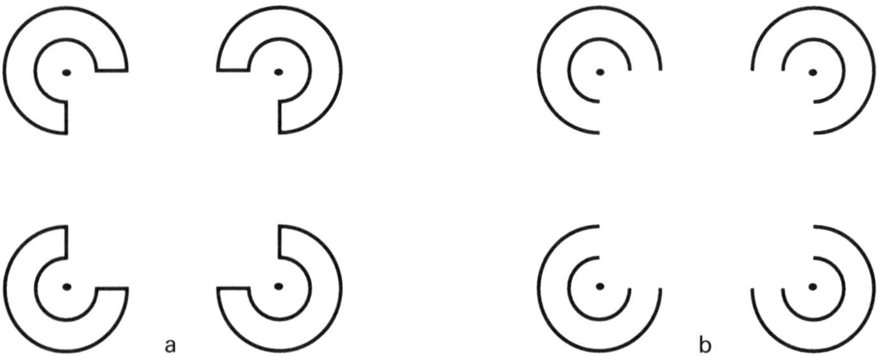

FIG. 6.15. (a) Blob outlines do not induce illusory contours. (b) Removing the short line segments in part (a) allows an ICO to emerge.

Kanizsa (1974) has discussed displays very similar to these in comparing his theory of IC perception with that of Gregory: "According to Gregory the sense data are used by the brain according to certain strategies, in order to decide which object has the highest probability of being present. But then, comparing the perceptual effects of Figures 12.26a and 12.26b [similar to our figures 6.15b and 6.15a, respectively], one should conclude that for the brain [a corner of the type in Figure 12.26b] is more probable than [a corner of the type in Figure 12.26a], a conclusion that seems to me rather implausible."

In our view it is not that the inducers in Figure 6.15a are more probable than those in Figure 6.15b, but that those in Figure 6.15a would be highly *improbable* if there were a surface (the potential illusory figure) in front, whereas those in Figure 6.15b would not. In other words, given that an ICO occurs in Figure 6.15b our theory helps us to understand why, despite the similarity between Figures 6.15a and 6.15b, an ICO does not occur in the former.

Genericity is also helpful in understanding the role of line segments that run along the length of an IC, as in Figure 6.12 discussed earlier. Consider Figure 6.16, in which the short line segments of Figure 6.15a have been moved away from the circular arcs. Most observers perceive an IC in this figure that is as strong or stronger than the one in Figure 6.15b. Note that the circular arcs appear to be partially occluded, whereas the line segments that lie along the IC do not. The line segments appear to be closer than the circular arcs, lying at the same depth as the IC. This is what would be predicted on the basis of the genericity arguments given above. The line segments are usually described by observers as entities of some sort attached to the edge of the illusory surface. Generally, in displays in which some of the inducers of an illusory figure are consistent with generic occlusion and some are not, the former are seen as partially occluded, while the latter are seen as unoccluded and are pulled forward to the same depth as the IC.

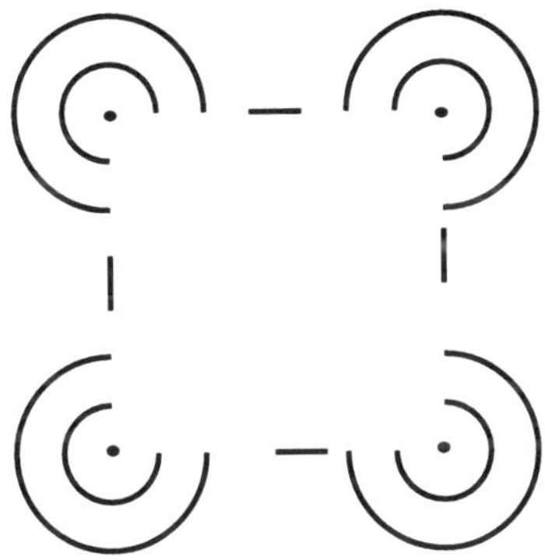

FIG. 6.16. A modification of Figure 6.15a that does induce a strong IC.

Kanizsa (1974) has argued that "closure" can explain the perception of ICs with line-end inducers. Supporters of this theory might claim that it can explain the effects seen in the displays in this section. However, we believe genericity to be a more satisfactory explanation, since it is a valid ecological constraint. It also predicts perceived depth relations, which closure does not.

Genericity can be applied in a similar way to obtain necessary conditions for the "neon color spreading" spreading effect (see Van Tuijl, 1975). Beginning with a display that produces neon color spreading using colored lines, more lines are added that intersect the original lines at their points of color change. The result is that the color spreading is greatly reduced and the perception of transparency disappears (see Albert and Hoffman, in press).

CONCLUSION

The reports of our visual systems are not unbiased accounts dictated by the state of the world. Rather these reports are the result of sophisticated inferences which have been wired into the billions of neurons which process vision. Many principles underlie these inferences. One of the most powerful and ubiquitous is the principle of genericity. As we have seen, genericity is integral to our perceptions of line drawings, the parts of objects, and subjective surfaces. Further study of genericity and of its interactions with other principles that shape our perceptions is a promising direction for research into human vision.

ACKNOWLEDGMENTS

We thank B. Bennett, M. Braunstein, J. Liter, C. Prakash, and S. Richman for useful discussions, and M. D'Zmura for comments on an earlier draft. The authors were supported by NSF grant DIR-9014278 and by ONR contract N00014–88–K–0354.

REFERENCES

Albert, M. K., & Hoffman, D. D. (in press). Generic visions: General position assumptions in visual perception. *Scientific American.*

Biederman, I. (1985). Human image understanding: Recent research and a theory. *Computer Vision, Graphics, and Image Processing, 32,* 29–73.

Biederman, I. (1987). Recognition-by-components: A theory of human image interpretation. *Psychological Review, 94,* 115–147.

Binford, T. O. (1981). Inferring surfaces from images. *Artificial Intelligence, 17,* 205–244.

Bonaiuto, P., Giannini, A. M., & Bonaiuto, M. (1991). Visual illusory productions with or without amodal completion. *Perception, 20,* 243–257.

Frisby, J. P., & Clatworthy, J. L. (1975). Illusory contours: Curious cases of simultaneous brightness contrast? *Perception, 4,* 349–357.

Gregory, R. L. (1970). *The intelligent eye.* New York: McGraw-Hill.

Grossberg, S., & Mingolla, E. (1985). Neural dynamics of form perception: Boundary completion, illusory figures, and neon color spreading. *Psychological Review, 92,* 173–211.

Guillemin, V., & Pollack, A. (1974). *Differential topology.* Englewood Cliffs, NJ: Prentice-Hall.

Hoffman, D. D., & Richards, W. A. (1984). Parts of recognition. *Cognition, 18,* 65–96.

Jepson, A., & Richards, W. (1992). *What makes a good feature?* (MIT AI Memo 1356).

Kanizsa, G. (1974). Contours without gradients or cognitive contours? *Italian Journal of Psychology, 1,* 93–112.

Kellman, P. J., & Shipley, T. F. (1991). A theory of visual interpolation in object perception. *Cognitive Psychology, 23,* 141–221.

Kennedy, J. M. (1978). Illusory contours not due to completion. *Perception, 7,* 187–189.

Kennedy, J. M. (1988). Line endings and subjective contours. *Spatial Vision, 3,* 151–158.

Kilpatrick, K. P. (1952). Elementary demonstrations of perceptual phenomena. In K. Kilpatrick (Ed.), *Human behavior from the transactional point of view* (pp. 1–15). Hanover, NH: Institute for Associated Research.

Koenderink, J. (1990). *Solid shape.* Cambridge: MIT Press.

Lowe, D. (1985). *Perceptual organization and visual recognition.* Boston: Kluwer.

Lowe, D. G., & Binford, T. O. (1981). The interpretation of three-dimensional structure from image curves. *Proceedings of IJCAI-7* (pp. 613–618). Vancouver, British Columbia, Canada:

Marr, D., & Nishihara, H. K. (1978). Representation and recognition of the spatial organization of three-dimensional shapes. *Proceedings of the Royal Society of London, B200,* 269–294.

Nakayama, K., & Shimojo, S. (1990). Towards a neural understanding of visual surface representation. *Cold Spring Harbor Symposium on Quantitative Biology, 40,* 911–924.

Nakayama, K., & Shimojo, S. (1992). Experiencing and perceiving visual surfaces. *Science, 257,* 1357–1363.

Purghe, F. (1991). Is amodal completion necessary for the formation of illusory figures? *Perception, 20,* 623–636.

Purghe, F., & Katsaras, P. (1991). Figural conditions affecting the formation of anomalous surfaces: Overall configuration versus single stimulus part. *Perception, 20,* 193–206.

Rock, I. (1987). A problem solving approach to illusory contours. In S. Petry & G. E. Meyer (Eds.), *The perception of illusory contours* (pp. 62–70). New York: Springer-Verlag.

Shipley, T. F., & Kellman, P. J. (1990). The role of discontinuities in the perception of subjective figures. *Perception and Psychophysics, 48,* 259–270.

Ullman, S. (1979). *The interpretation of visual motion.* Cambridge, MA: MIT Press.

van Tuijl, H.F.J.M. (1975). A new visual illusion: Neonlike color spreading and complementary color induction between subjective contours. *Acta Psychologica, 39,* 441–445.

Witkin, A. P., & Tenenbaum, J. M. (1983). On the role of structure in vision. In J. Beck, B. Hope, & A. Rosenfeld (Eds.), *Human and machine vision* (pp. 481–543). New York: Academic Press.

7 Empirical Meaningfulness, Measurement-Dependent Constants, and Dimensional Analysis

Ehtibar N. Dzhafarov
University of Illinois at Urbana–Champaign

ABSTRACT

The "empirical meaningfulness" analysis in theory of measurement imposes a priori restrictions on statements involving a given set of quantities, by striking down as "empirically meaningless" those of logically possible statements whose truth value (true or false) is not invariant under mutual substitutions of "admissible" measurements of the quantities involved. However, any logically unambiguous statement that is "empirically meaningless" by this invariance criterion can be equivalently reformulated to become, by the same criterion, "empirically meaningful." This is achieved by explicating in the statement all its measurement-dependent constants, whose numerical values covary with choices of measurements within a specified class. Provided that the basic, nonderivable laws of a given area (such as mechanics or psychophysics) can be formulated in some specific measurements (such as mass in grams or in absolute threshold units), a simple algorithm described in this chapter determines the set of measurement-dependent constants that ensure the invariance of these basic laws under any specified class of transformations of these measurements: the choice of this class, and thereby of the measurement-dependent constants, is subject to no substantive constraints. The only context in which the invariance considerations may be restrictive is that of deciding whether a given statement is logically derivable from a given list of basic laws: if it is, then one should be able to make it invariant under the same class of transformations with the aid of the same set of measurement-dependent constants. Dimensional analysis in physics, for instance, can determine that a statement is not derivable from a given set of physical laws (such as the gravitation law and the second law of motion) by demonstrating that it cannot be made dimensionally homogeneous (invariant under scaling transformations) if one only utilizes the dimensional constants that have been explicated in these basic laws themselves, when presenting them in a dimensionally homogeneous form. Outside the context of derivability, however, the requirement of dimensional homogeneity does not restrict the class of possible laws of physics, as their dimensional homogeneity can always be achieved by an appropriate choice of dimensional constants.

INTRODUCTION

The main points of this chapter can be summarized as follows.

(1) Any grammatically correct sentence (statement, law) considered "empirically meaningless" by traditional criteria (based on the idea of *invariance under mutual substitutions of measurement functions from a given class of functions*) is logically *equivalent* to a sentence that, by the same criteria, is "empirically meaningful." The difference therefore is in *form* rather than *content,* having nothing to do with empirical truth or falsity. By a universal algorithm called *covariant substitution* any logically complete sentence can be (re)formulated in an "empirically meaningful" form: this is achieved by *explicating,* within the sentence, all its *measurement-dependent constants,* whose values covary with choices of measurement functions. Confusions and ambiguities in the content of a sentence may only arise when its formulation is *logically incomplete,* and they can always be resolved by standard logical analysis. Once a sentence is logically complete, it can be classified as *logically false* (i.e., stating something that can be shown to be false by mathematical means, like the sentence "this length is 5, in all possible units"), *logically true* (like the sentence "this length is 5, in some units"), or *empirical* ("this length is 5 m").

(2) The key concept in the analysis of sentences involving measurement functions is that of the measurement-dependent constants. In physics all or most of measurement-dependent constants are commonly known *dimensional constants*. By appropriate choice of those, *any* physical sentence can be written in a *dimensionally homogeneous* form, which is the physics version of "empirical meaningfulness." This is done by what I call Bridgman's algorithm, a particular case of covariant substitutuion. Dimensional analysis is an algebraic technique determining whether and how a sentence containing a given list of variables can be written in a dimensionally homogeneous form *using only those dimensional constants that have been explicated (by Bridgman's algorithm) in other, more basic sentences, from which the sentence in question is assumed to be logically derivable*. The sentence is determined to be *nonderivable* from these basic sentences when the list of their dimensional constants is not sufficient to write this sentence in a homogeneous form. Outside the context of derivability, dimensional analysis does not and cannot tell which sentences may and which may not be empirically true, which is why dimensional considerations cannot impose any restrictions upon possible *fundamental laws* of an area (such as mechanics, material science, or psychophysics), or upon its *situational sentences,* describing

specific circumstances to which the fundamental laws are to be applied to derive a given sentence.

(3) A decision on "empirical meaningfulness" or "meaninglessness" of a sentence containing measurement functions cannot be based on the *automorphisms of the empirical operations* used in constructing these measurement functions. Insofar as we are able to *empirically distinguish* between different measurement functions of one and the same quantity, at least one of these measurement functions must be defined *empirically*—by a set of *nonumerical* (or prenumerical) relations. A specific measurement function, such as length in meters, cannot be, however, defined within a relational system that contains automorphisms other than identity (e.g., the traditional system for length, with order and concatenation): in such a system no *ostensive* proposition can be formulated for its elements, which means that no specific length can be identified as "the length such that [. . . empirical relations involving this length. . .]." To *empirically* define a specific measure, one should be able to *empirically* ("qualitatively") refer to a specific length, which means that additional empirical relations should be appended to those already in the system, reducing the group of its automorphisms to identity. It is only an artefact of legitimate but arbitrary formalization (axiomatization) choices that some empirical relations (e.g., order and averaging in constructing an "interval scale" for temperature) are explicitly included in the *formal system,* whereas other relations (such as, "to be below the temperature of freezing water") are not, being instead used as elements of an *interpretation* of the formal system. One can always formalize a relational system so that its only automorphism is identity, forming thereby an *empirically complete relational system. Empirically incomplete relational systems,* those with nontrivial automorphism groups, can always be viewed as *groups* of complete systems.

The views presented here considerably overlap with those of three authors: P. W. Bridgman (1922) on dimensional analysis in physics, J. Michell (1986, 1990) on what can be loosely characterized as the emphasis on logical derivability, and W. W. Rozeboom (1962) on the critical importance of measurement-dependent constants in formulation of scientific propositions.

A BIT OF INFORMAL LOGIC: VARIABLES, CONSTANTS, AND SENTENCES

I begin by mentioning, without elaboration, some general logical concepts used throughout this paper. A *sentential function* (or *predicate*), such as "$x + f(y) = 6$, where $x, y \in Re$ and $f \in F$," is a (grammatically correct) formulation relating *constants* and *variables*. All variables considered here are either *variable functions* (in our example, f) or *variable quantities* (x and y). Variables of both kinds assume their *values* in certain sets ($x, y \in Re, f \in F$; note that the value of a variable function is a fixed function). All nonvariable terms in a sentential function are constants: fixed quantities (in our example, 6), fixed sets (*Re, F*), fixed relations and functions (. . . + . . . =), and fixed logical terms ("where," "and"). In this chapter, however, the term "constant" is used exclusively to designate fixed numbers.

A sentential function becomes a *sentence* (or *statement*) when all its variables are bound by *logical quantifiers,* such as "for all," "for some," "for precisely three," etc., referring to the variables' possible values. Sentences, but not sentential functions, can be evaluated in terms of their *truth values:* TRUE or FALSE. The following two sentences are obtained by quantifying, in two different ways, the sentential function above: (1) "for all $x, y \in Re$ and all $f \in F$, $x + f(y) = 6$" and (2) "for all $x, y \in Re$ there is an $f \in F$, such that $x + f(y) = 6$." If *Re* designates the class of reals, and *F* the class of positive linear functions, then the first sentence is false, the second is true.

Using variable functions and, especially, binding them by quantifiers "for any function," "there exists a function," etc., is not as common as doing the same with variable numbers. The main reason for this is that in many cases (though not always) a variable function can be *parametrized,* i.e., expressed through a *fixed function* of several *variable numbers*. The class of positive linear functions *F*, for example, can be parametrized by conventional coefficients $a \in Re^+$ (positive reals) and $b \in Re$, so that the sentential function above can be written as "$x + (ay + b) = 6$, where $x, y, b \in Re$ and $a \in Re^+$." The two sentences previously formed then become (1) "for all $x, y \in Re$, all $b \in Re$, and all $a \in Re^+$, $x + (ay + b) = 6$," and (2) "for all $x, y \in Re$ there are $b \in Re$ and $a \in Re^+$ such that $x + (ay + b) = 6$."

The order in which different quantifiers enter in a sentence is important in determining which variable numbers (or functions) depend on which. In the sentence "for all $x, y \in Re$ there is an $f \in F$, such that $x + f(y) = 6$" the choice of function *f* depends on the values of *x, y,* whereas in the sentence "there is an $f \in F$ (such that) for all $x, y \in Re : x + f(y) = 6$" it does not. In this chapter whenever it is not obvious or immaterial, interdependences between different variables and functions will be explicitly stated ("an $f \in F$ depending on *x* and *y*") or indicated by subscripts: "for all $x, y \in Re$ there is an $f_{xy} \in F$ such that $x + f_{xy}(y) = 6$," or "for all $x, y \in Re$ there are $b_{xy} \in Re$ and $a_{xy} \in Re^+$, such that $x + (a_{xy}y + b_{xy}) = 6$".

Since explicitly written quantifiers make formulations cumbersome, the usual convention is that if a variable number (or a variable function) is not explicitly bound by quantifiers, it is treated as bound by the generality quantifier "for all," referring to the class of the variable's possible values. I will only use this *generality convention* with respect to those variables that are not focal for the analysis. If, as it will usually be the case, the analysis focuses on variable functions (or their parameters), then our two example sentences may be written as (1) "for all $b \in Re$ and all $a \in Re^+$, $x + (ay + b) = 6$," and (2) "there are b_{xy}

$\in Re$ and $a_{xy} \in Re^{+}$, such that $x + (a_{xy}y + b_{xy}) = 6$" (the missing quantifiers are "for all $x,\ y \in Re$").

In this paper, sentences whose truth value (TRUE or FALSE) can be ascertained by logical (mathematical) means only, are called *logical sentences* (they are either *logically true* or *logically false*). Valid mathematical theorems are logical (and logically true) in this sense, and both example sentences above are logical, one being logically true, another logically false. Sentences that are not logical are called *empirical* (being *empirically true* or *empirically false*). If $\mathscr{A}$ denotes the class of all geometric points that can be placed on this page, then the following sentence is empirical (and probably empirically true): "for all $\mathscr{x}, \mathscr{y} \in \mathscr{A}$, distance between $\mathscr{x}$ and $\mathscr{y}$ in meters < 6." Note, however, that this sentence is empirical not because it refers to an empirical operation (measurement of distance in meters), but because its truth or falsity cannot be ascertained by purely logical or mathematical means. The sentence "there are $\mathscr{x}, \mathscr{y} \in \mathscr{A}$, such that distance between $\mathscr{x}$ and $\mathscr{y}$ in meters < 6" is logical, and logically true (because one can always find such pairs $\mathscr{x}, \mathscr{y} \in \mathscr{A}$, namely, $\mathscr{x} = \mathscr{y}$, for which the distance in meters is zero).

This following comment might seem superfluous, but is essential in the present context: all functions entering sentences under consideration are assumed to be *well defined*. This means that when one says $\log_e(x)$ there is an *effective mathematical procedure* to compute $\log_e(x)$ given x from its domain; when one says "numerical value t of temperature $\mathscr{t}$ in degrees Celsius," there is an *effective empirical procedure* that allows one, given temperature $\mathscr{t}$, to arrive at the number t. "Effective," here, means performable in a countable (generally infinite) number of steps defined by induction.

MEASUREMENT FUNCTIONS

The example with "numerical value t of temperature $\mathscr{t}$ in degrees Celsius," is that of a *measurement function* (MF). In general, a MF $x = x(\mathscr{x})$ is an effective empirical procedure by which a number x is assigned to any "instance" (or "magnitude") of an "empirical quantity" $\mathscr{x}$. Avoiding philosophical discussions, I will assume that the meanings of the terms "quantity" (like mass) and "magnitude" thereof (a value of mass) are understood. Different MFs measuring one and the same quantity are defined by their mathematical relations to each other, *conversion functions,* usually forming an N-parametric mathematical group of strictly increasing functions. The class of the "interval-scale" temperature MFs, for example (call this class TEMP), satisfies the following proposition: if MF $t(\mathscr{t})$ $\in$ TEMP, then for any real b and any positive real a, $t'(\mathscr{t}) = [at(\mathscr{t}) + b] \in$ TEMP. Here, the conversion functions form a two-parametric group of positive linear transformations.

Obviously, the conversion functions associated with a given class of MFs do not effectively define this class, unless at least one of the MFs is defined by some independent empirical procedure (an *anchoring* MF, e.g., temperature in degrees Celsius), allowing one to compute its values directly from an empirical quantity, rather than from other MFs for this quantity. Thus, an effective definition of the class TEMP of temperature MFs would be: $t(\mathcal{t}) \in$ TEMP, if and only if for some real b and some positive real a, $at(\mathcal{t}) + b = \bar{t}(\mathcal{t})$, where $\bar{t}(\mathcal{t})$ is the anchoring MF (the description of the empirical procedure follows).

Now we are ready to form various sentences involving temperature MFs. The most celebrated example of "empirical meaninglessness," given in virtually any treatise on the subject, is of the form

$$\text{if } \mathrm{B}(\mathcal{t}_1, \mathcal{t}_2) \text{ then } t(\mathcal{t}_1)/t(\mathcal{t}_2) = 2, \tag{7.1}$$

where $\mathrm{B}(\mathcal{t}_1, \mathcal{t}_2)$ can be replaced by some nonnumerical relation involving $\mathcal{t}_1$ and $\mathcal{t}_2$, such as "$\mathcal{t}_1$ and $\mathcal{t}_2$ are temperature magnitudes of objects $\mathbb{O}_1$ and $\mathbb{O}_2$, respectively [an identification of $\mathbb{O}_1$ and $\mathbb{O}_2$ in nontemperature terms follows]." Sentential function 7.1 is not, strictly speaking, a sentence, and it can only be considered a sentence under the convention of generality (free variables treated as bound by generality quantifiers). Then the explicit form of this sentence is

$$\begin{gathered}\text{for } \textit{all} \text{ MFs } t \in \text{TEMP and } \textit{all } \mathcal{t}_1, \mathcal{t}_2\text{:}\\ \text{if } \mathrm{B}(\mathcal{t}_1, \mathcal{t}_2) \text{ then } t(\mathcal{t}_1)/t(\mathcal{t}_2) = 2.\end{gathered} \tag{7.2}$$

This sentence can be trivially shown to be false by reductio ad absurdum, so in the terminology of this chapter the sentence is logical, and its truth value is FALSE. No empirical knowledge of temperature MFs or of the empirical relation $\mathrm{B}(\mathcal{t}_1, \mathcal{t}_2)$ is involved in this derivation; the latter is based exclusively on the specific combination of conversion functions with logical quantifiers (which is why the sentence is logical). The key words here are "specific combination of . . . with logical quantifiers," as can be seen from the fact that the following sentence is *logically true:*

$$\begin{gathered}\text{for } \textit{all } \mathcal{t}_1, \mathcal{t}_2 \textit{ there is} \text{ a MF } t \in \text{TEMP (depending on } \mathcal{t}_1, \mathcal{t}_2\text{) such that}\\ \text{if } \mathrm{B}(\mathcal{t}_1, \mathcal{t}_2) \text{ then } t(\mathcal{t}_1)/t(\mathcal{t}_2) = 2.\end{gathered} \tag{7.3}$$

Since 7.3 and 7.2 include one and the same sentential function, there can be nothing illegitimate in computing ratios of the temperature MFs per se; it is simply that some sentences about such ratios turn out to be logically false. The following sentence is yet another way of quantifying sentential function 7.1:

$$\begin{gathered}\textit{there is} \text{ a MF } t \in \text{TEMP such that for } \textit{all } \mathcal{t}_1, \mathcal{t}_2\text{:}\\ \text{if } \mathrm{B}(\mathcal{t}_1, \mathcal{t}_2) \text{ then } t(\mathcal{t}_1)/t(\mathcal{t}_2) = 2.\end{gathered} \tag{7.4}$$

This sentence is empirical, with its truth value depending on $\mathrm{B}(\mathcal{t}_1, \mathcal{t}_2)$. It would be (empirically) true, for instance, if $\mathrm{B}(\mathcal{t}_1, \mathcal{t}_2)$ describes such temperature pairs $(\mathcal{t}_1, \mathcal{t}_2)$ that when water of temperature $\mathcal{t}_1$ is mixed with equal amount of freezing water without heat loss, the eventual temperature of the mixture is $\mathcal{t}_2$. For this $\mathrm{B}(\mathcal{t}_1, \mathcal{t}_2)$, temperature in degrees Celsius provides one possible solution.

With one important exception, considered in the next section, *all* examples of "empirically meaningless" propositions given in the literature (e.g., Falmagne & Narens, 1983; Pfanzagl, 1968; Roberts, 1979; Suppes, 1959; Suppes & Zinnes, 1963) are simply logically false sentences. In all such examples a sentential function, like 7.1, is *implicitly* interpreted under the generality convention, like 7.2, and the resulting sentence is shown to be false by reductio ad absurdum. Then, however, this logical falsity is attributed to the "inadmissibility" of the operations with MFs contained in the sentence, rather than to the logical structure of the sentence (and most importantly, its logical quantification). There is no big harm in calling things differently, and mental translation of "empirical meaninglessness" into logical falsity is not a demanding exercise. This terminology, however, is potentially misleading, because it might suggest that some empirical considerations, in addition to simple logical principles, are being involved—when in fact they are not.

MEASUREMENT-SPECIFIC SENTENCES AND INVARIANCE UNDER SUBSTITUTIONS OF MFS

I mentioned that there was an important exception to the rule that all "empirically meaningless" sentences are simply logically false when formulated unambiguously. This exception relates to *measurement-function-specific* (MF-specific) sentences, those referring to uniquely specified MFs (such as length in meters or Celsius temperature). Returning to the putative ratio of temperature MFs (from the class TEMP), an example might be (I begin using the generality convention here and omit the quantifiers "for all t_1, t_2")

$$\text{if } \mathrm{B}(t_1, t_2) \text{ then } \bar{t}(t_1)/\bar{t}(t_2) = 2, \tag{7.5}$$

where $\bar{t}(t)$ denotes a specific temperature MF belonging to TEMP, say Celsius temperature. To be well defined, this MF should either be anchoring itself (i.e., defined through an effective empirical procedure), or be reducible to an anchoring MF by a conversion $a\bar{t}(t) + b$. Assume for simplicity that $\bar{t}(t)$ (Celsius temperature) is an anchoring MF of the class TEMP.

According to the position considered (Falmagne, 1992; Falmagne & Narens, 1983; Narens & Mausfeld, 1992), sentence 7.5 is "empirically meaningless" because its truth value is not preserved under direct replacements of $\bar{t}(t)$ by other MFs from the class TEMP. Indeed, if one substitutes Fahrenheit temperature $\bar{\bar{t}}(t)$ for $\bar{t}(t)$, then the sentence

$$\text{if } \mathrm{B}(t_1, t_2) \text{ then } \bar{\bar{t}}(t_1)/\bar{\bar{t}}(t_2) = 2, \tag{7.6}$$

cannot be true if 7.5 is true (and vice versa). Now we have a serious discrepancy between "empirical meaninglessness" and logical falsity: both 7.5 and 7.6 are empirical sentences, and one of them may very well be empirically true. This approach has been criticized in the literature by pointing out its logical arbitrariness (Guttman, 1971; Michel, 1986, 1990). My analysis goes one step farther: I will show that insofar as the *content* (rather than a specific *form*) of a MF-specific sentence is concerned, the substitution criterion is not restrictive at all—*any such sentence can be equivalently reformulated so that its truth value will be preserved under substitutions of MFs within any class of MFs, however broad or arbitrary.* The dichotomy of form and content is not as vague or philosophical as it may appear. I call a characterization of a sentence *content related* if, and only if, whenever it holds for the sentence, it also holds for all sentences logically equivalent to it; if a characterization holds for a sentence but does not hold for at least one of its equivalents, then the characterization is *form related.* Logical truth value (TRUE or FALSE) is content related, by definition, and so are logical derivability and informal characterizations like "profound," or "interesting." "Empirical meaningfulness," by contrast, is only form related.

A systematic demonstration of this claim for sentence 7.5 involves two steps. First, we construct a sentence equivalent to 7.5 but referring to the *entire* class TEMP of MFs:

$$\text{for any MF } t \in \text{TEMP there is a real number } c_t \text{ such that}$$
$$\textit{if } \mathrm{B}(\ell_1, \ell_2) \text{ then } \frac{t(\ell_1) - c_t}{t(\ell_2) - c_t} = 2,$$
$$\text{where } c_t = 0 \text{ when } t \text{ is } \bar{t} \text{ (Celsius temperature).} \tag{7.7}$$

It is easy to verify that this sentence has precisely the same truth value as 7.5, that is, they are indeed logically equivalent (interdeducible). For the moment I will leave open the question of what is the general algorithm by which this sentence is derived from 7.5. The second step consists in constructing direct logical *specializations* of this general sentence to specific MFs, such as Celsius $\bar{t}(\ell)$ and Fahrenheit $\bar{\bar{t}}(\ell)$:

$$\text{if } \mathrm{B}(\ell_1, \ell_2) \text{ then } \frac{\bar{t}(\ell_1) - 0}{\bar{t}(\ell_2) - 0} = 2 \quad \text{and} \quad \text{if } \mathrm{B}(\ell_1, \ell_2) \text{ then } \frac{\bar{\bar{t}}(\ell_1) - 32}{\bar{\bar{t}}(\ell_2) - 32} = 2. \tag{7.8}$$

(One could even attach subscripts $\bar{t}$ and $\bar{\bar{t}}$ to 0 and 32, respectively, to indicate that they are "measured in" °C and °F.) These two sentences are both equivalent to 7.5, they are both equivalent to 7.7 of which they are specializations, and they have a common "form" *up to a measurement-function-dependent* (MF-dependent) constant c_t. This constant is a mathematical function (or "reduction") of the conversion coefficients a, b that define the class TEMP, $c_t = -b/a$, and its sole purpose is to ensure the generalizability of 7.5 to the entire class TEMP.

One can clearly see now that the "empirical meaninglessness" of 7.5 is due to the fact that the constant 0, subtracted from the Celsius temperature values, has been overlooked, and the question of whether this constant is or is not MF dependent has not been raised. De facto the decision has been (unknowingly) made in favor of the MF independence of 0, resulting in the unjustifiable replacement of 7.5 with 7.6. To appreciate the peculiarity of the situation, note that this mistake would probably be avoided if the initial observations that led to 7.5 were made in °F rather than °C. Then the initial sentence would have the form of 7.8 (right), rather than 7.5, and it would be easy to realize that 32 is "measured in °F" and hence must change when we switch to other MFs.

The concept of a MF-dependent constant generalizes the familiar notion of a *dimensional constant*. It is a well-known fact that physical laws preserve their truth values (and their "forms") under changes of measurement units only if the numerical values of all dimensional constants in these laws are changed "correspondingly." In the abstract measurement literature, sentences containing MF-dependent constants were considered by Pfanzagl (1968) under the name of "meaningfully parametrized relations." The notion seems to have escaped the attention of later writers (e.g., Roberts, 1979, pp. 79–80, treats dimensional constants as essentially a nuisance for otherwise straightforward theory), but "meaningfully parametrized relations" have been reintroduced in Luce, Krantz, Suppes, and Tversky (1990, chap. 22). Using this language, the point of this section, systematically developed in the next one, is that any sentence can be "meaningfully parametrized," with respect to any class of MFs that includes those contained in the sentence. "Meaningful parametrization" of a sentence, therefore, is logically ensured and has nothing to do with its empirical content.

COVARIANT SUBSTITUTION

A MF-specific sentence, like 7.5, can be rewritten in an infinity of equivalent forms, containing different numerical constants. For example, every occurrence of $\bar{t}(\mathcal{t})$ in 7.5 can be replaced with $1 \cdot \bar{t}(\mathcal{t})^1 - 0$. How can one know which of these constants are and which are not MF dependent? Is the choice determined by a substantive theory of temperature? Is it guaranteed that MF-dependent constants can always be found, allowing one to rewrite sentences like 7.5 in a measurement-function-class (MF-class) form like 7.7? Precisely how do the MF-dependent constants, if found, covary with MFs within a given class, like TEMP? Are there any restrictions on the possible classes of MFs? The answers to all such questions are contained in the algorithm for the procedure I call *covariant substitution*. The essence of the algorithm is this. Let $S(\bar{x}(\mathcal{x}))$ be an arbitrary MF-specific sentence, referring to a well-defined MF (or a vector of MFs) $\bar{x}(\mathcal{x})$. Let all explicit numerical constants in this sentence be treated as "pure numbers." Let X be an arbitrary class of MFs that contains $\bar{x}(\mathcal{x})$, such that any MF $x(\mathcal{x}) \in X$ can be converted into $\bar{x}(\mathcal{x})$ by a one-to-one parametrizable conversion function $\bar{x} = f(x)$. The parameters $\mathbf{C}$ of the conversion functions are called *conversion coefficients*. The algorithm shows how one constructs from $S(\bar{x}(\mathscr{x}))$ a MF-class sentence (referring to *all* MFs from X) of the form

$$\text{for any } x(\mathscr{x}) \in X \text{ there is a vector } \mathbf{c} \text{ (depending on } x\text{) such that}$$
$$S^*(x(\mathscr{x}), \mathbf{c}),$$
$$\text{where } \mathbf{c} = \mathbf{c_0} \text{ when } x \text{ is } \bar{x}. \tag{7.9}$$

The components of vector **c** are MF-dependent constants: the algorithm shows that they are functions (or "reduced forms") of the conversion coefficients: $\mathbf{c} = \mathbf{c(C)}$. The specialization of 7.9 back to the original MF $\bar{x}(\mathscr{x})$ yields the sentence $S^*(\bar{x}(\mathscr{x}), \mathbf{c_0})$. This sentence is identical with the original $S(\bar{x}(\mathscr{x}))$, except that now all MF-dependent constants in it have been *explicated*. A specialization to another MF $\bar{\bar{x}}(\mathscr{x})$ from X will have the form $S^*(\bar{\bar{x}}(\mathscr{x}), \mathbf{c_1})$: not only $\bar{\bar{x}}(\mathscr{x})$ substitutes for $\bar{x}(\mathscr{x})$ in the original sentence, but also all MF-dependent constants change their numerical values "accordingly."

The algorithm is as follows.

Step 1. Formulate the MF-specific sentence, $S(\bar{x}(\mathscr{x}))$. Take $\bar{x}(\mathscr{x})$ to be the anchoring MF(s). Treat all explicit numerical constants in $S(\bar{x}(\mathscr{x}))$ as "pure numbers."

Illustration. Let mass $\mathscr{m}$, distance $\mathscr{l}$, and force $\mathscr{f}$ be measured in well-defined specific measures $\bar{m}(\mathscr{m})$, $\bar{l}(\mathscr{l})$, and $\bar{f}(\mathscr{f})$, say, kg-m-N. Then the following is a MF-specific form of Newton's law:

$$B(\mathscr{m}_1, \mathscr{m}_2, \mathscr{l}, \mathscr{f}) \quad \text{iff} \quad \frac{\bar{m}_1\,\bar{m}_2}{\bar{f}\,\bar{l}^2} = \Gamma \qquad (\text{where } \Gamma^{-1} \approx 6.673 \cdot 10^{-11}). \tag{7.10}$$

The two explicit numerical constants (Γ and exponent 2) are treated as "pure numbers." [Hereafter, I omit arguments of MFs in all examples, writing m or $\bar{m}$ instead of the explicit $m(\mathscr{m})$ and $\bar{m}(\mathscr{m})$. Note that different subscripts, like in $\bar{m}_1$ and $\bar{m}_2$, refer to different arguments, rather than different MFs: $\bar{m}_1 = \bar{m}(\mathscr{m}_1)$ and $\bar{m}_2 = \bar{m}(\mathscr{m}_2)$.]

Step 2. Define the class(es) X of MFs related to $\bar{x}(\mathscr{x})$ through a certain (parametrizable) group of conversion functions: $x(\mathscr{x}) \in X$ if and only if for some vector of constants **C**, $\bar{x}(\mathscr{x}) = g(x(\mathscr{x}), \mathbf{C})$. Constants **C** are conversion coefficients.

Illustration. Consider the following class MASS* of mass MFs: $m(\mathscr{x}) \in$ MASS* if and only if $\bar{m} = c_m m^{a_m}$, for some positive constants c_m and a_m. In traditional terminology, mass is measured "on a log-interval scale" (Stevens, 1974). Classes LENGTH* and FORCE* are defined analogously, with conversion functions $\bar{l} = c_l l^{a_l}$ and $\bar{f} = c_f f^{a_f}$, respectively.

Step 3. Substitute $g(x(\mathscr{x}), \mathbf{C})$ for $\bar{x}(\mathscr{x})$ in $S(\bar{x}(\mathscr{x}))$ and simplify the expression algebraically to reduce the number of constants to a minimum. Denote the resulting vector of constants by **c:** these are MF-dependent constants. Express **c** as a function of conversion coefficients: $\mathbf{c} = \mathbf{c(C)}$.

Illustration. By substitution, algebra, and renaming,

$$\mathrm{B}(m_1, m_2, l, f) \quad \text{iff} \quad \frac{[c_m m_1^{a_m}][c_m m_2^{a_m}]}{[c_f f^{a_f}][c_l l^{a_l}]^2} = G_{lmf}\frac{(m_1 m_2)^{a_m}}{f^{a_f} l^{2a_l}} = \Gamma. \quad (7.11)$$

The MF-dependent constants **c** here are (G_{lmf}, a_l, a_m, a_f), expressed through the conversion coefficients $\mathbf{C} = (c_l, c_m, c_f, a_l, a_m, a_f)$ as $(G_{lmf}, a_l, a_m, a_f) = (c_l^{-2}c_m^2c_f^{-1}, a_l, a_m, a_f)$.

Step 4. The proposition constructed at Step 3 is a sentential function, not a sentence. To make it a sentence, first, affix to it the quantifier(s) "for any $x \in X$ there are constants **c** (depending on x)." Next, determine conversion coefficients $\mathbf{C_0}$ in the equation $g(\bar{x}(\mathscr{x}), \mathbf{C_0}) = \bar{x}(\mathscr{x})$, and compute $\mathbf{c_0} = \mathbf{c}(\mathbf{C_0})$: these are the values of **c** when $x = \bar{x}$. Suffix the specifying proposition "where $\mathbf{c} = \mathbf{c_0}$ when $x = \bar{x}$" to the sentence. This is the resulting MF-class sentence, 7.9, containing MF-dependent constants **c** and their special values $\mathbf{c_0}$ for the original MF $\bar{x}(\mathscr{x})$. This sentence is logically equivalent to (interdeducible with) the original sentence $S(\bar{x}(\mathscr{x}))$.

Illustration. For LENGTH*-MASS*-FORCE* classes the resulting MF-class sentence, logically equivalent to 7.10, is

for any $l \in$ LENGTH*, $m \in$ MASS*, $f \in$ FORCE*
there are positive reals G_{lmf}, a_l, a_m, a_f, such that

$$\mathrm{B}(m_1, m_2, l, f) \quad \text{iff} \quad G_{lmf}\frac{(m_1 m_2)^{a_m}}{f^{a_f} l^{2a_l}} = \Gamma$$

where $G_{lmf} = a_l = a_m = a_f = 1$ when (l, m, f) is $(\bar{l}, \bar{m}, \bar{f})$. (7.12)

(If one omits the anchoring proposition "where $\mathbf{c} = \mathbf{c_0}$ when $x = \bar{x}$" from the sentence, the result is still a valid MF-class generalization, only it will now be logically weaker than, rather than equivalent to, the MF-specific sentence it is derived from.)

Step 5. To specialize the MF-class sentence 7.9 to any MF $\bar{\bar{x}}(\mathscr{x})$ from X, compute $\mathbf{c_1}$ as $\mathbf{c}(\mathbf{C_1})$, where $\mathbf{C_1}$ satisfies $g(\bar{\bar{x}}(\mathscr{x}), \mathbf{C_1}) = \bar{x}(\mathscr{x})$, and substitute $\bar{\bar{x}}(\mathscr{x})$ and $\mathbf{c_1}$ for $x(\mathscr{x})$ and **c**, respectively, in the sentential function $S^*(x(\mathscr{x}), \mathbf{c})$ of 7.9. (I omit an illustration since it is obvious.) This concludes the algorithm for covariant substitution.

It is clear now why the "direct substitution criterion" of meaningfulness does not work for MF-specific sentences: a correct substitution of one MF for another should be preceded by the explication of all MF-dependent constants and followed by changing their numerical values. When this is done, however, the substitution criterion becomes expressly nonrestrictive: *any* MF can be generalized to *any* class of MFs and thereby substituted for by *any* other MF from this class. To emphasize this fact, the classes of MFs for mass-length-force in our illustration have been chosen broader than the traditional "ratio scales" (obtained by putting $a_l = a_m = a_f = 1$). Perhaps the most important characteristic of MF-dependent constants is that they are merely mathematical reductions, $\mathbf{c} = \mathbf{c}(\mathbf{C})$, of conversion coefficients **C** that define the class *X* of MFs to which a given MF-specific sentence is being generalized. MF-dependent constants, therefore, have no substantive ("qualitative") meaning: they are not given by the theory of the quantities being measured, and their sole purpose is to ensure the generalizability of sentences from specific MFs to classes of MFs. The conversion coefficients **C**, obviously, vary from one class of MFs to another, and so do MF-dependent constants $\mathbf{c} = \mathbf{c}(\mathbf{C})$; in addition, for a given class of MFs, the MF-dependent constants will generally vary with the MF-specific sentence to be generalized.

DIMENSIONAL ANALYSIS IN PHYSICS: DERIVABILITY-FROM

Luce (1978) suggested that dimensional analysis in physics is a particular case of the "empirical meaningfulness" analysis. To the extent one subscribes to this point of view, my previous analysis should leave one wondering: if "empirical meaningfulness" either is a terminological replacement for logical falsity, or has nothing to do with the empirical content of a sentence altogether, why is then dimensional analysis clearly sound and useful?

The point of resemblance between dimensional analysis and the "empirical meaningfulness" analysis lies in the fact that usually physical sentences are written in a *dimensionally homogeneous* form. This means that a complete logical formulation of a physical sentence has the form of sentence 7.9, in which the classes *X* of MFs for all physical quantities $\mathscr{x}$ are defined as follows: $x(\mathscr{x}) \in X$ if and only if $Cx(\mathscr{x}) = \bar{x}(\mathscr{x})$ for some positive real *C* and some anchoring MF $\bar{x}(\mathscr{x})$ (defined by an effective empirical procedure). The conversion coefficients *C* are referred to as conversion *factors,* and MF-dependent constants *c* are called *dimensional* constants. The similarity conversions are rarely mentioned explicitly, due to their universal use, and the affix proposition (the first line) of sentence 7.9 is usually omitted. One simply says then that physical sentences hold "for all units of measurement" (provided one remembers that all dimensional constants are *unit covariant*). This is taken in the representational theory of measurement to constitute the essence of the restrictive power of dimensional analysis. According to this position, dimensional analysis operates by striking down formulations that are not dimensionally homogeneous (in Pfanzagl's terms, are not "meaningfully parametrized") and thereby cannot be "true laws of physics;" the main problem to be solved, therefore, is *why* physical sentences are dimensionally homogeneous (Krantz, Luce, Suppes, & Tversky, 1971, chap. 10). In some form or another this idea permeates the whole issue of "empirical meaningfulness" from its outset. This is certainly what Suppes (1959) meant by referring to "systematic language of physics" and quoting from Newton's *Principia,* or what Falmagne (1992) meant by saying that "only meaningful statements have reached posterity" and quoting from Galileo's *Dialogues.*

It must be clear from the algorithm for covariant substitution that *any sentence can be written in a dimensionally homogeneous form, by introducing appropriate dimensional constants:* homogeneity or inhomogeneity of the form of a sentence, therefore, has nothing to do with its content and cannot serve as a selection criterion for "possible laws." This is clearly explained in Bridgman's classical book (1922), and I will refer to covariant substitution restricted to similarity conversions as *Bridgman's algorithm.* As an example, consider the traditional classes MASS, LENGTH, TIME, FORCE of MFs m, l, t, f, related to the respective anchoring functions by positive similarities: $\bar{m} = c_m m$, $\bar{l} = c_l l$, $\bar{t} = c_t t$, $\bar{f} = c_f f$. The following two sentences are MF-specific versions of the gravitation law and the second law of motion:

$$B^{GRV}(m_1, m_2, l, f) \quad \text{iff} \quad \frac{\bar{m}_1 \bar{m}_2}{\bar{f}\,\bar{l}^2} = \Gamma;$$

$$B^{MTN}(m, l, t, f) \quad \text{iff} \quad \frac{\bar{m}}{\bar{f}} \cdot \frac{d^2\bar{l}}{d\bar{t}^2} = 1. \tag{7.13}$$

By Bridgman's algorithm these sentences are generalized to the conventional MF classes as

for any $l \in$ LENGTH, $t \in$ TIME, $m \in$ MASS, $f \in$ FORCE
there are positive reals G_{lmf}, A_{ltmf}, such that

$$B^{GRV}(m_1, m_2, l, f) \quad \text{iff} \quad G_{lmf}\frac{m_1 m_2}{f\,l^2} = \Gamma;$$

$$B^{MTN}(m, l, t, f) \quad \text{iff} \quad A_{ltmf}\frac{m}{f} \cdot \frac{d^2 l}{dt^2} = 1$$

$$\text{where } G_{lmf} = A_{ltmf} = 1 \text{ when } (l, t, m, f) \text{ is } (\bar{l}, \bar{t}, \bar{m}, \bar{f}). \tag{7.14}$$

To derive 7.14, every MF in 7.13 has been multiplied with its own conversion factor, and these factors have algebraically "coalesced" (using Bridgman's language) into two dimensional constants of a monomial structure: $G_{lmf} = c_l^{-2} c_t^0 c_m^2 c_f^{-1}$, $A_{ltmf} = c_l^1 c_t^{-2} c_m^1 c_f^{-1}$. In the technical language of dimensional analysis, called dimensional algebra, the same fact is expressed by introducing dimensional symbols L, T, M, F for the four basic quantities and presenting the dimensionality of G_{lmf} and A_{ltmf} as $L^2T^0M^{-2}F^1$ and $L^{-1}T^2M^{-1}F^1$, respectively (the exponents being those for the corresponding conversion factors multiplied by -1). According to Bridgman, this is the essential logic of where physics takes its dimensional constants from: all dimensional constants are "coalesced" conversion factors. Physical theory does not play any role here, in agreement with the general position of this chapter: all MF-dependent constants are merely mathematical reductions of conversion coefficients, and they can be computed for any MF-specific sentence generalized to any class of MFs. Thus, from the point of view of dimensional analysis, any sentence involving gravitation forces, masses, and distances could be the true gravitation law (even if it contained, say, masses added to distances), and any such sentence could be made dimensionally homogeneous by an appropriate choice of dimensional constants.

To understand the aspect of physical sentences that dimensional analysis does address, consider the following MF-specific sentences:

$$\begin{aligned} B_1(l, t) &\quad \text{iff} \quad \bar{t} + \bar{l} = 100, \\ B_2(l, t) &\quad \text{iff} \quad \bar{l} \cdot \bar{t}^{-1} = 100, \\ B_3(m, l, t) &\quad \text{iff} \quad \bar{m}^{-1} \cdot \bar{l}^{-2} \cdot \bar{t}^{1} = 100. \\ B_4(m, l, t) &\quad \text{iff} \quad \bar{m}^{-1} \cdot \bar{l}^{3} \cdot \bar{t}^{-2} = 100. \end{aligned} \qquad (7.15)$$

Again, all these sentences can be put in a dimensionally homogeneous form by Bridgman's algorithm, each with its own dimensional constants; none of these sentences taken in isolation can be struck down as "impossible" or "meaningless." Suppose, however, that one asks whether these sentences can be *mathematically derived* from Newton's gravitation law and the second law of motion (sentences 7.14 or 7.13). In classical physics, this is what one would expect if, for example, $B_1(l, t)$ were describing period of revolution (t) of two celestial bodies about their gravitation center as a function of distance (l) between them (ignoring, for simplicity, sentences specifying initial conditions). Here is where dimensional analysis comes into operation, in the context of judging *derivability* of a given sentence from other sentences. It is easy to prove that since sentences 7.13 could be put in the dimensionally homogeneous form 7.14 by means of two dimensional constants, G_{lmf} and A_{ltmf}, any consequent of these sentences should be presentable in a dimensionally homogeneous form by means of dimensional constants that are functions of G_{lmf} and A_{ltmf}; moreover, these functions can only be monomials of the form $G^{a}_{lmf}A^{b}_{ltmf}$ (since all dimensional constants are monomials over conversion factors).

By simple algebra one can show now that the dimensional constants associated with $B_1(l, t)$ in 7.15, c_l and c_t (dimensional formulae $L^{-1}T^{0}M^{0}F^{0}$ and $L^{0}T^{-1}M^{0}F^{0}$), cannot be expressed as two monomials over G_{lmf} and A_{ltmf}; hence this sentence is not derivable from 7.13–7.14. This has nothing to do with the addition operation specifically, as can be seen from the fact that the same conclusion (nonderivability) applies also to $B_2(l, t)$, whose homogeneous reformulation requires one dimensional constant, c_{lt} ($L^{-1}T^{1}M^{0}F^{0}$). Both $B_1(l, t)$ and $B_2(l, t)$ might very well be empirically true (e.g., B_1 could describe spring length changing under an external force, B_2 could be stating the constancy of the speed of light, in some units)—dimensional analysis only tells us that they are *not derivable* from two *particular* sentences. The comparison of $B_3(m, l, t)$ with $B_4(m, l, t)$ is instructive, too. Both numerical expressions are monomial triples, their homogeneous formulations require one dimensional constant each, c_{ltm} and c'_{ltm} ($L^{2}T^{-1}M^{1}F^{0}$ and $L^{-3}T^{2}M^{1}F^{0}$, respectively). $B_3(m, l, t)$, however, is struck down as nonderivable, whereas $B_4(m, l, t)$ is not, because c_{ltm} cannot be presented as $G^{a}_{lmf}A^{b}_{ltmf}$ but $c'_{ltm} = G_{l}^{-1}{}_{mf}A^{l}_{ltmf}$ (note that *not being nonderivable* in the considered sense is necessary but not sufficient for being de facto derivable).

This is the entire essence of dimensional analysis. (Algebraic techniques involved, however, based on the Vaschy-Buckingham Pi theorem, are quite a bit more powerful than in the illustrations given; see, e.g., Kurth, 1972; Langhaar, 1951.) It follows that any attempt to theoretically restrict (based on dimensional considerations only) a class of possible laws in a given area *without the context of derivability* is doomed to failure (see Palacios, 1964, chap. 7.4, "First Rule"). For instance, dimensional analysis cannot restrict the class of the *basic* laws of an area, because by definition they are not supposed to be derivable from other laws. In particular, if one wishes to present them in dimensionally homogenous forms, one is not restricted by any phsycial principle as to the number and character of the dimensional constants one has to introduce (explicate). The same clearly applies to psychophysical laws, such as Weber's law or near-miss to Weber's law (cf. Narens & Mausfeld, 1992).

It seems quite obvious that the differences between the dimensional analysis of the four sentences in 7.15 cannot be accounted for on the basis of the "empirical meaningfulness" analyses. If "100" is considered to be a "pure number" (as intended), all four sentences are "empirically meaningless" by the direct substitution criterion. If "100" is dimensioned, then $B_1(l, t)$ is "meaningless," and the remaining three sentences are "meaningful" (in Pfanzagl's terms, "meaningfully parametrized"). Without elaborating, if one sets up the "structure of physical quantities with basis," along the lines suggested by Krantz et al. (1971, chap. 10) and Luce et al. (1990, chap. 22), one will see that the truth value of B_2, B_3, B_4, but not of B_1, is preserved under "similarities of the structure." This is definitely not what dimensional analysis is about.

The restrictive power of dimensional analysis (its ability to detect nonderivable sentences) is due to the fact that the basic laws of (some areas of) physics, derivability from which is being tested, happen to be such that, when written in a dimensionally homogeneous form, the number of the resulting dimensional constants is less than maximal (the maximum number equals that of basic dimensions, e.g., it is 4 in the LENGTH-TIME-MASS-FORCE).[1] Specifically, the basic laws of some areas of physics can be decomposed into fewer sentences than there are basic dimensions, each sentence containing a single monomial over basic quantities, and thereby yielding a single dimensional constant by Bridgman's algorithm (Palacios, 1964). If the number of the MF-dependent constants in all derivations equaled or exceeded that of basic dimensions, then there would

[1]A precise formulation should refer to the rank of the dimensional matrix associated with the constants, rather than their number. Those familiar with dimensional algebra might find useful the following theorem I state here without proof. Let Γ be a set of formulae containing variables $V_1, \ldots, V_m$ in specific units. Let $C_1, \ldots, C_k$ be the minimal set of dimensional constants that have to be introduced to write Γ in a dimensionally homogeneous form (this set is found by Bridgman's algorithm). Then the Pi theorem restricts the class of formulae $F(V_1, \ldots, V_m) = 0$ (in specific units) derivable from Γ if, and only if, the rank of the dimensional matrix for $C_1, \ldots, C_k$ is less than that for $C_1, \ldots, C_k, V_1, \ldots, V_m$.

be no advantage in using dimensionally homogeneous formulations over MF-specific ones: in the imaginary world with such a structure of physical laws, physicists could very well have adopted fixed "scientific" units of measurement for all physical sentences. Remaining in our own world, the real reason why only dimensionally homogeneous sentences "have reached posterity" (Falmagne, 1992) is not their "empirical meaningfulness," but their optimality with respect to derivability decisions involving relatively few MF-dependent constants. Note that sentences of physics are almost never derived directly from fundamental laws: in addition, one should also include "situational sentences" specifying boundary conditions and intervening external forces. These situational sentences bring in their own dimensional constants (again, by unrestricted application of Bridgman's algorithm), which in many cases are sufficient to annul the restrictive power of dimensional analysis.

It should also be clear why physical MFs are not embedded into classes of conversion functions broader than positive similarities. Physical theory itself does not limit MFs to particular classes. We have seen, for example, that the law of gravitation can be trivially presented in MFs defined up to power conversion functions (sentences 7.10, 7.12), as well as traditional similarity conversions (7.13, 7.14). However, in 7.14 the number of MF-dependent constants in the law reduces to just one, G_{lmf}, due to the algebraic "coalescing" of the conversion factors. By contrast, in 7.12 the three "dimensional exponents" remain separate, and their number equals that of the basic quantities. As a result, in deciding whether a given sentence is or is not derivable from the gravitation law, writing them in a "power-homogeneous" form 7.12 would provide no additional advantage over usual dimensionally homogeneous formulations 7.14. Luce et al. (1990), discussing power transformaton groups in the context of "real unit structures," point out that these transformations are "just how far the dimensional structure of physics can be generalized" (p. 124). It seems that the generalization could very well go much farther, but there is no useful purpose in its going even this far. This seems to explain why "at present there are no substantive examples of such a generalization" (ibid).

COMPLETE EMPIRICAL RELATIONAL SYSTEMS

To be well defined, any class of MFs should contain at least one anchoring MF, defined through an effective empirical procedure: it would do little good to know that different MFs for length are interrelated by positive similarities if none of them could be computed independently, "from empirical objects." H. Helmholtz (see Menger, 1959) has shown that empirical measurement procedures for quantities like length or mass can be described by a set of a few operations whose basic properties are formalized in a set of axiomatic sentences. Suppes and

Zinnes (1963) called such a theoretical construct (a set of magnitudes with relations defined through their axiomatic properties) an *empirical relational system* (ERS). For example, the "ratio-scale" representation of length is traditionally associated with the ERS $\mathcal{L}_1 = \{\mathscr{L}, l_1 \leq_l l_2, l_1 \oplus_l l_2 = l_3\}$, involving a linear ordering $\leq_l$ of length magnitudes $\mathscr{L}$ and a concatenation operation $\oplus_l$ with sum-like properties. A representation-uniqueness theorem tells us that there exists such a MF $\bar{l}(l)$ mapping $\mathscr{L}$ onto Re^+ that $l_1 \leq_l l_2$, iff $\bar{l}(l_1) \leq \bar{l}(l_2)$ and $l_1 \oplus_l l_2 = l_3$ iff $\bar{l}(l_1) + \bar{l}(l_2) = \bar{l}(l_3)$; the same holds for, and only for, any MF $l(l) = C\bar{l}(l)$, $C > 0$, a member of the class I have referred to as LENGTH. In the language of algebra, the ERS $\mathcal{L}_1$ is isomorphically mapped onto a numerical relational system (NRS) $L_1 = \{Re^+, l_1 \leq l_2, l_1 + l_2 = l_3\}$, the isomorphisms being defined up to positive scaling.

Consider now a "qualitative" sentential function (a predicate containing no MFs) $P(l_1, l_2, \ldots)$ that can be expressed through the defining predicates of $\mathcal{L}_1$ exclusively, $\leq_l$ and $\oplus_l$ (interconnected by logical and mathematical terms). Let such a predicate be called "empirically definable in $\mathcal{L}_1$" (Luce et al., 1990, chap. 22). A necessary condition for this is that the following sentence be logically true:

$$\begin{aligned} &\text{for any } l_1, l_2, \ldots, l_1^*, l_2^*, \ldots, \text{ and for any } C > 0: \\ &\quad \text{if } l_1^* = \bar{l}^{-1}[C\bar{l}(l_1)],\ l_2^* = \bar{l}^{-1}[C\bar{l}(l_2)], \ldots, \\ &\quad\quad \text{then } [P(l_1, l_2, \ldots) \text{ iff } P(l_1^*, l_2^*, \ldots)]; \end{aligned} \tag{7.16}$$

$\bar{l}$ stands here for a specific MF $\in$ LENGTH. The transformation $\bar{l}^{-1}[C\bar{l}(l)]$ mapping $\mathscr{L}$ onto itself is an automorphism of $\mathcal{L}_1$; the automorphisms form a group, with the identity (or trivial automorphism) corresponding to $C = 1$. Consider now the numerical sentential function $P^*(\bar{l}_1, \bar{l}_2, \ldots)$ obtained by a direct substitution of $\bar{l}(l_1)$ for l_1, $\bar{l}(l_2)$ for l_2, etc., accompanied by a direct substitution of $\leq$ for $\leq_l$ and $+$ for $\oplus_l$. The numerical predicate P^* can be referred to as *representing* an empirically definable (in $\mathcal{L}_1$) predicate. For the moment, I leave open the question of whether the predicate $P^*(\bar{l}(l_1), \bar{l}(l_2), \ldots)$ is itself empirically definable in $\mathcal{L}_1$, *when viewed as a "qualitative" predicate over* $(l_1, l_2, \ldots)$. Obviously, the truth value of any sentence formed from $P(l_1, l_2, \ldots)$ by quantification or specialization (on particular values of $l_1, l_2, \ldots$) should coincide with that of the sentence formed from $P^*(\bar{l}_1, \bar{l}_2, \ldots)$ by the same quantification or specialization. Since this must also be true for any MF $C\bar{l}(l)$, one comes to the following conclusion: if a sentence $S(\bar{l}(l))$ does not preserve its truth value under direct substitutions of $C\bar{l}(l)$ for $\bar{l}(l)$, then it must contain a predicate that does not represent an empirically definable predicate in $\mathcal{L}_1$. Such a sentence then can be labeled "empirically meaningless" with respect to $\mathcal{L}_1$. This is the essence of "empirical meaningfulness" understood on a "qualitative" level: invariance under mutual substitutions of MFs is justified as a *necessary condition* for "definability" in terms of a particular ERS.

As an example, consider the following two MF-specific sentences (to be read with the generality convention in mind):

$$B_1(\ell_1, \ell_2, \ell_3) \quad \text{iff} \quad \bar{l}_1 + \bar{l}_2 = \bar{l}_3 \tag{7.17}$$

$$B_2(\ell_1, \ell_2, \ell_3) \quad \text{iff} \quad \bar{l}_1 \cdot \bar{l}_2 = \bar{l}_3 \tag{7.18}$$

Sentence 7.18 does not pass the direct substitution test under similarity conversions, and one concludes that $\bar{l}_1 \cdot \bar{l}_2 = \bar{l}_3$ does not represent an empirically definable predicate in $\mathcal{L}_1$. Sentence 7.17 does pass the test, and one concludes that $\bar{l}_1 + \bar{l}_2 = \bar{l}_3$ *might* represent an empirically definable predicate in $\mathcal{L}_1$; in this case it obviously represents $\ell_1 \oplus_l \ell_2 = \ell_3$, which is empirically definable de facto. Clearly, the "empirical meaninglessness" of 7.18 thus understood does not refer to the sentence in isolation, but only in its relation to a particular ERS. In particular, it has nothing to do with truth or "scientific significance" of this sentence—the "meaninglessness" here is void of negative connotations, being a purely technical characterization (see, e.g., Narens, 1985, p. 155). Sentence 7.17, for example, is "empirically meaningless" with respect to the ERS $\mathcal{L}_0 = \{\mathscr{L}, \ell_1 \leq_l \ell_2\}$, that can be isomorphically mapped onto $L_0 = \{Re^+, l_1 \leq l_2\}$ (the MFs $l(\ell)$ here are defined up to arbitrary strictly increasing conversion functions). Obviously, one can find an infinity of ERSs in which a given sentence is "empirically meaningless," and this should not concern a researcher any more than an abstract algebraic exercise. The "empirical meaninglessness" of 7.18 in $\mathcal{L}_1$ simply indicates that the factual empirical procedures that led to its formulation cannot be formalized by the axioms of $\mathcal{L}_1$. On this note the discussion might have ended, perhaps with pointing out, in addition, that the literature regrettably abounds with misleading statements suggesting that "empirical meaninglessness" indicates things like "concepts that have neither empirical nor qualitative interpretations in the substantive domain" (Narens & Mausfeld, 1992, p. 467).

The issue is somewhat deeper, however: when applied to numerical statements involving well-defined MFs, the notion of "empirical meaningfulness" cannot serve even the limited technical purpose just discussed. The reason for this is in that *no specific MF (such as length in meters) can be defined within an ERS that has nontrivial automorphisms (equivalently, an ERS that isomorphically maps onto a given NRS by more than one MF).* Thus, the MF $\bar{l}$ in sentences 7.17 and 7.18 is an empirical predicate, "$\bar{l}(\ell) = \bar{l}$," that is not empirically definable in $\mathcal{L}_1$ (or $\mathcal{L}_0$, or any other ERS whose isomorphisms onto a given NRS consist of more than one MF). Therefore, by the very fact of formulating sentences invoking this MF (whether these sentences are "meaningful" or "meaningless" in $\mathcal{L}_1$ or $\mathcal{L}_0$), one guarantees that ERSs like $\mathcal{L}_1$ and $\mathcal{L}_0$ cannot formalize the factual empirical procedures involved.

Indeed, in the language consistent with $\mathcal{L}_1$, the empirical predicate "$\bar{l}(\ell) = \bar{l}$" is defined as "$\ell \circledR_l \ell_0 = \bar{l}$," where ℓ_0 refers to some standard length ("yardstick") and $\circledR_l$ is the "empirical ratio" operation effectively defined through $\oplus_l$ and $\leq_l$ by the standard Archimedean algorithm (parallel concatenations of "the measured" and the "yardstick"; see, e.g., Narens, 1985). This semiformal definition of "$\bar{l}(l) = \bar{l}$" presumes that l_0 is "known and fixed"; that is, it can be uniquely identified by a designatory sentential function "such length that [a unique description follows]." It is obvious, however, that the description in brackets cannot be written in terms of operations $\oplus_l$ and $\leq_l$ only. Put formally, the predicate $\mathcal{I}_0(l)$ that says "l is the standard length l_0" is not empirically definable in $\mathcal{L}_1$. A formal proof consists in observing that the sentence

$$\text{for any } l_1, l_2 \text{ and any } C > 0, \text{ if } l_2 = \bar{l}^{-1}[C\bar{l}(l_1)] \text{ then } [\mathcal{I}_0(l_1) \text{ iff } \mathcal{I}_0(l_2)] \quad (7.19)$$

is logically false (compare this with 7.16). There is nothing surprising or contradictory in the fact that $\mathcal{L}_1$ does not allow one to effectively identify any member of the class LENGTH of its own isomorphisms. Indeed, the class of all MFs isomorphically mapping $\mathcal{L}_0 = \{\mathcal{L}, l_1 \leq_l l_2\}$ onto $L_0 = \{\mathcal{L}, l_1 \leq_l l_2\}$, too, contains any specific MF one can think of, say, the meter measure, but here it is quite obvious that the language of $\mathcal{L}_0$ is too limited to single out and identify this measure. One might say that "from the point of view" of $\mathcal{L}_1$ different MFs from the class LENGTH are not just intersubstitutable, they are indistinguishable. Insofar as one can effectively distinguish between different specific MFs within a class of "admissible transformations" for a given ERS, a relevant formalization of the empirical procedures requires a *completion* of this ERS by some additional predicates, that would reduce its automorphisms to identify.

For length MFs such a construction was proposed by D. Hilbert in his classical axiomatization of Euclidean geometry (Hilbert, 1902); in Krantz et al. (1971, chap. 2) this construction is considered under the name of Archimedean ordered rings. The ERS in question is $\mathcal{L}_2 = \{\mathcal{L}, l_1 \leq_l l_2, l_1 \oplus_l l_2 = l_3, l_1 \otimes_l l_2 = l_3\}$, where $\otimes_l$ is an operation with multiplication-like properties. This ERS can be isomorphically mapped onto $L_2 = \{Re^+, \bar{l}_1 \leq \bar{l}_2, \bar{l}_1 + \bar{l}_2 = \bar{l}_3, \bar{l}_1 \cdot \bar{l}_2 = \bar{l}_3\}$, the MF $\bar{l}(l)$ being defined *uniquely*. Equivalently put, the group of the automorphisms of this ERS is reduced to identity. I will call such an ERS *complete*. Observe that $\mathcal{L}_2$ is equivalent to the ERS $\mathcal{L}_2^* = \{\mathcal{L}, l_1 \leq_l l_2, l_1 \oplus_l l_2 = l_3, \mathcal{I}_0(l)\}$, where $\mathcal{I}_0(l)$ says "l is the standard length l_0 defined as [a unique identification of l_0 in nonlength terms follows]." One recognizes here the common practice of complementing descriptions of the empirical operations of "comparing the measured with a yardstick" (which is what $\leq_l$ and $\oplus_l$ provide) by a definition of the "yardstick" itself. A NRS uniquely isomorphic to $\mathcal{L}_2^*$ is, for example, $L_2^* = \{Re^+, \bar{l}_1 \leq \bar{l}_2, \bar{l}_1 + \bar{l}_2 = \bar{l}_3, \bar{l} = 1\}$.

It must be quite clear that *all* predicates involving length values are de facto empirically definable in $\mathcal{L}_2^*$ (equivalently, $\mathcal{L}_2$). Indeed, any predicate $P(l_1, l_2, \ldots)$ can be defined by the proposition "$P(l_1, l_2, \ldots)$ iff $P^*(\bar{l}_1, \bar{l}_2, \ldots)$," where P^* is as explained earlier. The predicate $\bar{l}(l) = \bar{l}$ is defined in $\mathcal{L}_2^*$ by the proposition "$\bar{l}(l) = \bar{l}$ iff $\mathcal{I}_0(l_0)$ and l ®$_l$ $l_0 = \bar{l}$," where ®$_l$ is the "empirical ratio" referred to earlier. Once $\bar{l}(l) = \bar{l}$ is empirically definable, $P^*(\bar{l}_1, \bar{l}_2, \ldots)$ is empirically definable too, because it only involves numerical operations on $\bar{l}(\ell)$. Thus, both predicates $\bar{l}_1 \cdot \bar{l}_2 = \bar{l}_3$ and $B_2(\ell_1, \ell_2, \ell_3)$ in 7.18 are de facto and a priori empirically definable in $\mathcal{L}_2^*$. In an incomplete ERS, like $\mathcal{L}_1$, even if a numerical predicate, like $\bar{l}_1 + \bar{l}_2 = \bar{l}_3$, represents an empirically definable predicate, it would not be empirically definable itself.

Without elaborating, it trivially follows from the definitions introduced by Narens (1981) that an *N*-point-unique ERS can be made complete by appending to its defining predicates *N* arbitrary "yardstick-predicates." For example, the "interval-scale" representation of temperature is sometimes associated with the ERS $\mathcal{T}_1 = \{\mathfrak{T}, t_1 \leq_t t_2, t_1 \therefore_t t_2 = t_3\}$, involving a linear ordering $\leq_t$ of temperature magnitudes $\mathfrak{T}$ and an operation $\therefore_t$ with averaging-like properties. It is isomorphically mapped onto NRS $T_1 = \{Re, t_1 \leq t_2, t_1 + t_2 = 2t_3\}$, the isomorphisms being defined up to positive linear conversions. To make $\mathcal{T}_1$ complete, one can append to it such predicates as $\mathcal{W}_f(t)$, standing for "water eventually freezes at t," and $\mathcal{W}_e(t)$, standing for "water eventually evaporates at t." The unique isomorphic mapping $\bar{t}(t)$ of $\mathcal{T}_2 = \{\mathfrak{T}, t_1 \leq_t t_2, t_1 \therefore_t t_2 = t_3, \mathcal{W}_f(t), \mathcal{W}_e(t)\}$ onto $T_2 = \{Re, \bar{t}_1 \leq \bar{t}_2, \bar{t}_1 + \bar{t}_2 = 2\bar{t}_3, \bar{t} \leq 0, \bar{t} \geq 100)$ is the Celsius MF. Any predicate involving temperature magnitudes is empirically definable in $\mathcal{T}_2$, and any sentence involving Celsius MFs is "empirically meaningful." I will not discuss here whether the predicates $\mathcal{W}_f$ and $\mathcal{W}_e$ are as "fundamental" as, say, the operation $\therefore_t$, but they are clearly as well defined empirically and as rigorous theoretically (the description of $t_1 \therefore_t t_2 = t_3$ is as follows: "if one mixes two equal amounts of some substance, with initial temperatures t_1 and t_2, and if a heat loss is prevented, then the eventual temperature of the mixture is t_3").

Once a complete ERS is constructed and represented by a well-defined (unique) MF, this MF, *or some transformation thereof,* can, of course, be embedded as an anchoring MF in any class of MFs interrelated by conversion functions: the algorithm of covariant substitution guarantees, as I have shown, that both the truth value and the form (up to MF-dependent constants) of any sentence involving this MF will be preserved under all possible substitutions within any such class. The specific empirical operations formalized in an ERS are, in fact, quite irrelevant insofar as it is complete. I suggest that the only goal of measurement is to construct a unique numerical *identification* of the magnitudes of a quantity being measured. Once such an identification is empirically available, the class of conversion functions with which it will be associated will be determined by the objective structure of the empirical laws of an area. If the only empirically available numerical identification $\bar{l}$ of length, for example, were logarithmically related to the conventional meter measure, all sentences of mechanics would still be formulated as we know them, for the MFs within the class LENGTH. The only difference would be, of course, that the class will now be defined as "all MFs $l(\ell)$ such that $l = C \exp(\bar{l})$, for some positive real C." In the section on dimensional analysis I have discussed the reasons why using classes like LENGTH is convenient and desirable.

I suggest that construction of empirically complete ERSs and investigation of their mutual relations could be the central subject in a new revision of theory of measurement. Incomplete ERSs and their automorphisms can be treated as groups of complete ERSs (extensive, set-theoretic approach), or their parts (intensive, logical approach). In such a theory, which I tentatively call the *constructive* theory of measurement, *measurement is understood as an effective algorithm by which one constructs within a set of linearly ordered objects an everywhere dense subset of standard objects. To measure an object in the set is to indicate a unique chain of steps (generally, countably infinite) of the algorithm that leads to a standard object which, in some well-defined sense, is "infinitely close" to the object being measured.* Such an approach would bring measurement procedures back into the measurement theory, while preserving most of the mathematical results established within its framework. The notion of "empirical meaningfulness," however, would have no useful purpose.

ACKNOWLEDGMENTS

I would like to acknowledge the prominent role of my long correspondence with R. Duncan Luce in the development of my views on theory of measurement issues, as well as excellent comments by Reinhard Niederée that forced me to make corrections in the earlier version of this text.

REFERENCES

Bridgeman, P. W. (1922). *Dimensional analysis.* New Haven: Yale University Press.

Falmagne, J-C. (1992). Measurement theory and research psychologist. *Psychological Science, 2,* 88–93.

Falmagne, J-C., & Narens, L. (1983). Scales and meaningfulness of quantitative laws. *Synthese, 55,* 287–325.

Guttman, L. (1971). Measurement as structural theory. *Psychometrika, 36,* 329–347.

Hilbert, D. (1902/1965). *Foundations of geometry.* La Salle, IL: Open Court.

Krantz, D. H., Luce, R. D., Suppes, P., & Tversky, A. (1971). *Foundations of measurement* (vol. 1). San Diego: Academic Press.

Kurth, R. (1972). *Dimensional analysis and group theory in astrophysics.* New York: Wiley.

Langhaar, H. L. (1951). *Dimensional analysis and theory of models.* Oxford: Pergamon Press.

Luce, R. D. (1978). Dimensionally invariant numerical laws correspond to meaningful qualitative relations. *Philosophy of Science, 45,* 1–16.

Luce, R. D., Krantz, D. H., Suppes, P., & Tversky, A. (1990). *Foundations of measurement* (vol. 3). San Diego: Academic Press.

Menger, K. (1959). Mensuration and other mathematical connections of observable material. In C. W. Churchman & P. Ratoosh (Eds.), *Measurement: Definitions and theories* (pp. 97–128). New York: Wiley.

Michell, J. (1986). Measurement scales and statistics: A clash of paradigms. *Psychological Bulletin, 100,* 398–407.

Michell, J. (1990, August). *Permissible statistics and the validity of inferences from measurements.* Paper presented at the Twenty-Fifth Annual Mathematical Psychology Meeting, Stanford University.

Narens, L. (1981). On the scales of measurement. *Journal of Mathematical Psychology, 24,* 249–275.

Narens, L. (1985). *Abstract measurement theory.* Cambridge, MA: MIT Press.

Narens, L., & Mausfeld, R. (1992). On the relationship of the psychological and the physical in psychophysics. *Psychological Review, 99,* 467–479.

Palacios, J. (1964). *Dimensional analysis.* London: Macmillan.

Pfanzagl, J. (1968). *Theory of measurement.* New York: Wiley.

Roberts, F. (1979). *Measurement theory.* Reading, MA: Addison-Wesley.

Rozeboom, W. W. (1962). The untenability of Luce's principle. *Psychological Review, 69,* 542–547.

Stevens, S. S. (1974). Measurement. In G. M. Maranell (Ed.), *Scaling: A sourcebook for behavioral scientists* (pp. 22–41). Chicago: Aldine.

Suppes, P. (1959). Measurement, empirical meaningfulness, and three-values logic. In C. W. Churchman & P. Ratoosh (Eds.), *Measurement: Definitions and theories* (pp. 129–143). New York: Wiley.

Suppes, P., & Zinnes, J. L. (1963). Basic measurement theory. In R. D. Luce, R. R. Bush, & E. Galanter (Eds.), *Handbook of mathematical psychology* (Vol. 1, pp. 1–76). New York: Wiley.

II COLOR

Michael D'Zmura
Geoffrey Iverson

INTRODUCTION

The section on color includes two chapters on the geometric properties of color space and two on the computational problem of color constancy. Concern with the geometric properties of color space dates back to Helmholtz (1891, 1892), who proposed the first line element model for color discrimination. Work by MacAdam (1942) in the mid-20th century showed how empirical data on small color differences can be used to determine the local metric structure of color space. Much of Indow's work has continued this line of research, through a focus on the scaling of relatively large color differences. His work on differences in appearance found among Munsell chips—a standard set of colored papers—also bears on the study of color constancy, which involves the (ideal) independence of surface color appearance from light source spectral properties.

Maloney, Wuerger, and Krauskopf use proximity judgments to measure the angles between lines in color space. By testing the additivity of angles formed between three coplanar lines, they can determine whether the color space is related linearly to the physical stimulus space. Their experimental results with isoluminant stimuli show that the angles are superadditive and, thus, that the color space revealed by the proximity task is not Euclidean. Maloney, Wuerger, and Krauskopf discuss applications of their technique to other domains, like binocular space.

Izmailov presents a four-dimensional color space that is based on multidimensional scaling of color difference judgments. The color points of real colors are located on a three-dimensional spherical surface in the four-dimensional space. He characterizes the color space by scaling two sets of data. First, he scales his own data on large color differences between colored light stimuli. Second, he analyzes Indow's surface color data on differences between Munsell chip colors. The axes of the color space are made to correspond to the responses of the standard color-opponent channels, while the spherical angles that describe a particular color correspond to hue, saturation and brightness.

Iverson and D'Zmura describe how bilinear models can be used to estimate the spectral properties of surfaces and light sources. Bilinear models describe the interaction between a visual system's fundamental spectral sensitivities and reflected lights. The authors point out a desirable property for a bilinear model: two different sets of lit surfaces always give rise to different quantum-catch data. Models with this property allow for unique recovery of light source spectra and surface reflectance functions. The authors use this property to describe fully the conditions on a trichromatic visual system and the reflected lights that it views for spectral recovery to be possible.

D'Zmura, Iverson, and Singer propose a maximum likelihood procedure for color constancy that generalizes gray-world methods. They formulate prior distributions on light source spectra and surface reflectance functions using stochastic linear models. These are used to model how chromatic data, received by a visual system, depend on the spectral properties of the light source. The estimation procedure determines the light source most likely to have shone on a set of surfaces to produce the received data. The results of computer simulation suggest that lights reflected from relatively few surfaces are needed to estimate light source spectral properties accurately.

REFERENCES

Helmholtz, H. von. (1891). Kürzeste Linien im Farbensystem. *Sitzungsberichte der königliche preussischen Akademie der Wissenschaften zu Berlin*, Dezember 17, 1071–1083.

Helmholtz, H. von. (1892). Versuch, das psychophysische Gesetz auf die Farbenunterschiede trichromatischer Augen anzuwenden. *Zeitschrift für Psychologie und Physiologie der Sinnesorgane, 3*, 1–20.

MacAdam, D. L. (1942). Visual sensitivities to color differences in daylight. *Journal of the Optical Society of America, 32*, 247–274.

8 A Method for Testing Euclidean Representations of Proximity Judgments in Linear Psychological Spaces

Laurence T. Maloney
Sophie M. Wuerger
John Krauskopf
Center for Neural Science and Department of Psychology, New York University

ABSTRACT

A new method is presented that uses proximity judgments to measure angles between lines in a Euclidean space known up to an invertible linear transformation. We use this method to test the hypothesis that human judgments of the proximity of colors are consistent with any Euclidean geometry on color-matching space.

INTRODUCTION

Psychological data are often described and interpreted by representing them spatially. Stimuli are modeled as points in the space of n-tuples $x = (x^1, \ldots, x^n)$ of real numbers $\mathbf{R}^n$, and aspects of human performance are then modeled by the metric, dimensional, or geometric properties of the space. Figure 8.1 illustrates the relations among mathematical structures commonly imposed on $\mathbf{R}^n$ (see Suppes, Krantz, Luce, & Tversky, 1989, chaps. 12–13).

The most restrictive and most familiar of the four structures is the Euclidean. An example of a *Euclidean space* is the real vector space $[\mathbf{R}^n, +]$ endowed with the inner product

$$\langle x, y \rangle = \sum_{i=1}^{n} x^i y^i. \tag{8.1}$$

The Euclidean inner product can be used to define a Euclidean norm

$$\|x\| = \langle x, x \rangle^{1/2}, \tag{8.2}$$

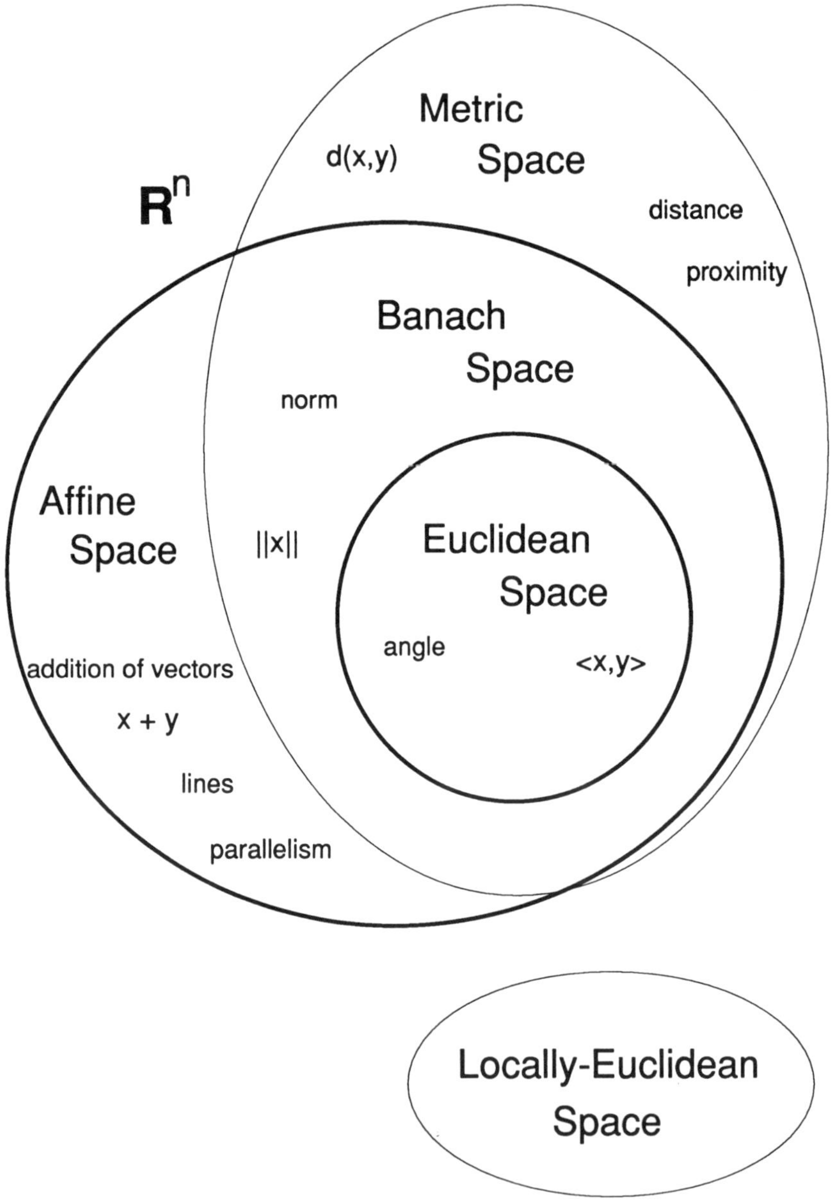

FIG. 8.1. Commonly used psychological representations and the relations among them. Every Euclidean space is a Banach (normed) space, every Banach space is both an affine space and a metric space. An affine space is a vector space with added mathematical structure. See text for further discussion. Locally Euclidean spaces are approximately Euclidean at a small enough scale, but need not be Euclidean globally.

and a Euclidean metric

$$d(x, y) = \|x - y\|. \tag{8.3}$$

We term this particular Euclidean space the *standard Euclidean space*. All possible Euclidean spaces on $\mathbf{R}^n$ can be generated from this one by invertible linear transformations A: such a space has inner product $\langle Ax, Ax \rangle$.

Every Euclidean space is also an *affine space* and a *metric space*. An affine space is a real vector space with lines and planes defined by means of the vector space operations. In an affine space, with no additional structure, we can meaningfully describe lines as intersecting, incident, or parallel. The angle between any two intersecting, noncoincident lines is, however, undefined. We can talk about the line joining two points, and, if three points a, b, and c are collinear, we can determine the ratio of the length ab and bc. The lengths ab and bc cannot be determined in isolation. The properties of an affine space that are defined are those that are preserved under an invertible linear transformation of the underlying vector space followed by a translation of the origin of the space.

If, in addition, an affine space has a metric consistent with the affine structure, then the space is *normed*. The magnitude of line segments is well defined, but the angle between any two intersecting line segments is not defined. The class of normed spaces that are complete[1] are the *Banach spaces*. The most familiar non-Euclidean norms are the Minkowski ρ-norms

$$\|x\| = \left(\sum_{i=1}^{n} (x^i)^\rho \right)^{1/\rho}, \tag{8.4}$$

but there are other norms as well.

Figure 8.1 indicates the relations among the four kinds of spaces, and the terms in small letters refer to concepts meaningful in each kind of space. In affine spaces, we can meaningfully describe lines and parallel lines, but we cannot talk about distance without also choosing a metric. If the metric is consistent with the vector space structure, then the space is a Banach space. Spaces that are locally Euclidean will be discussed in the section "Potential Applications."

[1]The technical assumption of completeness is defined in Young (1988).

In this chapter we are interested primarily in two of these models, affine spaces and Euclidean space. We describe a method for testing whether judgments of proximity on stimuli modeled as points in an affine space $\mathbf{R}^n$ can be accounted for by imposing a Euclidean metric on $\mathbf{R}^n$. That is, we are testing whether *any* Euclidean interpretation of the affine space is consistent with the proximity data. Equivalently, we are testing whether there is any linear transformation of the affine space that would make the proximity data consistent with standard Euclidean space. If the affine space can be interpreted as Euclidean, then the method determines the Euclidean geometry up to a rotation, reflection, and overall scaling of the space.

Geometric Idea

In this section we describe a method for measuring angles between pairs of lines in a Euclidean space when the space is only known up to a linear transformation (denoted ψ). The method assumes that the observer can judge the proximity of pairs of stimuli and that the proximity judgments are determined by the Euclidean metric. It further assumes that, given any two stimuli x and y in the Euclidean space, we can generate the stimuli with coordinate $(1 - \alpha)x + \alpha y$ for any real α.

Two key observations are that (1) judgments of orthogonality of lines in a normed linear space determine the space up to a similarity transformation (rotation/reflection and overall scaling) (Suppes et al., 1989, pp. 31ff), and (2) proximity judgments in a Euclidean space suffice to determine when intersecting lines are orthogonal (Young, 1988, Section 3.2).

Figure 8.2a corresponds to the space of possible stimuli, known to the experimenter. It is an affine space. Figure 8.2b represents the internal psychological space, which is assumed to be Euclidean. Each point or line labeled with a lowercase letter in Figure 8.2a is mapped by an unknown linear transformation Ψ to the corresponding point or line labeled in uppercase. For example, A_1 is the internal representation of stimulus a_1.

Consider two lines λ_1 and λ_2 in a plane in the experimenter's space (Figure 8.2a), and suppose that the experimenter chooses a stimulus a_1 on line λ_1. The observer then considers the stimuli corresponding to the line λ_2 and judges the stimulus on λ_2 that is most proximate to a_1. This most similar stimulus is denoted by b_2. In the Euclidean space, there must a unique nearest point on the line, and the line joining the internal representations A_1 and B_2 corresponding to the stimuli must be orthogonal to the line Λ_2 in Figure 8.2b. Note that, in the experimenter's space, the lines will typically not be orthogonal. The unknown linear transformation Ψ not only alters the angles between the two lines. It also may scale distances along the two lines differently. One orthogonality judgment cannot determine both the angle between the lines λ_1 and λ_2 and the relative scaling. With two such judgments, it is possible to determine both the angle and the relative scaling. We next demonstrate how to estimate the angle.

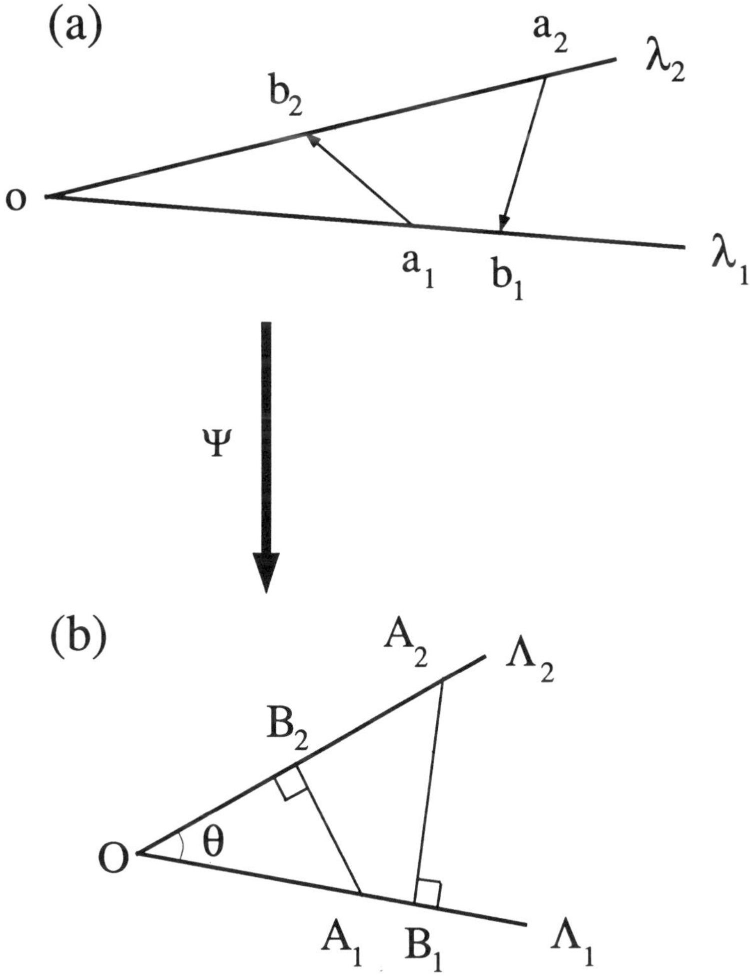

FIG. 8.2. The lines and points shown are those used in determining the angle θ between the lines Λ_1 and Λ_2. (a) corresponds to the affine space representing stimuli available to the experimenter. (b) is a second space that is Euclidean. The Euclidean metric of this space is assumed to determine the subjects judgments of proximities of pairs of stimuli. The two spaces are related by an unknown linear transformation Ψ that maps the points and lines with lowercase letters in (a) to the corresponding uppercase points in (b). The text describes the role of the points and lines shown.

Figure 8.2a illustrates (from the experimenter's point of view) two point-to-line proximity judgments of the sort just described: b_1 is the point on λ_1 most proximate to a_2 (fixed) on λ_2. Similarly, b_2 is the point on λ_2 most proximate to a_1 (fixed) on λ_1.

Note that the distances OA_1, OA_2, OB_1, and OB_2, are all unknown to the experimenter.

$$\cos(\theta) = \frac{OB_2}{OA_1} = \frac{OB_1}{OA_2}. \tag{8.5}$$

The quantities on the right-hand side are unknown. The distances oa_1, oa_2, ob_1, and ob_2, are all known to the experimenter. As a consequence of the linearity of the transformation from Figure 8.2a to Figure 8.2b,

$$\frac{ob_1}{oa_1} = \frac{OB_1}{OA_1} \tag{8.6}$$

and

$$\frac{ob_2}{oa_2} = \frac{OB_2}{OA_2} \tag{8.7}$$

From Equation 8.5

$$\cos^2(\theta) = \frac{OB_2}{OA_1}\frac{OB_1}{OA_2} = \frac{OB_2}{OA_2}\frac{OB_1}{OA_1} = \frac{ob_2}{oa_2}\frac{ob_1}{oa_1}. \tag{8.8}$$

This formula permits us to compute the cosine squared of the unknown angle θ in terms of quantities known to the experimenter.

It is also possible to determine the cosine and, therefore, the (unsigned) angle θ between the two lines as follows. The points o, a_1, and b_1 are collinear and, if they are distinct, one is between the other two (on the line λ_1). The point o is between a_1 and b_1 precisely when the point O is between A_1 and B_1. The same is true for the points on the lines λ_2 and Λ_2. If θ is greater than 90° then the point o will lie between a_1 and b_1 on λ_1 and also between a_2 and b_2 on λ_2. If $\theta < 90°$ then the point o will not lie between a_1 and b_1 on λ_1 and will not lie between a_2 and b_2 on λ_2. (It is not possible for the point o to lie between the two points on one line but not the other). We can therefore determine the magnitude of θ between 0° and 180° be examining the betweenness relations of the points o, a_i, b_i, $i = 1, 2$. The sign of θ tells us whether the angle runs clockwise or counterclockwise and cannot be determined from this data. In conclusion, two point-to-line "orthogonality judgments" determine the (unsigned) angle θ between two lines.

Once we know the angle between the two lines and the relative scale factor, the linear transformation Ψ is determined up to rotation, reflection, scaling, and translation. It should be noted that the derivation of the angle does not depend on prior knowledge of the transformation Ψ. We would get the same estimate θ for the angles in Figure 8.2b even if we transformed Figure 8.2a linearly before doing the computation above. It is only assumed that Ψ is linear. The relative scale factor is not invariant under such a linear transformation of Figure 8.2a. Such a transformation may scale the two lines in Figure 8.2a differently. The measured relative scale factor would then include this new additional scaling.

In order to determine Ψ when the linear spaces are three dimensional, only five orthogonality judgments are needed. We choose three noncoplanar lines λ_1, λ_2, and λ_3 that intersect in a common point o, and measure the angle and relative scale factor between λ_1 and λ_2, and the angle and relative scale factor between λ_2 and λ_3, using the procedure described above. These two angle/scale measurements require four orthogonality judgments. The relative scale factor between the third pair of lines, λ_1 and λ_3 is then known (it is the product of the other two scale factors). Only one additional orthogonality judgment is needed to determine the angle between λ_3 and λ_1. Five orthogonality judgments permit us to measure the three angles and two scale factors that determine Ψ up to rotation, reflection, scaling, and translation.

Test of Consistency

If proximity measurements are controlled by an underlying Euclidean space that is a linear transformation of the stimulus space ("the Euclidean hypothesis"), we can determine the angles between any pair of intersecting lines. Suppose that, for any three *coplanar* lines in the space λ_1, λ_2, and λ_3 that intersect in a common point, we measure the angle θ_{12} between λ_1 and λ_2, the angle θ_{23} between λ_2 and λ_3, and the angle θ_{13} between λ_1 and λ_3. If the Euclidean hypothesis holds, then $\theta_{13} = \theta_{12} + \theta_{23}$ necessarily. We can therefore test the Euclidean hypothesis by testing additivity of the angles between all pairs of three lines that are coplanar and intersect in a common point.

POTENTIAL APPLICATIONS

The test described above is applicable when (1) the stimuli are drawn from an affine space, and (2) the observer can judge the proximity of pairs of stimuli. The first condition is effectively satisfied if, given any two stimuli with coordinates x and y in $\mathbf{R}^n$, and any real number α in [0, 1], it is possible to generate the 'mixture' stimulus with coordinates $\alpha x + (1 - \alpha)y$. There may be many possible proximity tasks on the stimuli. It is possible that all, some, or none of the proximity judgments considered on a given affine space are consistent with underlying Euclidean metrics. We return to this point in the discussion. We next describe three potential application areas for the test above.

(1) *Proximity tasks*. Tasks in which subjects judge the psychological proximity of pairs of objects can be represented as distances in a space by means of

multidimensional scaling (see Davies & Coxson, 1982; Romney, Shepard, & Nerlove, 1972; Shepard, Romney, & Nerlove, 1972). The scaling procedure assigns spatial coordinates (in $\mathbf{R}^n$) to each object. The distance between the points in $\mathbf{R}^n$ assigned to a pair of objects is assumed to control the subject's judgment of the proximity of the objects. Distance in $\mathbf{R}^n$ is often computed using a Euclidean metric. For any two points $(x^1, \ldots, x^n)$ and $(y^1, \ldots, y^n)$ the distance between them is,

$$d(x, y) = \left[\sum_{i=1}^{n} (x^i - y^i)^2 \right]^{1/2}.$$

To apply the test above, at a minimum the stimuli must be continuously parameterized. In addition, the test can be applied only when the experimenter can meaningfully generate stimuli that lie along lines in the MDS solution. In principle this can be done (at least locally) even if the map between stimulus space and MDS solution is nonlinear but smooth.

(2) *Color matching and color proximity.* In a typical color-matching experiment, an observer is asked to judge whether two lights have the same color appearance. If the two lights are presented on identical patches of retina in the same adaptational state, then physically identical lights are judged to have the same color appearance. Many other pairs of lights, though, that have distinct spectral power distributions, will also be judged identical in color appearance. We can model performance in such an experiment by assigning to each light a color code (an n-tuple of numbers) so that two lights match precisely when they have the same code. Experimental results indicate that the mapping from the space of spectral power distributions of lights to $\mathbf{R}^n$ (*color-matching space*) is a linear transformation between a linear function space containing the spectral power distributions of possible lights (*light space*) and the real vector space $\mathbf{R}^n$. For trichromatic human observers, n may be taken to be 3. The necessary and sufficient conditions for this representation to hold are given in Krantz (1975). The real vector space need not have a metric. We apply the method to a color proximity task below.

(3) *Binocular space.* Luneburg (1947, 1950) has advanced the hypothesis that the representation of visual space based solely on disparity cues can be modeled as a hyperbolic geometry imposed on $\mathbf{R}^3$ (for reviews see Indow, 1982; Suppes et al., 1989, pp. 135–153). Such a geometry is locally Euclidean: a sufficiently small region around a point in the space is approximately Euclidean, and, by varying the size of the region we can make the approximation as good as we would like it to be. The method can potentially be used to test the hypothesized locally Euclidean properties of binocular space by measuring angles in a sufficiently small region of the space.

We next illustrate the method by testing whether a candidate color proximity judgment can be modeled by means of an inner product. This experiment as well as similar experiments using other color proximity tasks and related experiments are described in greater detail in Wuerger, Maloney, and Krauskopf (1993).

METHODS

Spatial Configurations and Tasks. The proximity task involved viewing three lights (a, b, b') and judging which of b or b' is "more proximate" to a (method of triads). We tested the Euclidean hypothesis employing the spatial configuration shown in Figure 8.3. Two small (2°) discs (b and b') are superimposed on a big (6°) disc (a) The observers were instructed to judge which of the two small discs was *least salient* when presented on the big disc (a).

Procedure. We wish to estimate the light b along a line λ in color space that is most proximate to a given *fixed light a*. We must estimate this most proximate point from 2-AFC judgments of the kind just described. We do this as follows: three different lights (b_1, b_2, b_3) that lie on the line λ are chosen. These lights are termed the primary comparision lights. Each primary comparision light (b_i, i =

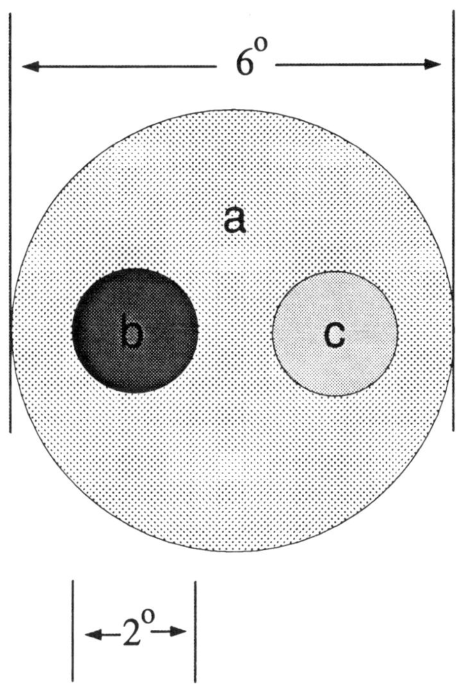

FIG. 8.3. At each trial three colored discs are briefly presented in the configuration shown. Two small (2°) discs (b and c) are superimposed on a big (6°) disc (a).

1, 3) is paired with nine different equally spaced[2] secondary comparison lights (c_j, j = 1, 9). These secondary comparison lights are also constrained to lie on the line λ. Each of these 27 pairs (three standards times nine comparisons) was presented 10 times in a randomized order. The light b was drawn from among the primary comparision lights b_i, and the light c from among the secondary comparision lights (c_j). The primary comparison light was as likely to be on the left as on the right.

The observer indicated by a button press (2-AFC) which of the two stimuli, b or c was more proximate ("less salient") to stimulus a.

Estimation Procedures. A psychometric function was plotted for each primary comparison light b based on estimates of the probability of selecting b as more proximate to a than the light c. If the primary comparison light b coincides with the most proximate point, then we expect to find a maximum relative frequency of 0.5 at $bv = c$. If the most proximate point is different from the primary comparison light b, then we expect to find a maximum relative frequency larger than 0.5 at some $c \neq b$. In either case, the location of the maximum relative frequency denotes the most proximate point which is estimated from the mean of a Gaussian curve fitted to the results. In each session three estimates (one for each b_i) of the most proximate point to a are obtained. Each condition is run three times, hence resulting in nine estimates of the most proximate point for each condition. The mean of these nine estimates is taken to be the estimate of the most proximate point.

Subjects. Four subjects participated in the experiment. Subject RP was naive as to the purpose of the experiment, subject JM was an experienced observer and aware of the purpose of the experiment. Subjects LM and SW are authors of this chapter. All subjects had normal vision or vision corrected to normal. All subjects had normal color vision.

Stimuli. The color space employed here is an affine transformation of the receptor space (LMS space; see MacLeod & Boynton, 1979, for details). We consider three lines (λ_1, λ_2, λ_3) We consider three fixed lights, a_1, a_2, and a_3. The origin o is defined as an equal-energy white light (40 cd/m^2), which is also the background color in all experiments. The line λ_i, i = 1, 2, 3, is the line containing a_i. The incremental cone coordinates (ΔL, ΔM, ΔS) of the three test stimuli with respect to the white point ($L = 0.66$, $M = 0.34$, $S = 0.017$) were as follows: for a_1: $\Delta L = 0$, $\Delta M = 0$, $\Delta S = -0.0029$; for a_2: $\Delta L = 0.0028$, $\Delta M = -0.0028$, $\Delta S = -0.0014$; for a_3: $\Delta L = 0.0056$, $\Delta M = -0.0056$, $\Delta S = 0$. The light a_3 along the L-2M line (λ_3) was chosen at 1/10 of the gamut available; the

[2]Stimuli that are equally spaced and *collinear* in the experimenter's affine space (Figure 8.2a) will also be equally spaced in color space.

test light along the S − (L + M) line constituted one fifth of the available gamut. The distance from the white point of each stimulus a_i (a = 1, 3) is five times greater than discrimination threshold in that color direction.

The line λ_3 (colors ranging from grey to red) is a line of constant luminance and constant S-cone-excitation (Krauskopf, Williams, & Heeley, 1982; MacLeod & Boynton, 1979). The line λ_1 (colors ranging from gray to greenish yellow) is a line of constant luminance and of constant L- and M-cone excitation. The line λ_2 (colors ranging from gray to orange) is a linear combination of the first two lines. The three lines are, therefore, coplanar in color space. In all the experiments the stimuli were presented as Gaussian pulses in time with a sigma of 160 msec and a total duration of 1000 msec. They appeared on a square background 512 min arc on a side.

Although all of the stimuli in this experiment are confined to an equiluminant plane, they need not appear *equally bright* to the observer. As the additivity law in brightness matching does not hold in general, there is no reason to expect them to be. Conversely, the set of all lights judged to be as bright as a given reference light is not, in general, a plane in color space (Wyszecki & Stiles, 1982, pp. 414–415). We return to this point below. In our experiments we fix a light (a_1) on line λ_1 and find by means of the 2-AFC procedure the light (b_2) on the line λ_2 which is judged as most similar to light a_1. Then we fix a light (a_2) on line λ_2 and find the most similar light (b_1) on the line λ_1. These two measurements permit us to estimate the angle θ_{12} using the procedure described above. The same sequence of measurements is conducted for the other two pairs of lines (λ_2 and λ_3, λ_1 and λ_3).

Data Analysis. We test whether angles are additive as predicted by a Euclidean geometry, that is, we test whether $\theta_{12} + \theta_{23} = \theta_{13}$. For instance, the angle θ_{12} is derived from the projection of a_1 on dimension λ_2 and the projection of a_2 on dimension λ_1. The key assumption is that the point on λ_2 which is most similar to the point a_1 defines the shortest distance in color space between A_1 and Λ_2. The most similar points are determined by the procedure as described in the method section. The angles θ_{13} and θ_{23} are computed in an analogous manner.

RESULTS

Table 8.1 shows the results for all four subjects. For each subject, the 9 (or 10) individual estimates of the three angles (θ_{12}, θ_{23}, θ_{13}) are tabulated as well as the mean and the standard error (in parentheses). The last column gives $\Delta\theta$, the difference $\theta_{12} + \theta_{23} - \theta_{13}$. In all but one case, the angle θ_{13} (angle between Λ_1 and Λ_3) is smaller than the sum of the angles θ_{12} (angle between Λ_1 and Λ_2) and θ_{23} (angle between Λ_2 and Λ_3). The sum $\theta_{12} + \theta_{23}$ exceeds θ_{13} by approximately 15%.

TABLE 8.1
Experimental Results for Four Subjects[a]

Subject	*Angle 12*	*Angle 23*	*Angle 13*	*Discrepancy*
JM	52	62	82	32
	52	61	93	20
	44	60	88	16
	53	61	88	26
	58	59	87	30
	46	58	86	18
	44	63	89	18
	53	62	87	28
	52	63	88	27
mean (sd)	50.4 (4.7)	61.0 (1.7)	87.6 (2.9)	23.9 (5.9)
LM	54	59	83	30
	46	60	83	23
	51	52	84	19
	53	54	85	22
	55	54	86	23
	49	54	81	22
	51	59	96	14
	53	57	86	24
	56	54	85	25
	52	56	85	23
mean (sd)	52.0 (3.1)	55.9 (2.9)	85.4 (4.3)	22.5 (4.09)
RP	42	65	86	21
	43	65	88	20
	36	72	92	16
	40	61	91	10
	29	62	92	−1
	41	63	90	14
	37	61	93	5
	47	60	92	15
	40	70	85	25
mean (sd)	39.4 (5.1)	64.3 (4.2)	89.9 (2.9)	13.89 (8.16)
SW	53	56	91	18
	47	57	90	14
	43	56	83	16
	44	58	93	9
	49	58	94	13
	54	66	83	37
	49	50	84	15
	48	60	84	24
	44	48	91	1
mean (sd)	47.9 (3.9)	56.6 (5.3)	88.1 (4.5)	16.33 (9.97)

[a]Angle 12 is the measured angled between lines Λ_1 and Λ_2. Angle 23 is the measured angled between lines Λ_2 and Λ_3. Angle 13 is the measured angle between lines Λ_1 and Λ_3. The three lines are coplanar and, in a Euclidean space, angle 12 and angle 23 must sum to angle 13. The last column gives the discrepancy $\Delta\theta$ between the sum of angles 12 and 23 and angle 13. The mean and standard deviation for the results of each subject are given as well.

Our null hypothesis is $\theta_{12} + \theta_{23} = \theta_{13}$ ($\Delta\theta = 0$). We reject the null hypothesis for each of the four subjects at a level of significant of $p < 0.01$ by means of a t-test. Note that 36 out of 37 of the $\Delta\theta$ are positive.

DISCUSSION

There is a long history of the use of similarity judgments in order to establish the structure of color space (Drösler, 1989; Ekman, 1954; Helmholtz, 1891; Indow, 1980; Krantz, 1967). Typically these approaches require judgments of the relative similarity of pairs of, in general, quite different lights. The data are then submitted to some method of multidimensional scaling to establish the number and nature of the fundamental coordinates of the Euclidean space which best encompass the judgments. The goodness-of-fit measures typically confound unpatterned error in the proximity data and patterned failures of the Euclidean model. As a consequence, multidimensional scaling methods provide no direct test as to whether a set of proximity data does, in fact, have a spatial representation. Tversky, Rinott, and Newman (1983) and Hutchinson and Tversky (1986) discuss methods for testing the Euclidean assumptions underlying multidimensional scaling.

In contrast, the method introduced here provides a straightforward test of the Euclidean hypothesis for proximity tasks on affine spaces. The observer need only choose the most proximate of a series of stimuli constrained to lie on a line in an affine space. The test is directly based on properties of Euclidean space. If the Euclidean hypothesis is satisfied, it requires only a small number of judgments to measure the angle between any pair of intersecting lines in the linear space.

The purpose of the reported experiment was to illustrate the method described above. We used it to test whether observers compare Euclidean distances in color space when judging the proximity of colored lights. For all four subjects, the angles derived under a Euclidean hypothesis are superadditive by about 15%. Hence, we reject the Euclidean hypothesis for our color proximity judgments. In brief, color space is not a Euclidean space with respect to the proximity task considered. Wuerger et al. (1995) report the results of two more experiments using other color proximity tasks. The Euclidean hypothesis was rejected for these tasks as well. We leave open the possibility that proximity judgments in one or more of these tasks is controlled by a non-Euclidean norm compatible with the affine structure of color space.

It should be noted that, if a color proximity task is describable by such a non-Euclidean norm, then the orthogonality relation between lines (based on finding the most proximal point) can be symmetric for some pairs of lines in the space, but not for all. That is, λ_1 may be judged perpendicular to λ_2, but λ_2 not be judged perpendicular to λ_1 when we reverse the roles of the lines. For the Minkowksi ρ-norms, with $\rho > 1$, the only mutually orthogonal pairs of lines are the axes of symmetry of the unit "ball" that are orthogonal in the Euclidean norm. All other pairs of lines fail to be mutually orthogonal. This property of the Minkowski ρ-norms holds for any norm and could provide the basis for testing the hypothesis that an affine space is normed (Suppes et al., 1989, p. 45).

Izmailov and Sokolov (1991) report that "the minimal dimensionality of the Euclidean space for equibright color discriminations is three." Their results are consistent with ours and with the Euclidean hypothesis. Their stimuli were not constrained to lie in a plane in color space. Rather, subjects adjusted spectrally narrowband lights to be equally bright. The equibright, spectrally narrowband stimuli were then rated for similarity and the numerical ratings interpreted as proximity data were averaged across trials and across subjects. They argue that a three-dimensional solution is preferable to a two-dimensional solution since "the two-dimensional solution has a smaller linear correlation between perceived color differences and interpoint distances" and since the deviations into the third dimension are patterned. They do not test the hypothesis statistically.

As noted in the description of the stimuli, sets of lights adjusted to be equibright are typically not coplanar in color space (Wyszecki & Stiles, 1982, pp. 413–420). We chose stimuli that *were* coplanar in color space, not equally bright. The observed failures of additivity cannot be explained by assuming that the stimuli were not coplanar without also abandoning the hypothesis that proximity is determined by a norm on color space.

The Izmailov & Sokolov results illustrate that multidimensional scaling is a powerful and underutilized technique for estimating a spatial representation of stimuli given a measure of proximity of the stimuli. It does not, however, lend itself to tests of the Euclidean assumptions underlying it.

Other Applications. The method presented here can be used to test whether any psychological representation in a affine space admits a Euclidean metric that can account for a particular kind of proximity judgment on the space. The test directly links judgments of proximity to fundamental properties of Euclidean space.

ACKNOWLEDGMENTS

The experiments were performed using the Postq experiment control system and a modified version of the VISTA board display device. Both Postq and the display device modifications were designed by Walter Kropfl of NYU whose excellent products and helpful consultations are gratefully acknowledged. We are grateful to Tarow Indow, Michael D'Zmura, Michael Landy, Kenneth Knoblauch, and Jan Drösler for helpful comments on previous drafts and presentations of this work. The research described in this chapter was supported in part by DFG grant Wu 204/1-1 from the German Research Foundation to Sophie Wuerger, AFOSR Grant F49620-92-J-0187 to Laurence T. Maloney, and the National Eye Institute grant EY06638 to John Krauskopf. Some of this work was presented at the annual meeting of the Association for Research in Vision and Ophthalmology, Sarasota, Florida, May 1993 and at the ECVP at Edinburgh, August 1993. The experiment reported is presented in fuller detail in Wuerger et al. (1993).

REFERENCES

Davies, P. M., & Coxson, A. P. M. (1982). *Key texts in multidimensional scaling*. Exeter, NH: Heinemann.

Drösler, J. (1989). *Farbadaptation und die Metrik des Farbraums*. Paper presented at teh '31 Tagung experimentell arbeitender Psychologen', Bamberg, Germany.

Ekman, G. (1954). Dimensions of color vision. *Journal of Psychology, 38*, 467–474.

Helmholtz, H. von (1891). Kürzeste Linien im Farbensystem. *Sitzungsbericht der Akademie zu Berlin*, 109–122.

Hutchinson, J. W., & Tversky, A. (1986). Nearest neighbor analysis of psychological spaces. *Psychological Review, 93*, 3–22.

Indow, T. (1980). Global color metric and color-appearance systems. *COLOR Research and Application, 5*, 5–12.

Indow, T. (1982). An approach to geometry of visual space with no a priori mapping functions: Multidimensional mapping according to Riemannian metrics. *Journal of Mathematical Psychology, 26*, 204–236.

Izmailov, Ch. A., & Sokolov, E. N. (1991). Spherical model of color and brightness discrimination. *Psychological Science, 2*, 249–259.

Krantz, D. (1967). Small-step and large-step color differences for monochromatic stimuli of constant brightness. *Journal of Optical Society, 57*, 1304–1316.

Krantz, D. H. (1975). Color measurement and color theory: I. Representation theorem for Grassmann Structures. *Journal of Mathematical Psychology, 12*, 283–303.

Krauskopf, J., Williams, D. R., & Heeley, D. M. (1982). Cardinal directions of color space. *Vision Research, 22*, 1123–1131.

Luneburg, R. K. (1947). *Mathematical analysis of binocular vision*. Princeton, NJ: Princeton University Press.

Luneburg, R. K. (1950). The metric of binocular visual space. *J. Opt. Soc. Am, 40*, 627–642.

MacLeod, D. I. A., & Boynton, R. M. (1979). Chromaticity diagram showing cone excitation by stimuli of equal luminance. *Journal of the Optical Society of America A, 69*, 1183–1186.

Romney, A. K., Shepard, R. N., & Nerlove, S. G. (1972). *Multidimensional scaling: Theory and applications in the behavioral sciences* (Vol. II). New York: Seminar Press.

Shepard, R. N., Romney, A. K., & Nerlove, S. B. (1972). *Multidimensional scaling: Theory and applications in the behavioral sciences* (Vol. I). New York: Seminar Press.

Suppes, P., Krantz, D. H., Luce, R. D., & Tversky, A. (1989). *Foundations of measurement: Vol. II. Geometrical, threshold, and probabilistic representations*. New York: Academic Press.

Tversky, A., Rinott, Y., & Newman, C. M. (1983). Nearest neighbor analysis of point processes: Applications to multidimensional scaling. *Journal of Mathematical Psychology, 27,* 235–250.

Wuerger, S. M., Maloney, L. T., & Krauskopf, J. (1995). Proximity judgments in color space: Test of a Euclidean color geometry. *Vision Research,* in press.

Wyszecki, G., & Stiles, W. S. (1982). Color science: Concepts and methods, quantitative data and formulae (2nd ed.). New York: Wiley.

Young, N. (1988). *An introduction to Hilbert space.* Cambridge, England: Cambridge University Press.

9 Spherical Model of Discrimination of Self-Luminous and Surface Colors

Chinghis Izmailov
Moscow State University, Russia

ABSTRACT

The metrical structure and dimensionality of color space were studied using estimates of large color differences. Multidimensional scaling of the data shows that the dimensionality of the space that provides a linear relationship between interpoint distances and chromatic differences is 4. However, color points do not completely fill in the four-dimensional space, but are located on a spherical surface. The perceived differences between colors are measured by Euclidean interpoint distances between points in the color space rather than by spherical distances. The phenomena of unique hue and color opponency were used to correlate the Cartesian axes with four neurophysiological channels of color vision. Three spherical angles at each point on the sphere correspond exactly to hue, saturation, and brightness of spectral lights. The colors of monochromatic lights are represented by a curve on this sphere. The subset of equibright colors is located on the spherical surface in a three-dimensional subspace. The same results were obtained by reanalyzing Indow and Kanazawa's (1960) data on color discrimination for Munsell colors. Functions are defined that relate Munsell color characteristics—hue, chroma, and value—with the three spherical angles of a color point in four-dimensional space.

INTRODUCTION

Constructing a uniform color space is a difficult problem that has yet to be solved. A traditional approach to this problem uses threshold measurement techniques that were developed primarily for application to the measurement of unidimensional sensory characteristics. Typically, discrimination of a single color characteristic is measured separately, and these measurements are then synthesized in a unified Euclidean model of color discrimination (Hurvich & Jameson, 1955; Judd & Wyszecki, 1963; Vos & Walraven, 1972; Wyszecki & Stiles, 1982). Yet this approach relies on assumptions about the dimensionality of color space that are often unjustified. Additional complications arise from the need to combine different kinds of experimental data in the framework of a unified model.

Quite a different approach to the study of color vision has been developed in the area of multidimensional scaling (MDS) (Indow, 1980; Izmailov, 1980; Izmailov, Sokolov, & Chernorizov, 1989; Shepard, 1962a, 1962b; Shepard & Carroll, 1966; Sokolov & Izmailov, 1988). This approach is based on the analysis of large, suprathreshold differences (or similarities) between colors. The method lets researchers investigate all color characteristics simultaneously. MDS lets one reconstruct spatial and metric structures of color difference without a prior assumptions (Kruskal, 1964; Shepard, 1962a, 1962b; Torgerson, 1958). One common restriction imposed in MDS in the analysis of color discrimination data is that a linear relationship holds between perceived color differences and interpoint distances in Euclidean space (Indow & Uchizono, 1960; Izmailov et al., 1989; Judd, 1967; Shepard & Carroll, 1966). This requirement of color space uniformity is the major criterion used to evaluate models (Judd & Wyszecki, 1963; Wyszecki & Stiles, 1982).

In earlier work, my colleagues and I considered a spherical model of uniform color space that was constructed using MDS analysis of large color differences (Izmailov, 1980; Izmailov & Sokolov, 1991; Izmailov et al., 1989; Sokolov & Izmailov, 1983, 1988). We tried to compare the spherical model obtained using colored light stimuli to that constructed by Indow and his colleagues on the basis of color differences between Munsell chip colors (Indow, 1980; Indow & Kanazawa, 1960). The general procedure for constructing a spherical model from data concerning the colors of equibright monochromatic lights is described in our earlier work (Izmailov & Sokolov, 1991).

METHODS

A detailed description of the apparatus and of the brightness-matching technique is presented in earlier papers (Izmailov, 1980; Izmailov et al., 1989). The apparatus, briefly, provided a visual display that was formed by an optical system comprising a visual photometer that included an objective, a photometric cube, and an eyepiece with Maxwellian view. The field of view comprised a circular test field of monochromatic light that subtended about 2° of visual angle that was surrounded by a dark annulus with outer diameter 6°.

Three subjects with normal color vision were asked to rate the color difference between successively presented pairs of lights using an integer ranging from 0 (the two stimuli in the pair were identical) to 9 (the two stimuli were maximally dissimilar). Maximum dissimilarity was not defined in the instructions to the subjects, who were free to use any number for any pair of nonidentical stimuli. There were 17 equibright color stimuli (16 monochromatic lights and 1 white light) paired in 136 combinations; each of the 136 combinations was presented 10 times in randomized order. Stimulus duration and interstimulus interval were 0.5 sec, and the interpair interval was 5 sec.

MDS—ANALYSIS OF LARGE COLOR DIFFERENCES

Estimates of the dissimilarity of each pair of light's colors were averaged across presentations and subjects and organized as a triangular matrix of color differences. A metric MDS procedure was applied to the matrix (Indow & Uchizono, 1960; Shepard & Carroll, 1966; Sokolov & Izmailov, 1983); it returned as solution a three-dimensional Euclidean space.

That the solution is three dimensional is interesting in light of the traditional view that only two dimensions are sufficient for describing equibright colors. This view is based mostly on color mixture data and on threshold discrimination data (Hurvich & Jameson, 1955; Judd & Wyszecki, 1963; Wyszecki & Stiles, 1982). Large color differences are supposed, by extension, to be described in a two-dimensional space also.

Let us first consider the formal foundation of the three-dimensional solution in terms of characteristic roots and coefficients of correlation. As can be seen in Table 9.1, not only the first and second dimensions but also the third dimension have large characteristic roots. The increase in the coefficient of correlation does

TABLE 9.1
Characteristic Roots and Coefficients of Correlations Describing the First Six Dimensions of Euclidean Space Obtained by MDS Analysis of Color Differences Between Equibright Light Stimuli (column a)[a] and Munsell Colors (column b) That Varied in Hue and Chroma[b]

	Characteristic Root		*Coefficient of Correlation*	
Dimension	*a*	*b*	*a*	*b*
1	14785	24978	0.782	0.776
2	9820	12556	0.977	0.966
3	2195	3068	0.993	0.974
4	1135	1127	0.993	0.976
5	698	540	0.996	0.976
6	498	394	0.996	0.977

[a]Indow and Uchizono (1960).
[b]Indow and Kanazawa (1960).

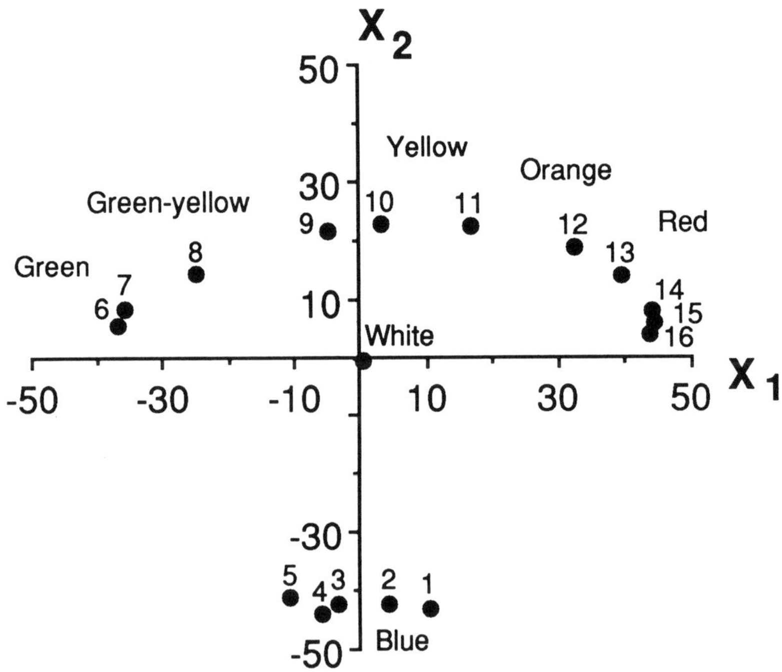

FIG. 9.1. Projections of the colors of monochromatic lights and a white light on the X_1X_2 plane formed of the first two dimensions. The white color is projected near the center of the X_1X_2 plane. Most saturated colors are projected far from the origin.

not end with the second dimension. While we could reject the third dimension if we had compelling reasons to do so, it will be shown that there are, on the contrary, strong reasons to consider the third dimension as a legitimate one in describing the data.

Figure 9.1 represents the traditional solution for spacing equibright colors in the X_1X_2 plane formed of the first two dimensions. However, there is a violation of color space uniformity, evident here as a considerable diminishing of linearity between perceived color differences and interpoint distances on the plane (Izmailov, 1980; Shepard, 1962a; 1962b; Shepard & Carroll, 1966). If the violation were due to measurement errors in the experimental data, then the distances between the color points and the X_1X_2 plane should be distributed randomly about zero. Figure 9.2 shows the projections of the color points onto the X_1X_3 plane of the same Euclidean space. It is evident from the data in Figures 9.1 and 9.2 that the spatial distribution of points along the third dimension is not random. There is a strict ordering in terms of color saturation. The most highly saturated colors (blue, green, and red), which are located far from the origin in the

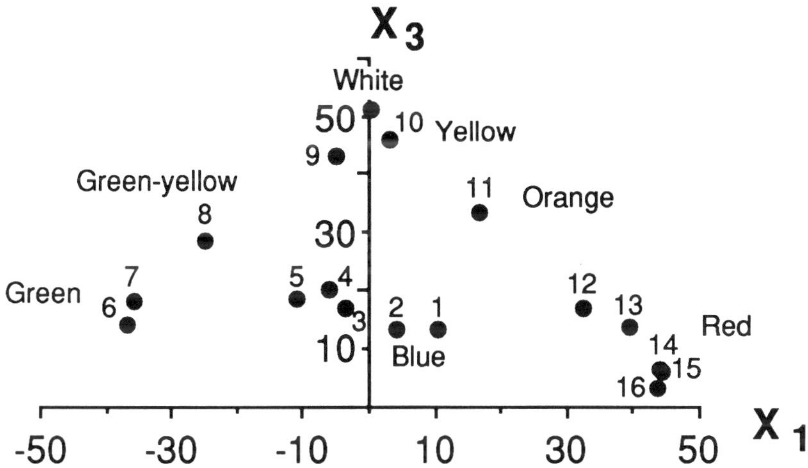

FIG. 9.2. Projection of the same colors on the X_2X_3 plane formed of the second and third dimensions. The white color point has the greatest height on this plane, while saturated colors have lesser heights.

chromatic plane X_1X_2, have small X_3 coordinates. The less saturated colors (green-yellow, yellow, and orange) are located closer to the origin of the X_1X_2 plane and have larger X_3 coordinates. The color with the greatest X_3 coordinate is white, which has small X_1 and X_2 values.

In other words, change in saturation corresponds to a special trajectory in the vertical plane of three-dimensional color space. Just as hue is characterized by azimuthal angle on the chromatic plane, saturation is described by vertical angle in three-dimensional space. The data show clearly that a third dimension is important for understanding color discrimination, particularly for describing changes in color saturation.

SPHERICAL PROPERTIES OF COLOR SPACE

An essential peculiarity of the obtained spatial structure is the following: the color points do not fill the three-dimensional space uniformly, rather they form a non-Euclidean surface of constant positive curvature within the three-dimensional space—namely a spherical surface. In order to prove this fact, it is necessary and sufficient to show that a geometric center can be found for a given configuration of color points.

Theoretically, the center must be equidistant from all color points. Yet the data are not free of error, so that one determines the center by finding the point for which the variance of the distances to all color points is minimized. I used an iterative procedure to minimize the variance of the radial distances by shifting the center point. A coefficient of variation is computed by taking the ratio of the standard deviation and the mean radius.

The goodness of fit to the spherical model is estimated in terms of both the coefficient of variance for radial distances and the coefficient of correlation between interpoint distances and initial color differences. A decrease in the size of the first coefficient, while maintaining the magnitude of the second one, represents a better fit. Values obtained from the given data are 0.078 and 0.995, respectively.

COLOR SPACE ROTATION

The interpoint distances, on which the resulting scaling solution is based, are invariant to all possible rotations. As a result, further motivation is needed for the particular directions that have been chosen for the cartesian axes of the solution space. In the spherical model of color discrimination, every axis is interpreted as opponent and so must be related to the color points in a certain way.

The well-known phenomenon of unique hue was used to orient the axes in the space. There are three hues in the spectrum seen when presenting monochromatic lights with wavelengths about 470 nm, 500 nm, and 575 nm and a further nonspectral hue (seen by mixing lights with dominant wavelengths 440 nm and 675 nm) that are considered as perceptually pure colors, namely "blue," "green," "yellow," and "red," respectively. Each such unique hue characterizes a single color characteristic in color-opponent theory (Hurvich & Jameson, 1955).

One axis in the solution space was chosen to be a red-green axis and was oriented to pass through the green point corresponding to wavelength 500 nm. Another axis was chosen to be a blue-yellow axis and was oriented to pass through the color points corresponding to wavelengths 470 nm and 580 nm. To obtain the frame of reference that represents the opponent properties of color vision in such a way, one rotates the initial configuration of points to reach as precise a localization of unique hue colors on the corresponding axes as possible.

The color-opponent spectral sensitivity functions that are derived from the final configuration are shown in Figure 9.3, and their similarity to corresponding functions, found empirically by using a hue cancellation technique (Hurvich & Jameson, 1955), helps to verify the solution.

NORMALIZATION OF THE COLOR SPHERE

A sphere has a constant radius for every point on its surface, speaking theoretically, but noise in the data cause radius values to fluctuate, and as a result the color sphere has a nonzero width which corresponds to the coefficient of variation (see Table 9.2). Since I do not deal with the interpretation of the radius

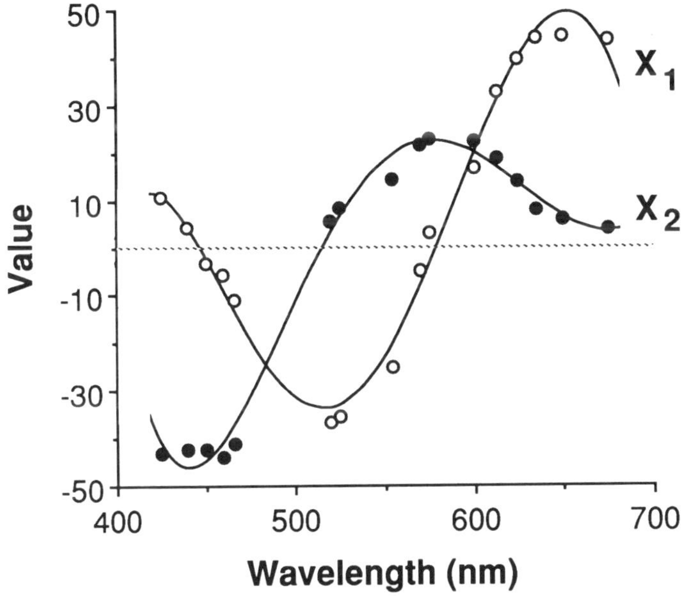

FIG. 9.3. The red-green (X_1) and blue-yellow (X_2) opponent functions derived from the spherical model of color discrimination. The curves are polynomials of degree 5 that fit the data points best.

TABLE 9.2
Coefficients of Variation Describing Sphericity of the Color Spaces Obtained by MDS Analysis of Color Differences Between Munsell Colors for the Six Original Paired Asymmetric Matrices (column a), for the Three Matrices Found by Averaging the Two Matrices for Each Subject (column b), the Two Matrices Found by Averaging Two Subjects' Matrices (1, 2, 3, 4 and 3, 4, 5, 6) (column c), and the Matrix Found by Averaging the Six Matrices of All Three Subjects (column d)

	Coefficients of Variance (%)			
Matrices of Color Differences	*a*	*b*	*c*	*d*
1	36.0	32.0		
2	34.0		29.0	
3	22.4	18.7		5.3
4	16.4		10.0	
5	16.3	12.7		
6	17.4			

values here, the sphere is reduced to a sphere of radius 1 by eliminating the fluctuations:

$$x_{1i}^2 + x_{2i}^2 + x_{3i}^2 = 1, \tag{9.1}$$

where

$$x_{ki} = X_{ki}/R_i, \qquad k = 1, 2, 3, \tag{9.2}$$

in which X_{ki} are the original coordinates and R_i is the radius of the ith color point.

ANALYSIS OF THE COLOR SPHERE

The presented model is a two-dimensional spherical surface in a three-dimensional Euclidean space. Each point on the surface corresponds to a particular color. Points that lie off of the spherical surface do not correspond to colors.

The colors of monochromatic lights are situated along a curvilinear path closed by purple colors. The pole of the sphere represents a "pure" white color. The azimuthal angle of a point codes its hue, while its vertical angle or elevation codes its saturation. The perceived difference between a pair of colors is determined by the central angle of the smaller of the two arcs determined by the great circle that passes through the two points. But quantitatively the difference is measured in terms of the line segment that joins the same points. This means that the structure of perceived color differences can be described in terms of a Euclidean metric:

$$d_{ij}^2 = \sum_{k=1}^{n} (X_{ki} - X_{kj})^2, \tag{9.3}$$

where d_{ij} is the distance between the ith and jth color points and n is the number of cartesian axes. This description holds true provided that each color point lies on the same spherical surface; i.e.,

$$S_i^2 = \sum_{k=1}^{n} X_{ki}^2 = \text{constant}. \tag{9.4}$$

THE BRIGHTNESS DIMENSION

For equibright colors, the number n of Cartesian axes Equation 9.3 is equal to 3. One possible way to visualize a new, second set of equibright colors at another level of luminance is as a concentric sphere of different radius. Yet another level of brightness would describe a third concentric sphere, etc.

We so arrive at the idea of Schrödinger (1920) and of Vos and Walraven (1972) that colors, varying in their chromatic characteristics as well as in brightness, fill an entire globe in three-dimensional Euclidean space.

Our experiments reject this idea.

What happens, in fact, when we add the dimension of brightness to the chromatic characteristics of the colors, is that we transform the spherical surface in three-dimensional Euclidean space to another spherical surface, this time in four-dimensional space.

Adding brightness variations increases the number of spherical axes. In addition to the two angles that are needed to characterize the hue and saturation of a color point, a third spherical angle is required to characterize its brightness.

This model has been confirmed by several independent experiments by my colleagues and me (Izmailov et al., 1989; Sokolov & Izmailov, 1983, 1988). Here I shall illustrate the spherical model of color space by applying it to Professor Tarow Indow's data on discrimination of Munsell colors.

REANALYSIS OF INDOW'S DATA

Our first step was to analyze the data of Indow and Uchizono (1960). Estimates of pair differences between 21 Munsell colors that varied in hue and chroma only were presented as a triangular matrix for one of four subjects.

Using MDS procedures, the coordinates of color points in 21-dimensional Euclidean space were obtained. Characteristic roots and coefficients of correlation were calculated for solution dimensionalities. As can be seen from Table 9.1 (columns marked b), the characteristic values stress the importance of the first and second dimensions. However, the third dimension also gives a prominent, additional contribution to the structure of the color points.

Goodness of fit to the spherical model of color vision was estimated for the Indow and Uchizono data as well as for the original color discrimination data. For points in three-dimensional space, the smallest coefficient of variance for radial distances of color points from the origin ("thickness" of the spherical layer) is 0.11 with the coefficient of correlation equal to 0.974.

Our second step was to apply this approach to Indow and Kanazawa's (1960) data. In this work, estimates of pair differences between 24 colors that varied in hue, chroma, and value were presented as pairs of asymmetric triangular matrices, one pair per subject. Each pair of matrices represented the data from a single subject; paired matrices differed in the order of stimuli. The first matrix represented estimates between *i* and *j* stimuli, while the second matrix represented estimates between *j* and *i* stimuli.

We used MDS to determine the coordinates of color points in four-dimensional space for the data of three subjects. For each of the subjects, we analyzed both the original pair of asymmetric matrices and also the matrices

TABLE 9.3
Coefficients of Correlation Describing Four-Dimensional Euclidean Space Obtained by MDS Analysis of Color Differences Between Munsell Colors for Separate Matrices (column a), Averaging for Each Subject (column b), Averaging for Two Subjects (column c), and Averaging for All Three Subjects (column d)

Matrices of Color Differences	*Coefficients of Correlation*			
	a	*b*	*c*	*d*
1	0.879	0.950		
2	0.896		0.971	
3	0.914	0.958		0.982
4	0.922		0.977	
5	0.923	0.964		
6	0.912			

found by averaging these. The goodness of fit by the spherical model of color vision to Indow and Kanazawa's data was determined. For the six original asymmetric matrices, the minimal coefficients of variance for radial distances ("thickness" of spherical layer for color points) varied from 16.3% to 36.0% (Table 9.2, column a), while the coefficients of correlation varied from 0.814 to 0.899 (Table 9.3, column a).

These results are bad from the point of view of one wishing to construct a spherical model. However, the data averaged for each subject give, in all cases, smaller coefficients of variation and larger coefficients of correlation (Table 9.2, column b and Table 9.3, column b). I suggest that the poor solutions obtained with the original pair of asymmetric matrices result from random errors in the initial estimates, and that averaging leads to better fits. In connection with this suggestion, we averaged data for two subjects (using two of the three possible groups of four matrices) and then for three subjects (all six matrices) and analyzed the resulting matrices the same way. Results are given in Tables 9.2 and 9.3 in columns c and d. The results show that averaging four matrices improves noticeably the characteristics of sphericity. Averaging all six matrices leads to a coefficient of variation equal to 5.3% and a coefficient of correlation equal to 0.982. These results agree with our work on the colors of monochromatic lights in showing a high degree of sphericity.

INTERCONNECTIONS BETWEEN MUNSELL AND SPHERICAL COLOR COORDINATES

I calculated the spherical coordinates representing hue, saturation and lightness of the 24 Munsell colors, using Equation 9.2 to normalize the four Cartesian

TABLE 9.4
Four Cartesian and Three Spherical Coordinates (in radians) of 24 Munsell Colors in Terms of the Spherical Model of Color Discrimination (Column R Lists the Radial Distance of Each Color Point from the Origin)

	Cartesian Coordinates					*Spherical Coordinates*		
Color	X_1	X_2	X_3	X_4	R	a_1	a_2	a_3
1	4.06	0.98	6.71	4.58	9.1	6.03	0.47	0.60
2	3.20	0.81	8.52	4.42	10.0	6.05	0.33	0.48
3	−1.46	2.93	8.47	4.53	10.0	4.23	0.34	0.49
4	−1.24	1.45	9.08	4.21	10.0	4.04	0.19	0.43
5	−5.50	0.23	7.37	4.33	10.0	3.18	0.57	0.53
6	−3.99	−0.16	7.86	3.61	9.6	3.09	0.44	0.43
7	−2.06	−4.27	7.16	4.66	9.8	2.02	0.51	0.58
8	−1.58	−3.13	8.24	4.38	9.9	2.05	0.36	0.49
9	3.09	−3.98	7.49	5.18	10.4	0.90	0.50	0.61
10	2.22	−3.33	8.60	4.63	10.5	0.96	0.38	0.49
11	2.92	0.98	6.40	5.26	8.9	5.92	0.35	0.69
12	1.93	−0.06	7.23	6.18	9.7	0.05	0.20	0.71
13	1.90	4.45	6.81	6.79	10.8	5.11	0.48	0.78
14	0.01	2.67	7.24	6.61	10.2	5.04	0.28	0.74
15	−4.45	0.66	6.85	6.77	10.6	3.30	0.45	0.78
16	−2.79	0.12	7.30	6.08	9.9	3.18	0.28	0.70
17	−3.03	−0.57	5.96	6.16	9.2	2.97	0.35	0.80
18	−2.45	−0.13	7.20	6.13	9.8	3.10	0.26	0.70
19	−3.13	−2.89	5.47	6.90	9.8	2.39	0.45	0.90
20	−2.21	−2.11	6.34	6.27	9.4	2.38	0.32	0.78
21	−1.59	−3.11	7.24	7.19	10.8	2.05	0.33	0.78
22	−1.27	−2.27	7.28	6.77	10.3	2.10	0.26	0.75
23	3.30	−0.17	6.27	6.82	9.8	0.06	0.35	0.82
24	2.32	−0.27	7.60	7.23	10.7	0.15	0.21	0.76

coordinates. Spherical coordinate values are listed in Table 9.4 in units of radians.

Hue. The hue of a color point is determined by its azimuthal angle in the chromatic plane x_1x_2 (a_1 in Table 9.4). All color models agreed with this description, which provides a linear relation between the Munsell hue coordinate and the value of the color point's azimuthal angle in the spherical model. The same result was demonstrated in the work of Indow (1980), in which the locations of Munsell colors in such chromatic planes were analyzed in detail.

Brightness. A relation between Munsell brightness value (V) and spherical angles (a_3 in Table 9.4) of color points on the achromatic plane x_3x_4 is not

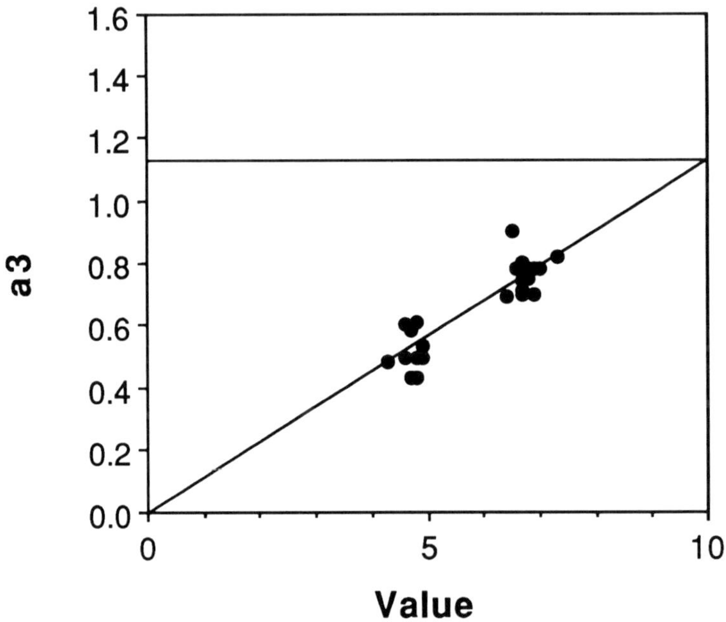

FIG. 9.4. Scatter diagram representing the relation between Munsell's scale of brightness (value) and the spherical coordinate a_3 obtained for 24 Munsell colors.

immediately evident. The situation is clarified by representing value and spherical angle in a scatter diagram (see Figure 9.4). Two clusters of points on this diagram mark two levels of brightness of the 24 Munsell colors (brightness values *V*5 and *V*7) (Indow & Kanazawa, 1960). Figure 9.4 shows a straight line drawn between the centers of the clusters and the origin of the scatter diagram: a linear model relates the two forms of brightness. Values of Munsell colors of reflected lights (abscissa) are related linearly to the brightnesses of self-luminous lights (ordinate). From this relationship, it follows that the maximal value of Munsell colors (10.0) has the intermediate value 1.1 rad on the scale of brightness of self-luminous colors.

Saturation. A scatter diagram for saturation is shown in Figure 9.5. The abscissa and the ordinate of the diagram represent Munsell chroma and the spherical coordinate a_2 for saturation, respectively, for the 24 Munsell colors. The linear relation between chroma and a_2 is very evident. The direct line that best approximates points on the scatter diagram intersects the vertical axis at a distance of 0.13 rad from the origin. This distance corresponds to nearly two JNDs in terms of the spherical model of color discrimination (Izmailov, 1980;

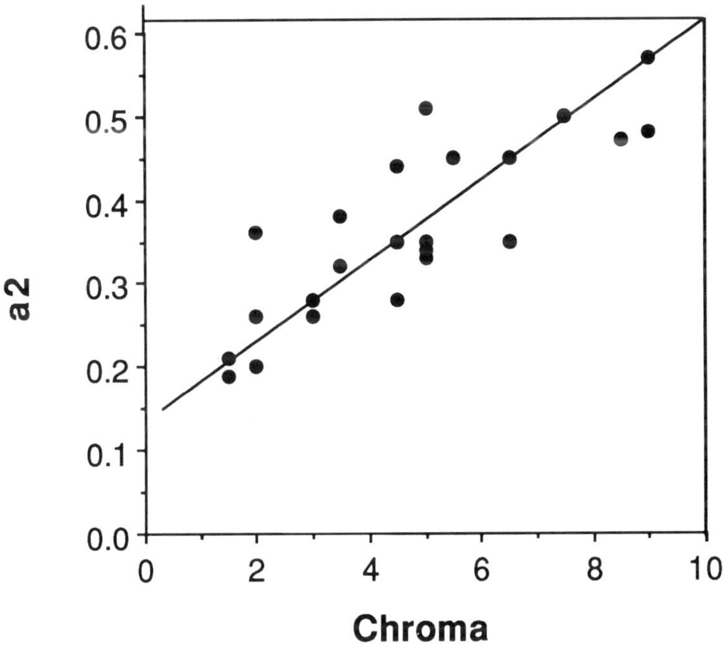

FIG. 9.5. Scatter diagram representing the relation between Munsell's scale of saturation (chroma) and the spherical coordinate a_2 obtained for 24 Munsell colors.

Sokolov & Izmailov, 1983). This small, constant shift of saturation may be due to experimental noise for a given set of stimuli evoked by influence of background and illuminance conditions.

CONCLUSION

These results represent the first step of analysis of the interrelation between the Munsell color system and the spherical model of color vision. They are far from complete. But even these results lead to positive conclusions about a color space that embeds self-luminous (aperture) colors as well as surface colors.

First, a general color space can be constructed for aperture and surface colors, which vary in hue, saturation, and brightness, by representing the three basic variables as three spherical angles of color points in this space. Euclidean distances between points in such a space coincides well with perceived color differences, both for measurements along separate color characteristic axes and for measurements between points that lie off the axes. The color space presents a solution to a problem with the cylindrical Munsell color solid, which, although the solid shows a good uniformity for each separate color variable, possesses substantial nonuniformity in interpoint distances when these points are allowed to vary in two *or more* color variables (Indow, 1980; Wyszecki & Stiles, 1982). Representing the Munsell variables as spherical coordinates provides total uniformity of color differences.

Second, relations between Munsell color characteristics and spherical angles are linear in first approximation. In this approximation, the set of surface colors does not fill the entire color space; rather it is a subset with the following limits. The hue scale for surface colors coincides with the azimuthal coordinate for aperture colors which subtends 360° (or 2π rad). Chroma has a range of 10 steps and is limited by a maximum of 0.5 rad (Figure 9.5). Value also has a range of 10 steps and is limited in the spherical model to 1.1 rad—the greater part of the spherical coordinate range (Figure 9.4). It may be of interest to note that, when represented in terms of the spherical model, one step in chroma is equivalent to two steps in value.

ACKNOWLEDGMENTS

The author gratefully acknowledges the help of Dr. E. Dzhafarov and Dr. M. D'Zmura in editing this manuscript.

REFERENCES

Hurvich, L. M., & Jameson, D. (1955). Some quantitative aspects of an opponent-colors theory: II. Brightness, saturation and hue in normal and dichromatic vision. *Journal of the Optical Society of America, 45,* 602–616.

Indow, T. (1980). Global color metrics and color appearance systems. *Color Research and Applications, 5,* 5–12.

Indow, T., & Kanazawa, K. (1960). Multidimensional mapping of Munsell colors varying in hue, and chroma, and value. *Journal of Experimental Psychology, 59,* 330–336.

Indow, T., & Uchizono, T. (1960). Multidimensional mapping of Munsell colors varying in hue and chroma. *Journal of Experimental Psychology, 59,* 321–329.

Izmailov, Ch. A. (1980). *Spherical model of color discrimination.* Moscow: Moscow State University.

Izmailov, Ch. A., Sokolov, E. N., & Chernorizov, A. M. (1989). *Psychophysiology of color vision.* Moscow: Moscow State University.

Izmailov, Ch. A., & Sokolov, E. N. (1991). Spherical model of color and brightness discrimination. *Psychological Science, 2,* 249–259.

Judd, D. B. (1967). Interval scales, ratio scales and additive scales for the sizes of differences perceived between members of a geodesic series of colors. *Journal of the Optical Society of America, 57,* 380–386.

Judd, D. B., & Wyszecki, G. (1963). *Color in business, science and industry.* New York: Wiley.

Kruskal, J. B. (1964). Nonmetric multidimensional scaling. A numerical method. *Psychometrika, 29,* 28–42.

Schrödinger, E. (1920). Grundlinien einer Theorie der Farbenmetric im Tagessehen. *Annalen der Physik (IV), 63,* 481–520.

Shepard, R. N. (1962a). The analysis of proximities: multidimensional scaling with unknown distance function: I. *Psychometrika, 27,* 125–140.

Shepard, R. N. (1962b). The analysis of proximities: Multidimensional scaling with unknown distance function: II. *Psychometrika, 27,* 219–246.

Shepard, R. N., & Carroll, J. D. (1966). Parametric representation of nonlinear data structures. In P. R. Krishnaich (Ed.), *International Symposium on Multivariate Analysis* (pp. 561–592). New York: Academic.

Sokolov, E. N., & Izmailov, Ch. A. (1983). The conceptual reflex arc and color vision. In H. G. Geissler (Ed.), *Modern Issues of Perception* (pp. 192–217). Berlin: VEB Deutscher Verlag.

Sokolov, E., & Izmailov, Ch. A. (1988). [A three-stage model of color vision]. *Sensory Systems, 2,* 400–407.

Torgerson, W. S. (1958). *Theory and method of scaling.* New York: Wiley.

Vos, J. J., & Walraven, P. L. (1972). An analytical description of the line element in the zone-fluctuation model of colour vision: I. Basic concepts. *Vision Research, 12,* 1327–1344.

Wyszecki, G., & Stiles, W. S. (1982). *Color science. Concepts and methods, quantitative data and formulae.* New York: Wiley.

10 Color Constancy: Spectral Recovery Using Trichromatic Bilinear Models

Geoffrey Iverson
Michael D'Zmura
Department of Cognitive Sciences & Institute for Mathematical Behavioral Sciences, University of California, Irvine

ABSTRACT

In recent work, we have shown that bilinear models of reflected lights can be used to provide accurate estimates of spectral properties of surfaces and light sources. We review that work here, paying special attention to trichromatic visual systems. We focus on the properties that a bilinear model must possess if it is to help recover spectral properties of lights and surfaces uniquely. Both numerical and analytical methods are discussed.

INTRODUCTION

The way that a material surface reflects lights of varying wavelength is described by a surface reflectance function. Such a reflectance function $R(\lambda)$ takes values between 0 and 1 and describes what proportion of incident light is reflected at each wavelength λ. Under simple viewing conditions, a light with spectral power $A(\lambda)$ that is incident on the surface will be reflected toward an observer with spectral power

$$L(\lambda) = R(\lambda)A(\lambda). \tag{10.1}$$

Variation in either surface composition or light source induces variation in reflected light. A remarkable fact of human vision is that a surface's color appearance remains stable under conditions of varying illumination (Beck, 1972; Katz, 1935; Land, 1983, 1986). This phenomenon of *color constancy* has prompted a number of theoretical schemes that estimate the spectral properties of a scene's illuminant and its surfaces (Brainard, Wandell, & Cowan, 1989; Brill, 1978, 1979; Buchsbaum, 1980; Drew & Funt, 1992; D'Zmura & Lennie, 1986; Forsyth, 1990; Hurlbert, 1986; Lee, 1986; Maloney, 1985; Maloney & Wandell, 1986; Rubner & Schulten, 1989; Sällström, 1973). In this chapter we sketch our own results on the matter (D'Zmura, 1992; D'Zmura & Iverson, 1993a, 1993b, 1993c, 1994; Iverson & D'Zmura, 1994).

We regard the first stage of a visual system as comprising p types of photoreceptors or light sensors at each image location. Each photoreceptoral type is associated with a spectral responsivity $Q_k(\lambda)$, $k = 1, 2, \ldots, p$. The spectral information that is available to a visual system concerning the light $L(\lambda)$ that impinges on its photoreceptors is provided by the responses q_k of those photoreceptors:

$$q_k = \int Q_k(\lambda)L(\lambda)\, d\lambda, \qquad k = 1, 2, \ldots, p. \tag{10.2}$$

If $L(\lambda)$ is the spectral power of light reflected at each wavelength λ from a surface with reflectance function $R(\lambda)$, when lit by a source with spectral power $A(\lambda)$, then we can combine Equations 10.1 and 10.2 to obtain

$$q_k = \int Q_k(\lambda)R(\lambda)A(\lambda)\, d\lambda, \qquad k = 1, 2, \ldots, p. \tag{10.3}$$

Equation 10.3 relates the visual system's data q_k to the physical properties of lights and surfaces. A visual system can exhibit color constancy by using the color data q_k to recover surface spectral properties. These spectral properties, which are represented by surface reflectance functions, do not depend on conditions of illumination, and recovering these reflectance functions is thus tantamount to color constancy. Unfortunately, Equation 10.3 makes it clear that, without further assumptions, there can be no color constancy: recovering functions of wavelength like $R(\lambda)$ using just a few data q_k is a hopeless task.

BASIS FUNCTIONS FOR REFLECTANCES AND ILLUMINANT SPECTRA

An important set of assumptions that has been brought to bear on the problem of color constancy is that typical lights and surfaces do not vary in an infinite number of ways; rather, illuminant spectral functions and surface reflectance functions each vary along only a few dimensions—comparable in number to the number p of photoreceptoral types (Lennie & D'Zmura, 1988). Consider, for instance, that typical photoreceptors are not sensitive to arbitrary spectral variation in either A or R. The light L that impinges on the photoreceptors may be decomposed as a sum of two components L_0 and L_1, such that for all $k = 1, 2, \ldots, p$,

$$\int Q_k(\lambda)L_0(\lambda)\, d\lambda = 0. \tag{10.4}$$

This suggests that a visual system need not attempt to reconstruct the physical spectra $R(\lambda)$ and $A(\lambda)$ but only those aspects of those functions which contribute to $L_1(\lambda)$, the part of the spectral function $L(\lambda)$ to which the photoreceptors are actually sensitive.

In light of this observation, it is useful to decompose each of $R(\lambda)$, $A(\lambda)$ into components, the first few of which dictate the responses of the photoreceptors. Specifically, the product $R(\lambda)A(\lambda)$ is decomposed as a sum

$$L(\lambda) = R(\lambda)A(\lambda) = \sum_{j=1}^{n} \sum_{i=1}^{m} r_j a_i R_j(\lambda) A_i(\lambda) + L_0(\lambda), \tag{10.5}$$

where $L_0(\lambda)$ satisfies Equation 10.4. The decomposition 10.5 is equivalent to replacing the actual surface reflectance by a weighted sum $\hat{R}(\lambda)$ of n components $R_j(\lambda)$, where

$$\hat{R}(\lambda) = \sum_{j=1}^{n} r_j R_j(\lambda), \tag{10.6a}$$

and replacing the actual illuminant spectral function $A(\lambda)$ by the weighted sum

$$\hat{A}(\lambda) = \sum_{i=1}^{m} a_i A_i(\lambda). \tag{10.6b}$$

The functions $R_j(\lambda)$, $j = 1, \ldots, n$, and $A_i(\lambda)$, $i = 1, \ldots, m$, constitute linear bases for surface reflectance functions and illuminant spectra, respectively. The weights r_j and a_i are called reflectance descriptors and illuminant descriptors, respectively.

Various basis functions $R_j(\lambda)$, $A_i(\lambda)$ have been proposed in the literature (Maloney, 1986). In Figure 10.1 (top panel), are shown the three CIE standard basis functions derived by Judd, MacAdam, and Wyszecki (1964) to account for the various phases of daylight. In the same figure (middle panel) are shown the three basis functions of Cohen (1964), which account for over 99% of the reflectance variation in a sample of 433 Munsell chips. An exemplary set of photoreceptoral sensitivities, those of Smith and Pokorny (1975), is displayed in the lowest panel of the figure.

More recently, Marimont and Wandell (1992) have argued persuasively that basis functions should be chosen so as to account for variation in the responses of a given set of photoreceptors (such as those of Smith & Pokorny). Unlike the reflectance and illuminant basis functions displayed in Figure 10.1, those of Marimont and Wandell are near zero at the extremes of the visible spectrum, where photoreceptors are quite insensitive. All authors agree that, for color

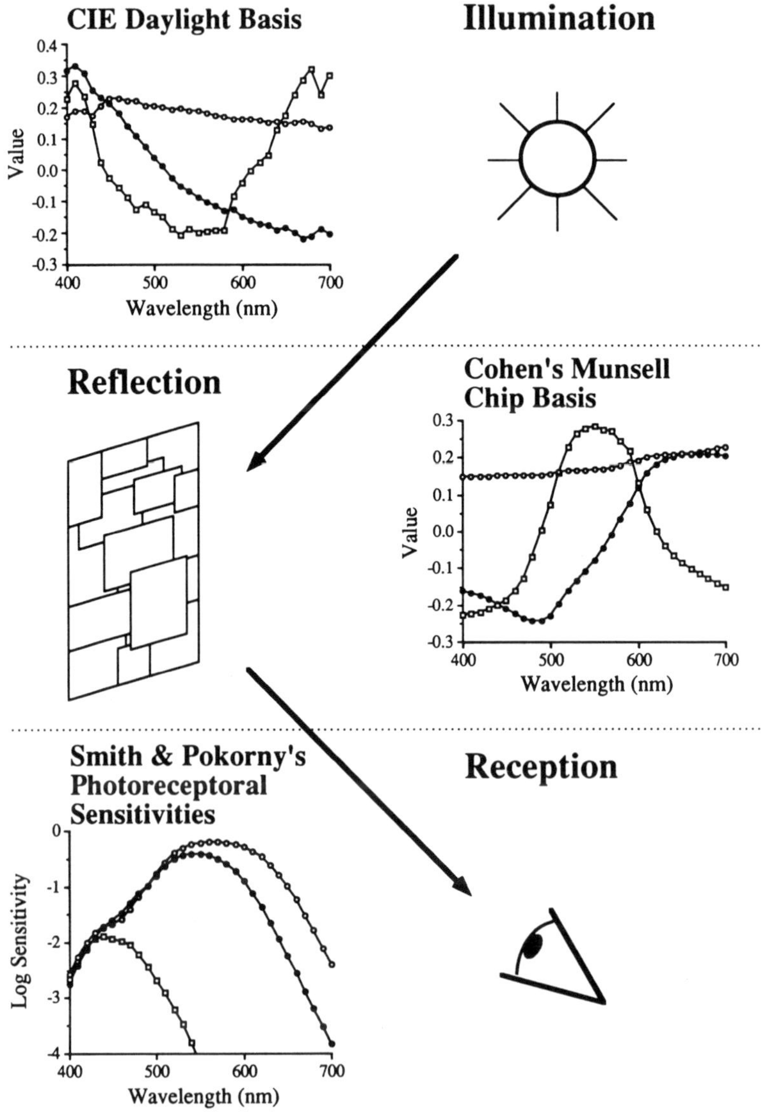

FIG. 10.1. Components of a trichromatic bilinear model. (Top) Model for illumination represented by three basic functions that describes the phases of daylight well (Judd, MacAdam, & Wyszecki, 1964). (Middle) Model for surface reflectance that describes the reflectance properties of Munsell chips well (Cohen, 1964). (Bottom) Linear model for human photoreception represented by the spectral sensitivities of Smith and Pokorny (1975).

vision under normal viewing conditions, at least three basis functions are required to model variation in both illumination and surface composition.

THE BASIC EQUATIONS FOR A BILINEAR MODEL

The use of a few basis functions to describe the dimensions along which reflectances and illuminant spectra can vary simplifies the computational problem of color constancy. The goal is now to recover a small, finite number of descriptors r_j and a_i from the photoreceptoral responses q_k. Toward this end, let us combine Equations 10.3–10.5 to express the dependence of data q_k on reflectance and illuminant descriptors:

$$q_k = \int Q_k(\lambda)\left(\sum_{j=1}^{n} \sum_{i=1}^{m} r_j a_i R_j(\lambda) A_i(\lambda) \right) d\lambda$$

$$= \sum_{j=1}^{n} \sum_{i=1}^{m} r_j b_{kji} a_i \tag{10.7}$$

where

$$b_{kji} = \int Q_k(\lambda) R_j(\lambda) A_i(\lambda)\, d\lambda. \tag{10.8}$$

The array b_{kji} of *pnm* numbers constitutes a *model* for a *p*-chromatic visual system. Note that according to Equation 10.7, the photoreceptoral responses q_k depend linearly on the vector $\mathbf{r} = [r_1, r_2, \ldots, r_n]^T$ of reflectance descriptors (here, as elsewhere, the superscript T denotes transpose). Likewise, photoreceptoral responses depend linearly on the vector $\mathbf{a} = [a_1, a_2, \ldots, a_m]^T$ of illuminant descriptors. We speak of Equation 10.7 as providing a *bilinear* model for a visual system.

The computational approach to the problem of color constancy is now seen to involve the following problem of bilinear algebra, namely, given sensory responses q_k satisfying Equation 10.7, how to extract the descriptors r_j and a_i. This task is made easier as more information is provided to the visual system. Suppose, for instance, that each of *s* surfaces is lit by a common light source. Then a visual system has available to it the *ps* photoreceptoral responses

$$q_{tk} = \sum_{j=1}^{n} \sum_{i=1}^{m} r_{tj} b_{kji} a_i, \qquad t = 1, \ldots, s, \quad k = 1, \ldots, p. \tag{10.9}$$

We shall also consider the situation in which each of *s* surfaces is lit, in turn, by *v* distinct lights. There are, in this case, *psv* sensor responses q_{tkw} with

$$q_{tkw} = \sum_{j=1}^{n} \sum_{i=1}^{m} r_{tj} b_{kji} a_{iw}, \qquad t = 1, \ldots, s, \quad k = 1, \ldots, p, \quad w = 1, \ldots, v. \tag{10.10}$$

A dominant theme of our work has involved the question, for which models b_{kji}, if any, can the nonlinear system of basic equations (10.10) be solved for the descriptors r_{tj} and a_{iw}? Note that there is an unavoidable ambiguity in any attempt to solve Equation 10.10. For if r_{tj}, a_{iw} satisfy Equation 10.10, then so do the *scaled* descriptors cr_{tj}, and $(1/c)a_{iw}$, for an arbitrary scalar c. If this arbitrary reciprocal scaling provides the only ambiguity in determining the descriptors r_{tj}, a_{iw} from Equation 10.10, then we say that recovery of descriptors is unique.

It is often convenient to organize the numerical array b_{kji} as a collection of matrices. This can be done in various ways, of which the following are the most important:

$$(\mathbf{B}_j)_{ki} = (\mathbf{B}'_i)_{kj} = (\boldsymbol{\beta}_k)_{ji} = b_{kji}. \tag{10.11}$$

For example, in terms of the $n \times m$ model matrices $\boldsymbol{\beta}_k$, the basic equations 10.10 read

$$\mathbf{R}\boldsymbol{\beta}_k\mathbf{A} = \boldsymbol{\Delta}_k, \quad k = 1, \ldots, p, \tag{10.12}$$

where the $s \times n$ reflectance matrix $\mathbf{R}$ holds the sn reflectance descriptors r_{tj}, and the $m \times v$ illuminant matrix $\mathbf{A}$ holds the mv descriptors a_{iw}. In the same way, the psv sensor responses q_{ktw} are grouped into p $s \times v$ matrices Δ_k.

Likewise, the basic equations 10.10 can be written in terms of the $p \times m$ matrices $\mathbf{B}_j$:

$$\sum_{j=1}^{n} r_{tj}\mathbf{B}_j\mathbf{A} = \mathbf{D}_t, \qquad t = 1, \ldots, s, \tag{10.13}$$

where the s response matrices $\mathbf{D}_t$ are of dimension $p \times m$.

In order for there to be any chance of solving either system 10.12 or 10.13 for the descriptor matrices $\mathbf{R}$ and $\mathbf{A}$, some restrictions on the model matrices $\boldsymbol{\beta}_k$(or $\mathbf{B}_j$) must be imposed. As will be evident shortly, much of our efforts have been concerned with finding conditions on the matrices $\boldsymbol{\beta}_k$ (or on the matrices $\mathbf{B}_j$) that are both necessary and sufficient for unique recovery of descriptors. Underlying such conditions is the assumption that the matrices $\boldsymbol{\beta}_k$ be of full rank. This condition is readily understood in the case of $m = n$, namely the case in which the matrices are square. We have supposed that the basis functions $R_j(\lambda)$, $A_i(\lambda)$ have been chosen to account accurately for all variation in the sensor responses. In particular, we require that there be no light for which some photoreceptor is insensitive to variation in surfaces lit by that light. That is, there is no vector $\mathbf{a}_0$

of illuminant descriptors such that, for all reflectance vectors **r** and some photoreceptoral type with index κ,

$$\mathbf{r}^{\mathrm{T}}\boldsymbol{\beta}_{\kappa}\mathbf{a}_0 = \mathbf{0}.$$

This means that all matrices $\boldsymbol{\beta}_k$ are nonsingular. For similar reasons we require the matrices $\mathbf{B}_j$ (or equivalently $\mathbf{B}_i'$) to be nonsingular when they are square. When a model's matrices are rectangular, the analogous requirement is that they be of full rank.

NECESSARY CONDITIONS FOR UNIQUE RECOVERY: FEASIBILITY CRITERION

The developments of the preceding section provide bilinear models to be used in recovering reflectance and illuminant descriptors from photoreceptoral data. Such models depend on the number p of photoreceptoral types and the dimensions n and m of the bases for reflectance and illumination, respectively. They take as input the data from v views (provided by different illuminants) of s surfaces. The meaning of each of these parameters is placed in Table 10.1 for ready reference. The building blocks for a model are the basis functions for reflectance, illumination, and the photoreceptoral spectral sensitivities (e.g., those shown in Figure 10.1).

In this and the following sections we shall be concerned primarily with distinguishing those bilinear models that work (viz., can be used to recover descriptors from photoreceptoral responses) from those that do not. We require already that a bilinear model's matrices be of full rank. This fact does not, of course, guarantee that unique recovery of descriptors obtains. A given model may allow unique recovery (a) always, (b) sometimes, or (c) never. In case (a), we say that a model allows perfect recovery; in case (b) partial recovery, and in case (c) we say that a model is a complete or total failure.

We come by our first criterion for successful recovery by comparing the number of data to the number of unknown descriptors: for a model to function adequately, sufficient data must be available. In particular, the number psv of sensor responses must compare favorably to the number $ns + mv$ of descriptors

TABLE 10.1
Bilinear Model Parameters

p	Number of photoreceptoral types
m	Dimension of the model for illumination
n	Dimension of the model for reflectance
v	Number of views provided by different illuminants
s	Number of different surfaces

to be recovered. Taking into account the ubiquitous reciprocal scaling we mentioned above, the following feasibility criterion must be satisfied if a given model is to be anything but a total failure:

$$psv \geq ns + mv - 1. \tag{10.14}$$

For reasons of linear independence of both illuminants and surfaces, $v \leq m$ and $s \leq n$, respectively, in Inequality 10.14.

PROBLEMS AND ENTAILMENTS

The criterion inequality 10.14 applies to all models that share some particular choice of the parameters p, m, n, v, and s. This generality suggests that we consider *color constancy* problems. A color constancy problem, denoted $(p\ m\ n\ v\ s)$, involves the question, for which bilinear models with parameters p, m, n, when provided v views of s surfaces, is unique recovery of the descriptors r_{tj}, a_{iw} always, sometimes, or never possible? If, for a given problem, there is a model that provides perfect recovery, then we say that the problem is a total success. If it can be shown that, for a given problem, no model is capable of perfect recovery, but there exists a model that allows unique recovery under some circumstances but not others, we say that the problem is a partial success (or partial failure). Finally, if all models for a problem are such that unique recovery is never possible, then we say that the problem is a total failure. Problems for which the feasibility criterion 10.14 is not met are total failures.

In previous work (D'Zmura & Iverson, 1993a, 1993b, 1994), we began the process of classifying problems as to their type (total success, partial success, and total failure). This task is made easier by taking into account the entailments recorded in Table 10.2 (taken from D'Zmura & Iverson, 1993b). The notation $(p\ m\ n\ v\ s) \Rightarrow (p'\ m'\ n'\ v'\ s')$ indicates that if the problem $(p\ m\ n\ v\ s)$ is a total success, then so is the problem $(p'\ m'\ n'\ v'\ s')$. Likewise, the notation $\sim(p\ m\ n\ v\ s) \Rightarrow \sim(p'\ m'\ n'\ v'\ s')$ indicates that if the problem $(p\ m\ n\ v\ s)$ is a total failure, then the same is true of $(p'\ m'\ n'\ v'\ s')$.

Problems can be partially ordered by entailment. Major interest centers on "root" problems. A root problem is one that is not entailed by any other problem. The problem (3 3 3 3 3) *is* entailed by the problem (3 3 3 2 3), for example (entailment (b) of Table 10.2), and so is not a root problem. Problems that are root problems include (31311), (33212), and (33322) (Brainard et al., 1989; Maloney & Wandell, 1986; Tsukada & Ohta, 1990). We shall discuss these and other trichromatic problems. Suffice it to note here that the problem (31311) is a total success, while the problems (33212) and (33322) are partial successes (D'Zmura & Iverson, 1993b).

TABLE 10.2
Entailments Among Color Constancy Problems[a]

Positive		
	$(p\ m\ n\ v\ s) \Rightarrow (p+1\ m\ n\ v\ s)$	(a)
	$(p\ m\ n\ v\ s) \Rightarrow (p\ m\ n\ v+1\ s)$	(b)
	$(p\ m\ n\ v\ s) \Rightarrow (p\ m\ n\ v\ s+1)$	(c)
	$(p\ m\ n\ v\ s) \Rightarrow (p\ n\ m\ s\ v)$	(d)
	$(m > v)\ \&\ (p\ m\ n\ v\ s) \Rightarrow (p\ m-1\ n\ v\ s)$	(e)
	$(v > m)\ \&\ (p\ m\ n\ v\ s) \Rightarrow (p\ m\ n\ v-1\ s)$	(f)
	$(n > s)\ \&\ (p\ m\ n\ v\ s) \Rightarrow (p\ m\ n-1\ v\ s)$	(g)
	$(s > n)\ \&\ (p\ m\ n\ v\ s) \Rightarrow (p\ m\ n\ v\ s-1)$	(h)
Negative		
	$\sim(p\ m\ n\ v\ s) \Rightarrow \sim(p-1\ m\ n\ v\ s)$	(i)
	$\sim(p\ m\ n\ v\ s) \Rightarrow \sim(p\ m\ n\ v-1\ s)$	(j)
	$\sim(p\ m\ n\ v\ s) \Rightarrow \sim(p\ m\ n\ v\ s-1)$	(k)
	$\sim(p\ m\ n\ v\ s) \Rightarrow \sim(p\ n\ m\ s\ v)$	(l)
	$(m \geq v)\ \&\ \sim(p\ m\ n\ v\ s) \Rightarrow \sim(p\ m+1\ n\ v\ s)$	(m)
	$(v \geq m)\ \&\ \sim(p\ m\ n\ v\ s) \Rightarrow \sim(p\ m\ n\ v+1\ s)$	(n)
	$(n \geq s)\ \&\ \sim(p\ m\ n\ v\ s) \Rightarrow \sim(p\ m\ n+1\ v\ s)$	(o)
	$(s \geq n)\ \&\ \sim(p\ m\ n\ v\ s) \Rightarrow \sim(p\ m\ n\ v\ s+1)$	(p)

[a] D'Zmura and Iverson (1993b).

NECESSARY AND SUFFICIENT CONDITIONS FOR PERFECT RECOVERY

The feasibility criterion 10.14, while providing a necessary condition for a model to allow perfect recovery, is, alas, not sufficient. As a simple example, we mention that the feasible problem (2 2 2 2 2) is a total failure (D'Zmura & Iverson, 1993a).

The task of discovering conditions that are both necessary and sufficient for a model to function flawlessly is not an easy one. Much of the difficulty arises from the fact that such conditions will surely involve the parameters p, m, n, v, and s in their expression. For example, if conditions on the model matrices $\boldsymbol{\beta}_k$ for a problem $(p\ m\ n\ v\ s)$ are known to guarantee perfect recovery, somewhat weaker conditions must be anticipated for the problem $(p\ m\ n\ v+1\ s)$. We give an illustration of this in the ninth section on "Trichromatic Problems."

Despite these reservations it is possible to give a model check algorithm which provides conditions for a given model to work flawlessly (D'Zmura & Iverson, 1993a, 1993b, 1994). The algorithm adapts itself to the parameters of a problem in a natural way. We do not attempt here to develop the algorithm in detail. Rather, we content ourselves with a simple illustration.

MODEL CHECK ALGORITHM FOR THE PROBLEM (3 3 3 2 3) INVOLVING TWO VIEWS OF THREE SURFACES

In earlier work (D'Zmura & Iverson, 1993a), we developed the model check algorithm for the problem (3 3 3 2 3) in terms of the model matrices $\mathbf{B}_j$, $j = 1, 2, 3$. For our present purposes, it is more convenient to work in terms of the matrices $\boldsymbol{\beta}_k$, $k = 1, 2, 3$.

Let us start by specializing Equation 10.12 to a trichromatic visual system ($p = 3$). The dependence of data on descriptors and bilinear model matrices is

$$\mathbf{R}\boldsymbol{\beta}_k\mathbf{A} = \Delta_k, \qquad k = 1, 2, 3,$$

where the reflectance matrix $\mathbf{R}$ and the bilinear model matrices $\boldsymbol{\beta}_1$, $\boldsymbol{\beta}_2$, and $\boldsymbol{\beta}_3$ are each 3×3, and the illuminant matrix $\mathbf{A}$ and the photoreceptoral response matrices $\boldsymbol{\Delta}_1$, $\boldsymbol{\Delta}_2$, and $\boldsymbol{\Delta}_3$ are each 3×2. All matrices are assumed to be of full rank.

There are several recovery algorithms that one might entertain for the problem (3 3 3 2 3).[1] We will describe one example, namely, the general linear recovery algorithm of D'Zmura and Iverson (1994). To explain the algorithm, let us first rewrite the above equation in the form

$$\boldsymbol{\beta}_k\mathbf{A} - \mathbf{R}^{-1}\,\boldsymbol{\Delta}_k = 0, \qquad k = 1, 2, 3, \tag{10.15}$$

which, when written out in detail, comprises a system of 18 linear equations in 15 unknowns (the nine entries of $\mathbf{R}^{-1}$ and the six entries of $\mathbf{A}$). Provided that the kernel of the 18×15 system matrix has dimension 1, recovery of the descriptors for the two illuminants and the (inverse) descriptors for the three surfaces is unique. (These descriptors, strung out as a 15-dimensional vector, constitute the 1-dimensional kernel.) When the dimension of the kernel of the system matrix exceeds 1, unique recovery fails and the model is at best a partial success.

For which models does the above recovery algorithm work for all legitimate data? Assume that there exist two sets of lit surfaces that give rise to identical quantum catch data. Suppose that the two matrices $\mathbf{R}$ and $\mathbf{S}$ of reflectance descriptors represent the two sets of surfaces and that the pair $\mathbf{A}$ and $\mathbf{Z}$ of illuminant descriptor matrices represents the lights that shine on the surfaces of $\mathbf{R}$ and $\mathbf{S}$, respectively. For unique spectral recovery to be possible, the only way that two different sets of surfaces lit by two different sets of lights can produce identical photoreceptoral responses is if the two sets of surfaces are identical and the two sets of lights are identical—identical up to some arbitrary scale factor. Specifically, $\mathbf{R} = c\mathbf{S}$ and $\mathbf{A} = (1/c)\mathbf{Z}$, for some constant c. If two, *truly* different

[1]One could use the two-stage linear recovery algorithm of D'Zmura (1992) and D'Zmura and Iverson (1993a, 1993b) or a nonlinear, least-squares procedure like that outlined by D'Zmura and Iverson (1993c).

sets of lit surfaces give rise to identical quantum catch data, then unique recovery is clearly not possible.

Thus, to resolve whether unique recovery is possible, we suppose that there exist two reflectance matrices **R, S** that represent two different sets of surfaces, and a corresponding pair of illuminant matrices **A, Z** that represent two sets of lights, such that the same data $\boldsymbol{\Delta}_k$, $k = 1, 2, 3$, can be expressed by Equation 10.12 both in terms of the pair (**R, A**) and the pair (**S, Z**):

$$\mathbf{R}\boldsymbol{\beta}_k\mathbf{A} = \mathbf{S}\boldsymbol{\beta}_k\mathbf{Z} = \Delta_k, \qquad k = 1, 2, 3.$$

It follows that

$$\boldsymbol{\beta}_k^{-1}\mathbf{S}^{-1}\mathbf{R}\boldsymbol{\beta}_k\mathbf{A} = \mathbf{Z}, \qquad \text{for } k = 1, 2, 3.$$

In terms of the matrix $\mathbf{E} = \mathbf{S}^{-1}\mathbf{R}$ and the matrices $\mathbf{G}_{21} = \boldsymbol{\beta}_2\boldsymbol{\beta}_1^{-1}$, $\mathbf{G}_{31} = \boldsymbol{\beta}_3\boldsymbol{\beta}_1^{-1}$, we obtain after some manipulation

$$[\mathbf{G}_{21}, \mathbf{E}]\boldsymbol{\beta}_1\mathbf{A} = [\mathbf{G}_{31}, \mathbf{E}]\boldsymbol{\beta}_1\mathbf{A} = \mathbf{0}, \tag{10.16}$$

where we have used the notation [**X, Y**] to denote the commutator $\mathbf{XY} - \mathbf{YX}$ (see Iverson & D'Zmura, 1994, for details). Unique recovery by the general linear recovery algorithm sketched above is guaranteed if and only if the only solutions of Equation 10.16 are of the form $\mathbf{E} = c\mathbf{I}$, $c \neq 0$ (**I** is the identity matrix).

Equation 10.16 means that the three rows of the matrix $[\mathbf{G}_{21}, \mathbf{E}]$ and those of the matrix $[\mathbf{G}_{31}, \mathbf{E}]$ are each perpendicular to the two independent columns of $\boldsymbol{\beta}_1\mathbf{A}$. In other words, the rows of $[\mathbf{G}_{21}, \mathbf{E}]$ and those of $[\mathbf{G}_{31}, \mathbf{E}]$ are mutually collinear. It follows that all 2×2 determinants formed from the 6×3 matrix

$$\begin{pmatrix} [\mathbf{G}_{21}, \mathbf{E}] \\ [\mathbf{G}_{31}, \mathbf{E}] \end{pmatrix} \tag{10.17}$$

vanish identically. Because $\mathbf{E} = c\mathbf{I}$ is always a solution of Equation 10.16, it is useful to decompose the matrix **E** as $\mathbf{E} = e_{11}\mathbf{I} + \boldsymbol{\epsilon}$. Since $[\mathbf{G}_{21}, \boldsymbol{\epsilon}] = [\mathbf{G}_{21}, \mathbf{E}]$ and $[\mathbf{G}_{31}, \boldsymbol{\epsilon}] = [\mathbf{G}_{31}, \mathbf{E}]$, it follows that each of the forty-five 2×2 determinants formed from the 6×3 matrix 10.17 is a quadratic polynomial in the eight variables $e_{11} - e_{22}$, $e_{11} - e_{33}$, and e_{ij}, $i \neq j$. Each such polynomial is a sum of the 36 monomials $(e_{11} - e_{22})^2$, $(e_{11} - e_{22})(e_{11} - e_{33})$, . . . , $e_{31}e_{32}$, e_{32}^2. Regarding each of these monomials as a different variable, we obtain 45 homogeneous linear equations in 36 variables. The model check algorithm computes the rank of the resulting 45×36 system matrix. If the rank of this matrix if full (i.e., 36) we are assured that the matrix $\boldsymbol{\epsilon}$ is null and that $\mathbf{E} = e_{11}\mathbf{I}$, in which case unique recovery of descriptors is guaranteed.

A natural generalization of this algorithm encompasses a wide array of other problems, and its application has met with considerable success (D'Zmura & Iverson, 1993a, 1993b, 1994). Even so, the model check algorithm does not yield much insight into the analytic structure of a model. To address this issue, we now turn to another approach. To keep the discussion both simple and pertinent, we consider only trichromatic problems.

TRICHROMATIC PROBLEMS INVOLVING ONE VIEW

There are two natural classes of color constancy problems to consider. Those involving two views, like (3 3 3 2 3) in the previous section, arise in cases where the illumination of a set of surfaces changes. This change can occur in time, so that two different illuminants shine in succession on a set of surfaces. This type of change was used by Edwin Land (1983, 1986) in his famous demonstrations of human color constancy. The change in illumination can also occur across space, as in the common situation where outdoor surfaces, partially in shadow, are lit simultaneously by bluish skylight and by the yellowish light from the solar disk. We shall treat these two-view problems in more detail in the next section.

The other natural class of problems involves a single view of a set of surfaces (Brainard et al., 1989; Maloney & Wandell, 1986). These problems are of the form $(p\ m\ n\ v\ s) = (3\ m\ n\ 1\ s)$, and we require, by the feasibility criterion 10.14, the inequality constraint $3s \geq ns + m - 1$, or equivalently $(3 - n)s \geq m - 1 \geq 0$. It follows that $n < 3$ (or $m = 1$ and $n = 3$). The fact that a trichromatic visual system (of the bilinear form considered here) cannot recover more than two reflectance descriptors (unless $m = 1$), was first discovered by Maloney and Wandell (1986). The result is, of course, a disappointment, for everyone agrees that our visual system is sensitive to at least 3 degrees of freedom in surface reflectance functions.

The only way to avoid this negative result is to accept the possibility $m = 1$. This was done by Brainard and colleagues (1989), who considered models of the form

$$\boldsymbol{\beta}_k = \mathbf{x}_k \mathbf{y}^{\mathrm{T}}, \tag{10.18}$$

where $\mathbf{x}_1$, $\mathbf{x}_2$, $\mathbf{x}_3$, $\mathbf{y}$ are nonnull three-dimensional vectors and the $\mathbf{x}_k$ are independent. Note that these model matrices $\boldsymbol{\beta}_k$ have a common two-dimensional kernel, consisting of all vectors orthogonal to $\mathbf{y}$. Such models are, in our terminology, singular. However, they are equivalent to nonsingular models for the problem (3 1 3 1 1) in which the matrices $\boldsymbol{\beta}_k$ are (independent) 3×1 vectors, and the vector of illuminant descriptors reduces to a scalar. Such models are sensitive to just 1 degree of freedom in illumination and thus cannot be regarded as satisfactory.

Because no trichromatic, bilinear model can recover three descriptors for both surface reflectances and illuminants, when provided a single view of any number of surfaces, Tsukada and Ohta (1990) and independently D'Zmura (1992) proposed that two or more views of a set of surfaces are required if one is to recover the requisite number of descriptors. Indeed, as D'Zmura (1992) pointed out, the very issue of color constancy does not arise unless the illumination changes.

TRICHROMATIC PROBLEMS INVOLVING TWO OR MORE VIEWS

The root problem (3 3 3 2 2) was first considered by Tsukada and Ohta (1990). They advanced a least-squares algorithm which they demonstrated could be used to recover the spectral characteristics of a color filter placed in front of an RGB camera. However, we know the problem (3 3 3 2 2) to be a partial success: there are certain choices of illuminants which cause recovery algorithms for this problem to fail (see D'Zmura & Iverson, 1993b). To be fair about the matter, we point out that these illuminants need not be physically realizable, depending on the detailed numerical structure of a model's matrices. A number of important and interesting features of the problem (3 3 3 2 2) remain unresolved. In contrast, the problems (3 3 3 3 3) and (3 3 3 2 3) are fully understood (D'Zmura & Iverson, 1993a, 1993b; Iverson & D'Zmura, 1994). For these two problems, Iverson & D'Zmura (1994) formulated two theorems that state necessary and sufficient conditons for unique recovery. We present these theorems below.

Let us consider first the simpler problem (3 3 3 3 3). We shall give conditions on the model matrices $\boldsymbol{\beta}_k$ that are both necessary and sufficient for unique recovery, at least for the broad class of so-called regular models. To frame these conditions in a satisfactory way, we use some familiar notions from linear algebra (see Iverson & D'Zmura, 1994, for details).

Suppose a model's matrices $\boldsymbol{\beta}_1$, $\boldsymbol{\beta}_2$, $\boldsymbol{\beta}_3$ are such that there exists a (nontrivial) subspace V and a subspace W of $\mathbb{R}^3$ such that $\boldsymbol{\beta}_k$ maps V into W for all $k = 1, 2, 3$. Then the model is said to be *reducible* (and the subspace V is the *reducing* subspace). For example, the (singular) models of the form 10.18 are reducible: V is the two-dimensional subspace that is orthogonal to the vector $\mathbf{y}$, and $W = \{0\}$. If $\mathbb{R}^3$ decomposes as a direct sum of a one-dimensional subspace V_1 and a two-dimensional subspace V_2, and the three matrices of a model are reduced simultaneously under both V_1 and V_2, then the model is *decomposable*. If a model is not reducible, it is *irreducible;* if a model is not decomposable, it is *indecomposable*. Note that for the class of nonsingular models, these notions can be reexpressed most effectively in terms of the matrices $\mathbf{G}_{kl} = \boldsymbol{\beta}_k\boldsymbol{\beta}_l^{-1}$, $k, l = 1, 2, 3$. If a nonsingular model is reducible, the matrices $\mathbf{G}_{kl}$ can, by an appropriate similarity transformation, be brought simultaneously to one of the two forms

$$\begin{pmatrix} x & x & 0 \\ x & x & 0 \\ x & x & x \end{pmatrix} \quad \text{or} \quad \begin{pmatrix} x & 0 & 0 \\ x & x & x \\ x & x & x \end{pmatrix};$$

where x denotes an entry which is not necessarily zero. Likewise, if a nonsingular model is decomposable, the matrices $\mathbf{G}_{kl}$ can be simultaneously brought to the following form by choosing an appropriate basis for the reflectances:

$$\begin{pmatrix} x & 0 & 0 \\ 0 & x & x \\ 0 & x & x \end{pmatrix}.$$

Finally, we say that a trichromatic model is *regular* if at least one of the matrices $\mathbf{G}_{kl}$ has distinct eigenvalues.

> **Theorem 10.1.** If a nonsingular trichromatic model for the problem (3 3 3 3 3) is regular and indecomposable, then it allows perfect recovery. Conversely, a regular, decomposable model is a total failure.

It is not within the scope of this chapter to prove this theorem; we give a detailed proof elsewhere (Iverson & D'Zmura, 1994). Note that Theorem 10.1 yields necessary and sufficient conditons for a regular trichromatic model using three (different) views of three (different) surfaces to guarantee unique recovery of all 18 descriptors. Note also that, for the problem (3 3 3 3 3), all models fall into one of two equivalence classes: those that work flawlessly, and those that never work. There are no partial successes.

There is some difficulty extending Theorem 10.1 to include nonregular models. To appreciate this, consider the following two examples:

Example (a).

$$\mathbf{G}_{21} = \begin{pmatrix} \lambda & 0 & 0 \\ 0 & \lambda & 0 \\ 0 & 0 & \lambda' \end{pmatrix}, \quad \lambda \neq \lambda'; \qquad \mathbf{G}_{31} = \begin{pmatrix} 1 & 1 & 1 \\ 0 & 1 & 0 \\ 0 & 0 & -1 \end{pmatrix}.$$

The model $\boldsymbol{\beta}_1 = \mathbf{I}$, $\boldsymbol{\beta}_2 = \mathbf{G}_{21}$, $\boldsymbol{\beta}_3 = \mathbf{G}_{31}$ is a total failure.

Example (b).

$$\mathbf{G}_{21} = \begin{pmatrix} \lambda & 0 & 0 \\ 0 & \lambda & 0 \\ 0 & 0 & \lambda' \end{pmatrix}, \quad \lambda \neq \lambda'; \qquad \mathbf{G}_{31} = \begin{pmatrix} 1 & 1 & 1 \\ 0 & 1 & 1 \\ 0 & 0 & -1 \end{pmatrix}.$$

The model $\boldsymbol{\beta}_1 = \mathbf{I}$, $\boldsymbol{\beta}_2 = \mathbf{G}_{21}$, $\boldsymbol{\beta}_3 = \mathbf{G}_{31}$ allows perfect recovery. Note that neither model is decomposable (although they are both reducible, with the same reducing subspace). It is worth mentioning here that we have examined many models that derive from empirically motivated choices of photoreceptors, reflectance bases, and illuminant bases and we have yet to meet a nonsingular or nonregular trichromatic model.

A result similar to Theorem 10.1 obtains for the more interesting problem (3 3 3 2 3). The needed criterion for (3 3 3 2 3) must be sharper than that in Theorem 10.1 for (3 3 3 3 3), because there is one less view. For (3 3 3 3 3), we used the notion of indecomposability; the more restrictive notion of irreducibility is the natural criterion for the problem (3 3 3 2 3). Again, Iverson and D'Zmura (1994) provide a detailed discussion and proof of the next result.

> **Theorem 10.2.** If a nonsingular, regular trichromatic model for the problem (3 3 3 2 3) is irreducible, then it allows perfect recovery. Conversely, a regular, reducible model is, at best, a partial success.

We have used the two theorems to test extensively whether bilinear models provide unique recovery for the problems (3 3 3 3 3) and (3 3 3 2 3). The results of these tests agree fully with the earlier results of the model check algorithm in showing that all empirically motivated bilinear models work perfectly for these problems.

CONCLUSIONS

In this chapter we have sketched results on the use of bilinear models by trichromatic visual systems to achieve color constancy. The groundbreaking work of Maloney and Wandell (1986) led to the disappointing conclusion that a trichromatic visual can recover no more than two spectral descriptors per surface reflectance. Because the roles of light sources and surface reflectances in a bilinear model are symmetric, we were led to consider the information added by shining two or more lights on two or more surfaces. This has led to positive results (D'Zmura, 1992; D'Zmura & Iverson, 1993a, 1993b; Iverson & D'Zmura, 1994). Indeed, we show elsewhere (D'Zmura & Iverson, 1994) that a trichromatic visual system can recover spectral descriptions of arbitrarily high dimension. For example, we have proven that all problems in the infinite chain of problems of form (3 c c c c), for $c \geq 2$, are total successes. However, it must be acknowledged that the need in these latter schemes to acquire sensory data from three or more independent views of a set of surfaces is not natural.

There are other color constancy schemes that recover three or more descriptors per surface reflectance. These schemes use additional assumptions, including (1) a known space-averaged reflectance function (the gray-world assumption) (Buchsbaum, 1980), (2) the presence of highlights (D'Zmura & Lennie, 1986), (3) the presence of interreflections (Drew & Funt, 1992), and (4) knowledge of how frequently surface reflectances and illuminant spectra are encountered (Brainard & Freeman, 1994; Trussell & Vrhel, 1991). We present work that draws on this last assumption in the chapter on probabilistic color constancy (D'Zmura, Iverson, & Singer, 1995).

ACKNOWLEDGMENTS

We thank Al Ahumada, Dave Brainard, Tarow Indow, Michael Landy, Larry Maloney, Jeff Mulligan, Misha Pavel, Brian Wandell, and Jack Yellott for helpful discussion. This work was supported by NEI EY10014 to M. D'Zmura and by NSF DIR-9014278 to the Institute for Mathematical Behavioral Sciences, University of California, Irvine, R. D. Luce, Director.

REFERENCES

Beck, J. (1972). *Surface color perception.* Ithaca, NY: Cornell University Press.

Brainard, D. H., & Freeman, W. T. (1994). Bayesian method for recovering surface and illuminant properties from photosensor responses. *Proceedings of the SPIE Symposium on Human Vision, Visual Processing, and Digital Display V, Vol. 2179,* 364–376.

Brainard, D. H., Wandell, B. A., & Cowan, W. B. (1989). Black light: How sensors filter spectral variation of the illuminant. *IEEE Transactions on Biomedical Engineering, 36,* 140–149.

Brill, M. H. (1978). A device performing illuminant-invariant assessment of chromatic relations. *Journal of Theoretical Biology, 71,* 473–478.

Brill, M. H. (1979). Further features of the illuminant-invariant trichromatic photosensor. *Journal of Theoretical Biology, 78,* 305–308.

Buchsbaum, G. (1980). A spatial processor model for object colour perception. *Journal of the Franklin Institute, 310,* 1–26.

Cohen, J. (1964). Dependency of the spectral reflectance curves of the Munsell color chips. *Psychonomic Sciences, 1,* 369–370.

Drew, M. S., and Funt, B. V. (1992). Variational approach to interreflection in color images. *Journal of the Optical Society of America A, 9,* 1255–1265.

D'Zmura, M. (1992). Color constancy: Surface color from changing illumination. *Journal of the Optical Society of America A, 9,* 490–493.

D'Zmura, M., & Iverson, G. (1993a). Color constancy: I. Basic theory of two-stage linear recovery of spectral descriptions for lights and surfaces. *Journal of the Optical Society of America A, 10,* 2148–2165.

D'Zmura, M., & Iverson, G. (1993b). Color constancy: II. Results for two-stage linear recovery of spectral descriptions for lights and surfaces. *Journal of the Optical Society of America A, 10,* 2166–2180.

D'Zmura, M., & Iverson, G. (1993c). Color constancy: Feasibility and recovery. *Investigative Ophthalmology and Visual Science, 34* (Suppl.), 748.

D'Zmura, M., & Iverson, G. (1994). Color constancy: III. General linear recovery of spectral descriptions for lights and surfaces. *Journal of the Optical Society of America A, 11,* 2389–2400.

D'Zmura, M., Iverson, G., & Singer, B. (1995). Probabilistic color constancy. In R. D. Luce, M. D'Zmura, D. D. Hoffman, G. Iverson, & A. K. Romney (Eds.), *Geometric representations of perceptual phenomena: Papers in honor of Tarow Indow's 70th birthday.* Hillsdale, NJ:

D'Zmura, M., & Lennie, P. (1986). Mechanisms of color constancy. *Journal of the Optical Society of America A, 3,* 1662–1672.

Forsyth, D. A. (1990). A novel algorithm for color constancy. *International Journal of Computer Vision, 5,* 5–36.

Hurlbert, A. (1986). Formal connections between lightness algorithms. *Journal of the Optical Society of America A, 3,* 1684–1693.

Iverson, G., & D'Zmura, M. (1994). Criteria for color constancy in trichromatic bilinear models. *Journal of the Optical Society of America A, 11,* 1970–1975.

Judd, D. B., MacAdam, D. L., & Wyszecki, G. (1964). Spectral distribution of typical daylight as a function of correlated color temperature. *Journal of the Optical Society of America, 54,* 1031–1040.

Katz, D. (1935). *The world of color* (R. B. MacLeod & C. W. Fox, Trans.). London: Kegan Paul, Trench, Trubner.

Land, E. H. (1983). Recent advances in retinex theory and some implications for cortical computations: color vision and the natural image. *Proceedings of the National Academy of Sciences USA, 80,* 5163–5169.

Land, E. H. (1986). Recent advances in retinex theory. *Vision Research, 26,* 7–21.

Lee, H.-C. (1986). A method for computing the scene illuminant chromaticity from specular highlights. *Journal of the Optical Society of America A, 3,* 1694–1699.

Lennie, P., & D'Zmura, M. (1988). Mechanisms of color vision. *Critical Reviews in Neurobiology, 3,* 333–400.

Maloney, L. T. (1985). *Computational approaches to color constancy* (Tech. Rep. No. 1985-01). Stanford: Psychology Laboratory.

Maloney, L. T. (1986). Evaluation of linear models of surface spectral reflectance with small numbers of parameters. *Journal of the Optical Society of America A, 3,* 1673–1683.

Maloney, L. T., & Wandell, B. A. (1986). Color constancy: A method for recovering surface spectral reflectance. *Journal of the Optical Society of America A, 3,* 29–33.

Marimont, D. H., & Wandell, B. A. (1992). Linear models of surface and illuminant spectra. *Journal of the Optical Society of America A, 9,* 1905–1913.

Rubner, J., & Schulten, K. (1989). A regularized approach to color constancy. *Biological Cybernetics, 61,* 29–36.

Sällström, P. (1973). Colour and physics: Some remarks concerning the physical aspects of human colour vision (Tech. Rep. No. 73-09). Stockholm: University of Stockholm Institute of Physics.

Smith, V. C., & Pokorny, J. (1975). Spectral sensitivity of the foveal cone photopigments between 400 and 500 nm. *Vision Research, 15,* 161–171.

Trussell, H. J., & Vrhel, M. J. (1991). Estimation of illumination for color correction. *1991 International Conference on Acoustics, Speech & Signal Processing* (pp. 2513–2516). Piscataway, NJ: IEEE Conference Publishing.

Tsukada, M., & Ohta, Y. (1990). An approach to color constancy using multiple images. *Proceedings of the Third International Conference on Computer Vision* (Vol. 3, pp. 385–393).

11 Probabilistic Color Constancy

Michael D'Zmura
Geoffrey Iverson
Benjamin Singer
Department of Cognitive Sciences and Institute for Mathematical Behavioral Sciences, University of California, Irvine

ABSTRACT

Specifying the frequency with which surface reflectance functions occur in the visual environment lets one use the chromaticities of reflected lights to provide maximum likelihood estimates of the spectral properties of a scene's illuminant. This approach to color constancy generalizes schemes that use the gray-world assumption. Monte Carlo simulation suggests that a trichromatic visual system needs the chromaticities of reflected lights from a random sample of relatively few surfaces to estimate accurately the correlated color temperature of daylight illumination.

INTRODUCTION

Color constancy refers to the perceptual stability of surface color appearance under conditions of changing, unknown illumination. This constancy can be posed as a computational problem: how can the visual system recover from photoreceptoral signals the spectral properties of the surfaces that it sees, physical properties that do not depend on the vagaries of illumination (Brill, 1978, 1979; Buchsbaum, 1980; Lennie & D'Zmura, 1988; Sällström, 1973)? One approach to the problem relies on finite-dimensional linear models of surface reflectance functions and of light source spectral functions (Cohen, 1964; Judd, MacAdam, & Wyszecki, 1964; Maloney, 1986). The linear models are used to construct some deterministic model of the change in reflected lights caused by changing illumination. One then inverts the image formation process to recover spectral descriptors of lights and surfaces. Schemes of this sort include two-stage linear recovery schemes (D'Zmura, 1992; D'Zmura & Iverson, 1993a, 1993b; Iverson & D'Zmura, 1994; Maloney & Wandell, 1986), a more general, one-stage linear recovery scheme (D'Zmura & Iverson, 1994) and various nonlinear recovery schemes (D'Zmura & Iverson, 1993c; Tsukada & Ohta, 1990). Variations on this approach use additional information such as highlights or interreflections to help recover spectral descriptions (Drew & Funt, 1992; D'Zmura & Lennie, 1986).

We present here a second approach, in which one uses the linear models for reflectance and illumination to construct a stochastic model of reflected lights. Such stochastic models help to generalize spectral recovery schemes, first formulated by Buchsbaum (1980), that use the gray-world assumption. The gray-world assumption, in its simplest form, holds that the space-averaged reflected light bears the chromaticity of the illuminant. This is true if the space-averaged reflectance is spectrally neutral (i.e., gray). One can assume, more generally, that the space-averaged reflectance function is known, but not necessarily gray, which suffices for illuminant chromatic properties to be determined from the space-averaged reflected light. Buchsbaum used the gray-world assumption to estimate the spectral properties of an unknown light source from the space-averaged reflected light, and then used this estimate to recover the reflectance properties of individual surfaces.

We examine explicit stochastic models of reflected lights which lead to a natural generalization of Buchsbaum's scheme. The scheme uses information about the entire distribution of the surface reflectance functions, not just the ensemble mean. This is done by using a statistical linear model for reflectance to help specify how likely it is that a particular reflectance function is encountered. Likewise, a statistical linear model for illuminant spectral functions specifies how likely it is that a particular light is shining on a scene. Knowledge of prior distributions on reflectances and illuminants is used to build a stochastic model for reflected lights.

We present a probabilistic color constancy scheme that uses the chromaticities of reflected lights in a maximum likelihood estimation procedure to determine the illuminant most likely to have shone on surfaces in a scene. Monte Carlo simulation of recovery, using probability distributions with realistically large dispersions for surface reflectance, shows that the chromaticities of relatively few reflected lights are needed to estimate light source spectral properties accurately.

RECOVERY METHOD

Suppose that the surface reflectance functions met by a trichromatic visual system are drawn randomly from some ensemble of reflectances (Hogg & Craig,

1978; Papoulis, 1984; Van Trees, 1968). The random variables $\mathbf{X}(A)$, $\mathbf{Y}(A)$, and $\mathbf{Z}(A)$, which represent the tristimulus values that characterize surfaces chromatically when viewed under some light source with wavelength dependence $A(\lambda)$, are then related to the random variable $\mathbf{R}(\lambda)$ that describes the distribution of reflectances as follows. For a trichromatic observer represented by color-matching functions $\bar{x}(\lambda)$, $\bar{y}(\lambda)$, and $\bar{z}(\lambda)$, we have[1]

$$\mathbf{X}(A) = \int \bar{x}(\lambda)A(\lambda)\mathbf{R}(\lambda)\, d\lambda, \tag{11.1a}$$

$$\mathbf{Y}(A) = \int \bar{y}(\lambda)A(\lambda)\mathbf{R}(\lambda)\, d\lambda, \tag{11.1b}$$

$$\mathbf{Z}(A) = \int \bar{z}(\lambda)A(\lambda)\mathbf{R}(\lambda)\, d\lambda. \tag{11.1c}$$

The tristimulus value distributions tell us how likely it is that one will record a particular tristimulus value (X, Y, Z) when the scene's illuminant is $A(\lambda)$. The tristimulus value distributions lead to a distribution of chromaticities that is represented by the random variables $\mathbf{x}(A)$ and $\mathbf{y}(A)$:

$$\mathbf{x}(A) = \frac{\mathbf{X}(A)}{\mathbf{X}(A) + \mathbf{Y}(A) + \mathbf{Z}(A)} \tag{11.2a}$$

$$\mathbf{y}(A) = \frac{\mathbf{Y}(A)}{\mathbf{X}(A) + \mathbf{Y}(A) + \mathbf{Z}(A)} \tag{11.2b}$$

The distribution of chromaticities tells us how likely it is that one will record a particular chromaticity (x, y) when the scene's illuminant is $A(\lambda)$.

Knowledge of the conditional probability densities $p((X, Y, Z)|A(\lambda))$ lets one use the chromatic properties of reflected lights to estimate scene illuminant properties. Suppose that s surfaces, drawn independently and at random from $\mathbf{R}(\lambda)$, are viewed under an unknown illuminant. The likelihood $L(A)$ of recording a particular set of s tristimulus values from the surfaces, given that the illuminant with spectral function $A(\lambda)$ is shone on the surfaces, is the joint probability

$$L(A(\lambda)) = \prod_{t=1}^{s} p((X, Y, Z)_t|A(\lambda)). \tag{11.3a}$$

[1]The chromatic response of a trichromatic visual system to a light is captured by three numbers (X, Y, Z) known as tristrimulus values. The tristimulus values are found by computing three integrals (as in Equation 11.1a–11.1c) each of which is an integral over the visible spectrum of the product of (1) the spectral function representing the light and (2) one of the visual system's color-matching functions. The three color-matching functions $\bar{x}(\lambda)$, $\bar{y}(\lambda)$, $\bar{z}(\lambda)$ represent the trichromatic visual system's spectral sensitivity. Chromatic information pertaining to the light can be represented separately from information on the light's intensity by projecting the tristimulus values onto a chromaticity plane. The chromaticity coordinates (x, y) that so arise characterize the light's chromatic properties in a way that is independent of the light's intensity. For an excellent general discussion of this and other matters in color vision, refer to Wyszecki and Stiles (1982).

It is often more practical to work instead with the log-likelihood function $L(A)$ given by

$$L(A(\lambda)) = \sum_{t=1}^{s} \ln p((X, Y, Z)_t | A(\lambda)). \tag{11.3b}$$

The maximum likelihood estimate $\hat{A}(\lambda)$ of the illuminant is determined by finding the maximum of the likelihood function (or the log-likelihood function) as a function of $A(\lambda)$, given some set of s input chromatic data (Hogg & Craig, 1978).

This estimation procedure supposes implicitly that each illuminant within the set of possible illuminants is equally likely. Yet one may have prior knowledge concerning the probability of encountering various illuminants. For instance, the locus of daylight chromaticities in the CIE 1931 chromaticity diagram is a one-dimensional locus, a curve parameterized by correlated color temperature (Judd et al., 1964; Wyszecki & Stiles, 1982). If we are certain that such a daylight (of unknown color temperature) is shining on the visible surfaces, then our estimate of illuminant chromaticity should fall on the daylight locus; i.e., the recovery procedure should not return a light with a chromaticity that lies off the daylight locus. We can express such prior knowledge through a distributuion that is concentrated on the daylight locus.

More generally, if the prior knowledge concerning illumination can be expressed as a probability distribution $p(A)$, then one can maximize the joint probability

$$l(A(\lambda)) = \prod_{t=1}^{s} p((X, Y, Z)_t | A(\lambda)) p(A(\lambda)), \tag{11.4a}$$

or taking logarithms,

$$\ell(A(\lambda)) = \sum_{t=1}^{s} \ln p((X, Y, Z)_t | A(\lambda)) + \ln p(A(\lambda)). \tag{11.4b}$$

The maximum a posteriori estimate $\hat{A}_{\text{map}}(\lambda)$ of the illuminant is determined by finding the maximum of the function $l(A(\lambda))$ as a function of $A(\lambda)$, given s chromatic data (Van Trees, 1968). A similar estimate is available if one must work with chromaticity distributions $p((x, y)|A(\lambda))$ rather than with tristimulus value distributions:

$$\ell(A(\lambda)) = \sum_{t=1}^{s} \ln p((x, y)_t | A(\lambda)) + \ln p(A(\lambda)). \tag{11.5}$$

To realize such an estimation procedure, one must specify prior distributions on the reflectances and illuminants. To do so, we consider statistical linear models, namely linear models with random coefficients. Following the notation of D'Zmura & Iverson (1993a, 1993b, 1993c), the random variable $\mathbf{R}(\lambda)$ that describes the distribution of (nonfluorescent) surface reflectance functions can be expressed as

$$\mathbf{R}(\lambda) = \sum_{j=1}^{n} \mathbf{r}_j R_j(\lambda), \tag{11.6}$$

where the n orthogonal functions $R_j(\lambda)$ are the basis functions for an n-dimensional linear model for reflectance, and the n random variables $\mathbf{r}_j$ are descriptors whose joint distribution specifies a statistical model for surface reflectance functions.

Likewise, the random variable $\mathbf{A}(\lambda)$ that describes the distribution of illuminant spectral functions can be written

$$\mathbf{A}(\lambda) = \sum_{i=1}^{m} \mathbf{a}_i A_i(\lambda), \tag{11.7}$$

where the m orthogonal functions $A_i(\lambda)$ are the basis functions for an m-dimensional linear model for illumination, and the m random variables $\mathbf{a}_i$ are descriptors whose joint distribution specifies a statistical model for illuminant spectral functions.

Let us consider the reflectance model of form given by Equation 11.6 in more detail. One issue that arises immediately is whether the model generates surface reflectance functions that are not physically realizable, e.g., take on negative values. (A real, nonfluorescent surface reflectance function $R(\lambda)$ must take on values between 0 and 1.) Note that a natural choice of model is one that uses normal distributions, of known means and variances, for the probabilistic descriptors $\mathbf{r}_j$, in combination with basis functions of approximately the form 1, sin, cos, . . . (Such a Fourier basis provides an average reflectance level and pairs of quadrature components that express chromatic variation.) The problem with using normally distributed descriptors is that there will be some nonzero probability of generating a reflectance function that is not physically realizable.

The same problem can occur with models for illumination. The use of the normal distribution for the probabilistic descriptors $\mathbf{a}_i$ will lead to illuminant spectral functions with negative values, a physical impossibility.

We demand that our models be physically realizable. In the case of reflectance, this has led us to consider random variables $\mathbf{r}_j$ whose joint distribution, in tandem with Fourier basis functions, produces physically realizable reflectance functions. For a three-dimensional statistical linear model, we use the basis functions

$$R_1(\lambda) = 1,$$
$$R_2(\lambda) = \sin \lambda,$$
$$R_3(\lambda) = \cos \lambda, \qquad (11.8)$$

where it is assumed that the interval of visible wavelengths (e.g., [400, 700] nm) has been shifted and scaled onto the interval $[0, 2\pi]$. The model (Equation 11.6) then has the form

$$\mathbf{R}(\lambda) = \mathbf{r}_1 1 + \mathbf{r}_2 \sin \lambda + \mathbf{r}_3 \cos \lambda. \qquad (11.9)$$

We choose the random variable $\mathbf{r}_1$ to be beta distributed. This is a natural choice. The beta distributions (Hogg & Craig, 1978) form a family with two parameters a, b: $\beta(x|a, b) = C_{a,b}x^{a-1}(1 - x)^{b-1}$, for x in [0, 1], where $C_{a,b}$ is a normalizing constant. The parameters (a, b) can be varied to alter the mean $a/(a + b)$ and the variance $ab/(a + b + 1)(a + b)^2$ of the distribution. A beta distribution is nonzero only on the interval [0, 1], so that the (constant) reflectance functions generated by the first term of the model alone are physically realizable. Having sampled the beta distribution to produce some particular value r_1, we then select the amplitude $(r_2^2 + r_3^2)^{1/2}$ of the sinusoidal component from a uniform distribution that is concentrated on the interval $[0, \min(r_1, 1 - r_1)]$. One can use a distribution on this interval other than the uniform; what is important is that using this interval ensures that the greatest possible saturations can be generated, given the constraint of physical realizability. We then choose the phase of the component at frequency one from the uniform distribution on $[0, 2\pi]$. The resulting distributions for $\mathbf{r}_1$, $\mathbf{r}_2$, and $\mathbf{r}_3$ are not independent in such a model.

In addition to producing only physically realizable reflectances, such a model has the desirable property of being able to produce skewed distributions for average reflectance. This property stems from use of the beta distribution, whose parameters can be chosen to produce an ensemble average of 0.25, for instance, yet continue to produce white surfaces—with reflectances near 1.0. Another desirable feature is the ease with which the model can be extended from three to some higher number of dimensions: the amplitude at any positive, integer-valued frequency[2] is chosen from a distribution that is concentrated on the largest possible interval that satisfies physical realizability, given the prior choice of amplitudes at lower frequencies.

---

[2]We have in mind models that provide truncated Fourier series expansions for reflectance functions. The positive integer value for frequency states how many wavelengths the particular sine and cosine components at that frequency possess on the visible spectrum.

SIMULATION

We simulated the recovery of illuminant spectral properties using the maximum likelihood estimation procedure and models described above. Our aims were several. First, we wanted to learn how the accuracy of illuminant estimation depends on the number of sampled surfaces. An estimation procedure is useful if only a few samples are needed to reach the required accuracy. Second, we wanted to see how the accuracy of recovery depends on the choice of surface reflectance model. In particular, we hoped to show that recovery is possible using models across a range of beta distribution parameters. Third, we wanted to discern and correct, if needed, bias in the estimation procedure.

To minimize computation time and memory requirements, we did not simulate tristimulus values of reflected lights and use these to estimate illuminant spectral functions (Equation 11.4b). Rather, we simulated chromaticities of reflected lights and used these to estimate illuminant chromaticities (Equation 11.5). Furthermore, we did not use a statistical linear model for illumination (Equation 11.7); rather we used a finite collection of daylight illuminants in our simulations. Computations were performed on a Digital Equipment Corporation 3000/400 workstation.

Our first step was to use the reflectance model of Equation 11.9 to compute discrete approximations to the conditional probability densities $p((x, y)|A(\lambda))$. These approximations were computed for each of the 56 CIE daylight illuminants with chromaticities listed by Wyszecki and Stiles (1982, Table IV(3.3.4)), which range in correlated color temperature from 4000 K (yellowish) to 25,000 K (bluish). In order of increasing temperature, the 56 particular phases of daylight range from 4000 to 8500 K in steps of 100 K, include 9000 K and 9500 K, range from 10,000 to 15,000 K in steps of 1000 K, and finally include 20,000 K and 25,000 K. Maximum likelihood estimation proceeded under the assumption that the 56 daylight illuminants were equally likely.

To approximate a conditional probability distribution, we calculated the chromaticities of 131,072 surface reflectances generated independently by the reflectance model. For each daylight, the frequency of each chromaticity's occurrence was recorded in the appropriate bin of a 100×100 array. The arrays' coordinates correspond to CIE 1931 standard observer x and y values that range from 0.0 to 1.0, so that chromaticities were represented to a precision of $(\Delta x, \Delta y) = (0.01, 0.01)$. These conditional probability densities were approximated, for each of the 56 daylight illuminants, for four choices of the beta distribution parameters: $(a, b) = (2.5, 2.5), (2.5, 4.0), (2.5, 5.5), (2.5, 7.0)$ with means 0.5, 0.385, 0.313, and 0.263, respectively.

Figure 11.1 shows the approximation to the conditional probability density $p((x, y)|(A(\lambda))$ for CIE standard daylight illuminant D_{65} (with correlated color temperature 6500 K) and for the reflectance model with beta parameters $(a, b) =$

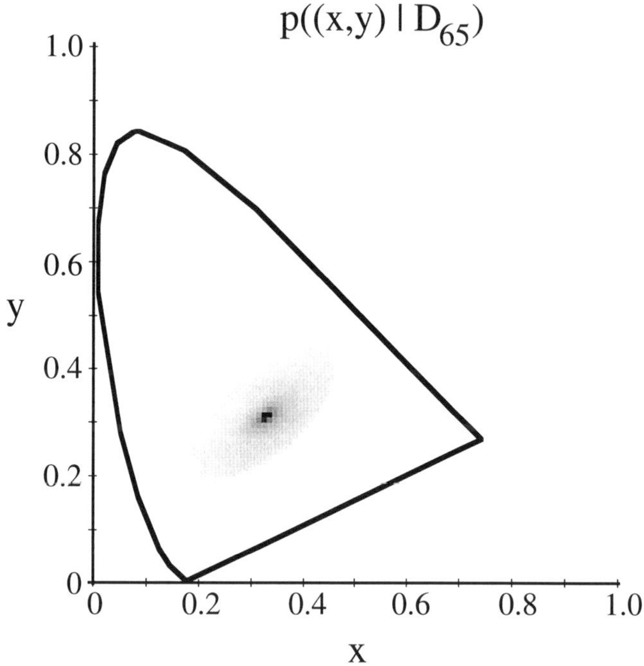

FIG. 11.1. Discrete approximation to $p((x, y)|A(\lambda))$ for CIE standard daylight illuminant D_{65} and for the reflectance model with beta parameters $(a, b) = (2.5, 4.0)$. Darker values represent higher frequencies. The solid curve in this chromaticity diagram (Wyszecki & Stiles, 1982) marks the spectrum locus, which is the locus described by the chromaticities of monochromatic lights as wavelength is varied.

(2.5, 4.0). The frequency of a particular chromaticity is represented by pixel value: more frequent chromaticities are darker.

We first investigated the accuracy of the estimation procedure. We chose the actual simulated illuminant to be D_{65} and then examined the frequency with which each of the 56 possible illuminants was recovered. In Figure 11.2 we show the results of recovery using the chromaticities of reflected lights from $s = 1, 2, 4, 8$, and 16 independently drawn surfaces. The graphs show the relative frequency with which each illuminant was recovered in 16,384 trials. As the number of surfaces increases, the estimates start to concentrate in the interior of the correlated color temperature interval 4000–25,000 K and then cluster increasingly about the correlated color temperature 6500 K of the true illuminant.

In the one- or two-surface cases, the distributions of estimates have peaks at the extreme yellow (4000 K) and blue (25,000 K) ends. These are caused by the relatively large variance in samples of very small size. For instance, single

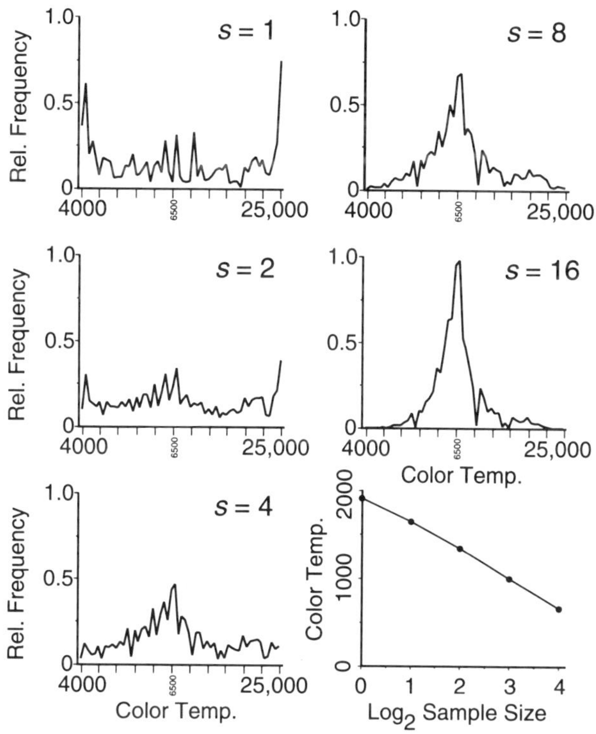

FIG. 11.2. Results of recovery using the chromaticities of reflected lights from $s = 1, 2, 4, 8$, and 16 surfaces with reflectances drawn from the three-dimensional statistical linear model with beta parameters $a = 2.5$, $b = 4.0$. The actual simulated illuminant was D_{65}. Each graph, for $s = 1$ through $s = 16$, shows the relative frequency with which each illuminant was recovered in 16,384 trials. The estimates cluster more tightly around the correlated color temperature of the actual illuminant as the number of surfaces increases. The tick marks along the horizontal axes mark every fifth illuminant correlated color temperature, namely 4000, 4500, 5000, 5500, 6000, 6500 (marked), 7000, 7500, 8000, 8500, 12,000, and 25,000 K in the set of 56 described in the text. Standard deviations of the distributions of estimates are plotted at bottom right.

surfaces ($s = 1$) that are more yellow or more blue than the most extreme light sources (at 4000 K and 25,000 K) lead to estimates that are the most extreme yellow or blue possible, respectively.

The graphs show that the distributions of estimates are fairly symmetric. This symmetry is found when the difference between illuminants is measured simply as a difference in list position, as in Figure 11.2, and is a pleasant surprise. Had we chosen a metric on the illuminants based on, for example, perceived differences between the colors of the illuminants, the distributions may well have lacked symmetry.

The bottom right panel of Figure 11.2 is a plot of the standard deviations of the distributions of estimates, in units of correlated color temperature, as a function of sample size. The graph documents the improvement in recovery with increasing sample size. The standard deviation is about 1000 K when $s = 8$ and about 700 K when $s = 16$. In the latter case, the bulk of the estimates lie between 5800 and 7200 K, which is a rather small chromatic interval about the true value 6500 K. The graph of standard deviation versus log sample size is nearly linear in the range examined; of course, this property cannot hold at larger sample sizes. The graphs of relative frequency suggest that the bias in the estimate is small; we examine this more carefully below.

We performed further simulations to find out how the standard deviations of the distributions of estimates depend on the beta distribution parameters. Again, the actual simulated illuminant was D_{65}. We determined the standard deviations of the distributions of relative frequencies with which each of the 56 daylight illuminants was recovered in 16,384 trials for the beta distribution parameters $(a, b) = (2.5, 2.5)$, $(2.5, 4.0)$, $(2.5, 5.5)$, $(2.5, 7.0)$. Figure 11.3 shows that the greatest accuracy is achieved in the case where the reflectance model uses the beta distribution with parameters $(a, b) = (2.5, 2.5)$, which has a mean of 0.5. This is the beta distribution, among these four, that generates the reflectance distribution with the greatest chromatic power. The beta distribution mean of 0.5

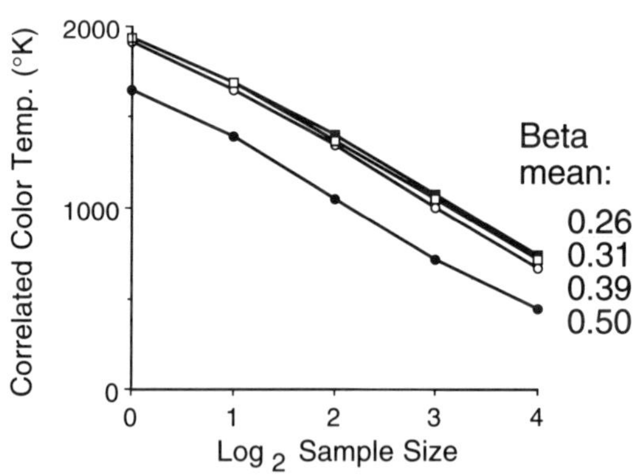

FIG. 11.3. Dependence of the accuracy of the estimate on beta distribution parameters. Plotted are the standard deviations of the distributions of estimates, for numbers of samples $s = 1, 2, 4, 8$, and 16, for four choices of the beta-distribution parameters: $(a, b) = (2.5, 2.5)$, (2.5, 4.0), (2.5, 5.5), (2.5, 7.0) with means 0.5, 0.385, 0.313, and 0.263, respectively.

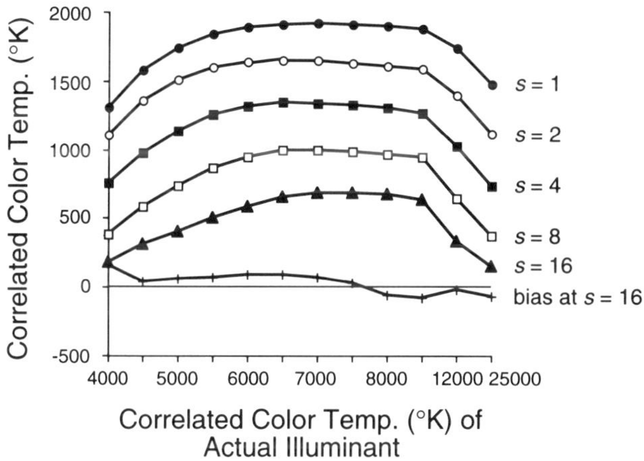

FIG. 11.4. Dependence of the accuracy of the estimate on actual illuminant and estimation bias. The standard deviations of the distributions of estimates for numbers of samples $s = 1, 2, 4, 8$, and 16 are plotted as a function of the correlated color temperature of the actual illuminant (top five graphs). Standard deviations and biases at 12,000 K and 25,000 K are scaled by 0.1 and 0.02, respectively. Tick marks along the horizontal axis are as in Figure 11.2. The bottommost graph shows the bias in the estimates when $s = 16$.

makes more frequent high-amplitude reflectances, which have an amplitude close to 0.5 at frequency one, the highest possible amplitude. The models with lesser means lead to poorer estimates, and one possible reason is that they do not provide as many telltale outlying chromaticities.

We investigated how recovery depends on the choice of illuminant. In the previous simulations, the illuminant was always D_{65}. We chose to simulate recovery for 12 illuminants across the range 4000–25,000 K, namely those with correlated color temperatures 4000, 4500, 5000, 5500, 6000, 6500, 7000, 7500, 8000, 8500, 12,000, and 25,000 K. For each of these we computed distributions of relative frequencies of illuminant recovery and computed their standard deviations, for numbers of samples $s = 1, 2, 4, 8$, and 16. The reflectance model used the beta distribution with parameters $(a, b) = (2.5, 4.0)$.

Figure 11.4 shows the results. The five topmost graphs show how the standard deviations of the distributions of estimates depend on the actual illuminant. Estimate accuracy is best near the yellow end of the interval and is fairly constant across the broad, intermediate range of temperatures. Color temperature changes much more rapidly as one approaches the blue end of the list of daylights, and the standard deviations at the blue extreme, expressed in terms of correlated color temperature, are much greater. To plot these in the same figure comfortably, the standard deviations in the estimates at 12,000 K are shown at 1/10 their true magnitudes, while the standard deviations in the estimates at 25,000 K are shown at 1/50 their true magnitudes.

Standard deviations decrease approximately linearly with log sample size across the range of daylight illuminants. The bottom graph plots the bias in the estimates found with 16 samples as a function of the actual illuminant's correlated color temperature. The estimates are nearly free of bias. With the exception of the estimates at 12,000 K and 25,000 K, which have again been scaled by 0.1 and 0.02, respectively, the bias exceeds 100 K in absolute value in only one instance, namely at 4000 K, the yellow end of the interval. Other than this, the highest bias is found at 6500 K, where the average estimate of the temperature was 6594 K, so providing a bias of 94 K.

We finally tested how well the probabilistic color constancy scheme recovers daylight illuminant chromaticities in the case where a five-dimensional statistical linear model for reflectance is used. The reflectance model used a beta distribution with parameters $(a, b) = (2.5, 4.0)$ for the frequency zero component and used uniform distributions on the (successive) remaining intervals for the amplitudes at frequencies 1 and 2. Figure 11.5 shows that the color constancy scheme works well to recover the actual illuminant, D_{65}, and that accuracy improves with increasing sample size in much the same way as shown before with the three-dimensional statistical linear models for reflectance.

DISCUSSION

We have described a probabilistic color constancy scheme that uses statistical linear models for reflectance and illumination to derive maximum likelihood estimates of the chromatic properties of a scene's illuminant. Monte Carlo simulation suggests that the scheme may be sufficiently powerful in practice to recover accurate estimates of daylight illumination chromaticity for a range of realistic reflectance models.

The statistical linear models that we developed here extend earlier work with linear models (Cohen, 1964; Dixon, 1978; Judd et al., 1964; Maloney, 1986; Parkkinen, Hallikainen, & Jaaskelainen, 1989). While empirical measurements of actual reflectances and illuminants guided the construction of the linear models, many color constancy schemes use the linear models in a way that (1) ignores the frequency with which reflectances and illuminants are encountered and (2) ignores the issue of whether estimates of reflectances and illuminants are physically realizable. Statistical linear models provide explicit information on the frequency with which reflectances and illuminants are encountered and can incorporate the constraint of physical realizability.

It is unfortunate that the statistical linear models that use normal distributions generate physically unrealizable reflectances and illuminants, because these models can lead to explicit formulas for estimation of the sort that have, so far,

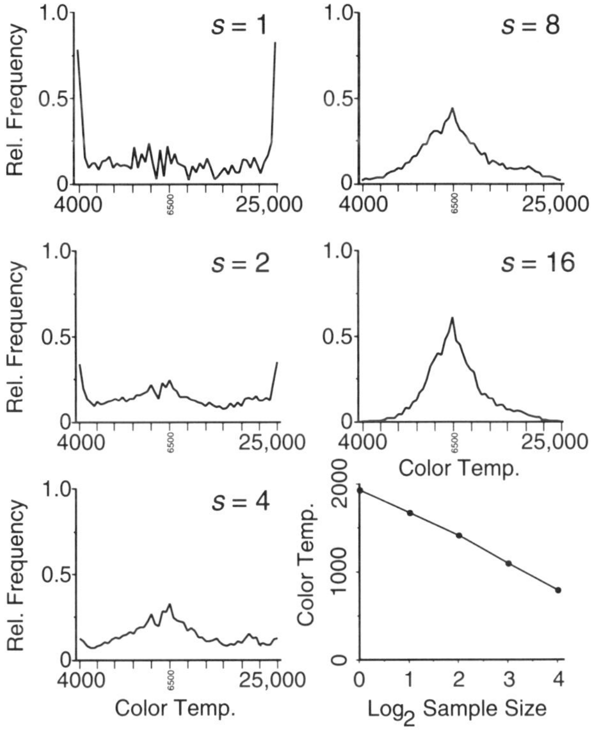

FIG. 11.5. Results of recovery using the chromaticities of reflected lights from $s = 1, 2, 4, 8$, and 16 surfaces with reflectances drawn from the five-dimensional statistical linear model with beta parameters $a = 2.5$, $b = 4.0$. The actual simulated illuminant was D_{65}. Accuracy of estimates improves with increase in sample size, but not as rapidly as in the case of a three-dimensional model.

eluded us in our work with physically realizable models that use the beta distribution. In groundbreaking work, Trussell and Vrhel (1991) considered maximum likelihood estimates of illumination in the case of normally distributed reflectances. They derived a nonlinear system of equations for maximizing the likelihood function, and used a numerical nonlinear procedure that was modified to avoid local extrema. In an attempt to avoid the consequences of physical unrealizability, they suggested that reflectance distributions should be heavily concentrated about the mean. Brainard and Freeman (1994) also have pursued estimation using statistical linear models based on normal distributions and have successfully estimated illuminant spectral functions in simulation. One hopes that future work will join the two approaches to produce realistic, physically realizable models for which one can obtain tidy analytic results.

Our estimation scheme requires that reflectance samples are drawn independently. While this is a good working assumption, there are certainly situations in which it does not hold true. For instance, any attempt to "coordinate colors" (as in the selection of clothing) implies a dependency in the sampling of surface reflectances.

While we have used a finite set of illuminants in our simulations, the statistical linear model for illumination represented by Equation 11.7 is more general. For instance, a model that may be more useful in practical situations is the one that uses the first several daylight basis functions of Judd and colleagues (1964) in combination with appropriate random variables for the coefficients. For instance, the first (average light) coefficient may be modeled in terms of a Rayleigh-distributed random variable (Papoulis, 1984), while the second (dependent) coefficient can be chosen from a distribution that is concentrated on the interval that ensures physical realizability; likewise for the third coefficient, and so on. Such a model could be used to help recover both the chromatic properties and the absolute intensity of a scene's illuminant. Note that this model generates illuminants that lie off the daylight locus, however. More sophisticated yet would be to parameterize the daylight model using color temperature to generate daylights that lie on the daylight locus with appropriate prior probabilities. The estimation procedure could then use tristimulus values to recover absolute spectral functions for daylight illuminants and restrict its estimates to the daylight locus. The ability to restrict one's estimates in this way is a decided advantage of the present scheme over gray-world schemes.

Having recovered an estimate of the illuminant, one can produce estimates of surface reflectance functions using deterministic methods (Maloney & Wandell, 1986) if the reflectance model is three dimensional. It is possible, however, that these estimates correspond to reflectance functions that are not physically realizable. One can extend the estimation scheme to incorporate the constraint of physical realizability using the methods of nonlinear programming (Gill, Murray, & Wright, 1981). Moreover, the results of Figure 11.5 show that the scheme works to recover the relative spectral functions of (three-dimensional) illuminants when four- or higher-dimensional models for reflectance are used. Such estimates by a trichromatic visual system are not possible using standard deterministic schemes, which can estimate the illuminant from a single view of coplanar array of matte surfaces only if a two-dimensional (Maloney & Wandell, 1986) model for reflectance proves accurate.

ACKNOWLEDGMENTS

We thank Al Ahumada, Dave Brainard, Tarow Indow, and Jack Yellott for helpful discussion and Al Ahumada, Don Hoffman, and Larry Maloney for their comments on an earlier version of this chapter. This work was supported by NEI

EY10014 to M. D'Zmura and by NSF DIR-9014278 to the Institute for Mathematical Behavioral Sciences, University of California, Irvine, R. D. Luce, Director.

REFERENCES

Brainard, D. H., & Freeman, W. T. (1994). Bayesian method for recovering surface and illuminant properties from photosensor responses. *Proceedings of the SPIE Symposium on Human Vision, Visual Processing, and Digital Display V, Volume 2179,* 364–376.

Brill, M. H. (1978). A device performing illuminant-invariant assessment of chromatic relations. *Journal of Theoretical Biology, 71,* 473–478.

Brill, M. H. (1979). Further features of the illuminant-invariant trichromatic photosensor. *Journal of Theoretical Biology, 78,* 305–308.

Buchsbaum, G. (1980). A spatial processor model for object colour perception. *Journal of the Franklin Institute, 310,* 1–26.

Cohen, J. (1964). Dependency of the spectral reflectance curves of the Munsell color chips. *Psychonomic Sciences, 1,* 369–370.

Dixon, E. R. (1978). Spectral distribution of Australian daylight. *Journal of the Optical Society of America, 68,* 437–450.

Drew, M. S., & Funt, B. V. (1992). Variational approach to interreflection in color images. *Journal of the Optical Society of America A, 9,* 1255–1265.

D'Zmura, M. (1992). Color constancy: Surface color from changing illumination. *Journal of the Optical Society of America A, 9,* 490–493.

D'Zmura, M., & Iverson, G. (1993a). Color constancy: I. Basic theory of two-stage linear recovery of spectral descriptions for lights and surfaces. *Journal of the Optical Society of America A, 10,* 2148–2165.

D'Zmura, M., & Iverson, G. (1993b). Color constancy: II. Results for two-stage linear recovery of spectral descriptions for lights and surfaces. *Journal of the Optical Society of America A, 10,* 2166–2180.

D'Zmura, M., & Iverson, G. (1993c). Color constancy: Feasibility and recovery. *Investigative Ophthalmology and Visual Science, 34* (Suppl.), 748.

D'Zmura, M., & Iverson, G. (1994). Color constancy: III. General linear recovery of spectral descriptions for lights and surfaces. *Journal of the Optical Society of America A, 11,* 2389–2400.

D'Zmura, M., & Lennie, P. (1986). Mechanisms of color constancy. *Journal of the Optical Society of America A, 3,* 1662–1672.

Gill, P. E., Murray, W., & Wright, M. H. (1981). *Practical optimization.* London: Academic.

Hogg, R., & Craig, A. (1978). *Introduction to mathematical statistics* (4th ed.). New York: MacMillan.

Iverson, G., & D'Zmura, M. (1994). Criteria for color constancy in trichromatic bilinear models. *Journal of the Optical Society of America A, 11,* 1970–1975.

Judd, D. B., MacAdam, D. L., & Wyszecki, G. (1964). Spectral distribution of typical daylight as a function of correlated color temperature. *Journal of the Optical Society of America, 54,* 1031–1040.

Lennie, P., & D'Zmura, M. (1988). Mechanisms of color vision. *Critical Reviews in Neurobiology, 3,* 333–400.

Maloney, L. T. (1986). Evaluation of linear models of surface spectral reflectance with small numbers of parameters. *Journal of the Optical Society of America A, 3,* 1673–1683.

Maloney, L. T., & Wandell, B. A. (1986). Color constancy: A method for recovering surface spectral reflectance. *Journal of the Optical Society of America A, 3,* 29–33.

Papoulis, A. (1984). *Probability, random variables, and stochastic processes* (2nd ed.). New York: McGraw-Hill.

Parkkinen, J. P. S., Hallikainen, J., & Jaaskelainen, T. (1989). Characteristic spectra of Munsell colors. *Journal of the Optical Society of America A, 6,* 318–322.

Sällström, P. (1973). Colour and physics: Some remarks concerning the physical aspects of human colour vision (Tech. Rep. No. 73-09). Stockholm: University of Stockholm Institute of Physics.

Trussell, H. J., & Vrhel, M. J. (1991). Estimation of illumination for color correction. *1991 International Conference on Acoustics, Speech & Signal Processing* (pp. 2513–2516). Piscataway, NJ: IEEE Conference Publishing.

Tsukada, M., & Ohta, Y. (1990). An approach to color constancy using multiple images. *Proceedings of the Third International Conference on Computer Vision, 3,* 385–393.

Van Trees, H. L. (1968). *Detection, estimation and modulation theory* (Part 1). New York: Wiley.

Wyszecki, G., & Stiles, W. S. (1982). *Color science. Concepts and methods, Quantitative data and formulas* (2nd ed.). New York: Wiley.

III SCALING

A. Kimball Romney

INTRODUCTION

The chapters in this section on scaling are concerned with problems in the organization of various bodies of data, mostly perceptual, through various scaling techniques, usually multidimensional. Scaling, with a history dating from the early days of psychophysics in the 19th century, has long played a key role in organizing data. However, the implementation of practical multidimensional solutions of significant size had to await the development of modern computers. As is fully documented by his bibliography in the Appendix, Indow has been one of the most seminal and prolific contributors to both theoretical developments and, more uniquely, to practical applications. An outstanding example of his ability to combine these two aspects of work is exhibited in his, the first, chapter of this volume.

This portion of the volume, devoted to the scaling of perceptual phenomena, contains six contributions. Each takes a different approach, but all deal with the challenge of organizing a considerable body of data into a coherent form, typically described as scales, for comparison with independent variables. The chapters range from being primarily concerned with some theoretical or practical problem in which scaling is a means to an end to a focus on purely methodological developments that would result in new or innovative methods of scaling. Brief comments on the various contributions reflect the general focus of each chapter.

Marks provides a detailed review of one of the standard methodological and analytic tools of modern psychophysical scaling, namely, cross-modality matching. He examines such foundational questions as, "What kinds of mental processes mediate cross-modality matching?" or "Does the existence of cross-modal similarities bear any consequences for the ways that people process perceptual information?" Marks brings a wealth of results to bear on these and other basic questions concerning the nature of cross-modality matching and its implications for perceptual processes. Scaling, including multidimensional, plays a critical role in the analysis, representation, and modeling of the results of several bodies of empirical data reported by Marks.

Baird notes that context effects, e.g., procedural variables, available response options, position of a standard, etc., influence the scaling results from fundamental sensory mechanisms. Baird attempts to understand the human judgment process so as to account for context effects and with a view to making the human being a better measuring instrument when applied to a variety of situations in psychophysical experiments. The judgment model, which looks at a subject on a trial-by-trial basis, is based on five assumptions. It provides accounts of standard psychophysical parameters, such as the exponent of the Stevens function and the shape of the Ekman function. In working out the implications of the model, he makes heavy use of both experimental- and simulation-based evidence. This model should have implications for experimentation, statistical analysis, and analytic description.

Eisler constructs a five-parameter psychophysical function for the subjective duration of time, and he applies it to the comparison of different subgroups of subjects. To avoid having subjects use numerals, as in magnitude estimation, his experimental trial consisted of first presenting a duration of sound that the subject was to match. To do that, the sound resumed after a short pause, and the subject was instructed to terminate the second sound when it was perceived to have lasted as long as the first sound. He shows that the psychophysical function for time is a power function with a discontinuity (at around 6 sec) and, accordingly, consists of two segments. He concludes with a description of the effect of stimulus and group differences in terms of differing parameter values.

Romney, Batchelder, and Brazill present a descriptive scaling method for obtaining a single spatial representation that contains the cognitive maps from a large number of subjects. The method is illustrated on judged similarity data for 21 common animals. The 125 subjects were each assigned to one of four data collection methods. The results of the scaling provide a spatial representation that includes the location of each animal for each subject. No constraints were placed on individual subjects concerning the configuration among judged items. The method serves three main purposes. First, it allows one to describe and test comparisons among individuals and subgroups; e.g., the study compares subjects assigned to different data collection methods. Second, it enables one to comprehend and examine very large data sets that otherwise would not be accessible in a single coherent view. Third, it provides an optimally aggregated representation that can be used to predict cognitive behaviors that relate to cognitive structure.

The chapters by Carroll and Chaturvedi and by Klauer and Carroll are both methodological in the sense that they present models that extend traditional multidimensional methods into new and novel contexts. Carroll and Chaturvedi generalize models for continuous parameters to discrete cases. Two models are developed in detail. The first is the more general CANDCLUS (canonical decomposition clustering), and SINDCLUS (separability-based INDCLUS [individual differences clustering) is an important special case of CANDCLUS. In applications the models provide various partitions of a set of items, e.g., judged similarities among kin terms collected with a pile-sorting task is an illustration used by the authors, for a number of individuals or groups. SINDCLUS provides estimates of weights for each of the various individuals or groups as well as each partition. In effect it provides insight into the extent that each group emphasizes a given partition. A summary statistic, in the form of variance accounted for (VAF), is used as a measure of fit for each group.

Klauer and Carroll present a family of discrete network models for representing proximity data. The concept of a minimum-path-length distance provides a bridge between interval-level proximity data and the discrete network representations consisting of connected and weighted graphs. They begin with an exploration of the formal foundations underlying network models as psychological representation of stimuli. They develop representation and uniqueness results for proximity measures with interval-scale as well as with ordinal-scale properties. They review various implementations of network models and propose the algorithms MAPNET and the closely related INDNET. They illustrate a discrete network model for the 15 English kin terms that fits the data well, accounting for 84% of variance. The results compare favorably to spatial models such as multidimensional scaling as well as to alternative discrete models such as tree structures, additive clusters, and multiple hierarchical structures.

12 Intermodal Similarity and Cross-Modality Matching: Coding Perceptual Dimensions

Lawrence E. Marks
John B. Pierce Laboratory and Yale University

ABSTRACT

In making cross-modality matches with respect to perceived intensity—for example, in matching the loudness of sound and the brightness of light—people recognize quantitative similarity in the face of qualitative difference. In fact, people recognize several kinds of auditory-visual similarity: between pitch and brightness (low pitch = dim; high pitch = bright) and pitch and size (low pitch = large; high pitch = small), as well as loudness and brightness (soft = dim; loud = bright): Such similarity relations can be characterized through a multidimensional spatial representation that is also multimodal, with different modalities "sharing" dimensions such as pitch, loudness, and brightness. Evidence that very young children, like adults, perceive similarity between pitch and brightness and between loudness and brightness, but not between pitch and size, accords with the view that pitch-brightness and loudness-brightness relations are intrinsic to perception, whereas the pitch-size relation may derive from experience. Auditory-visual similarities express themselves widely—in language as well as perception, and in functional interactions as well as structural relations: Similarities appear in synesthetic perception, in cross-modal matches, in judgments of cross-modal metaphors, and in response-time interactions of perceptual and linguistic stimuli. Even when they are initially perceptual, intermodal similarities infiltrate language, so that semantic codes may even come to dominate perceptual ones.

CROSS-MODALITY MATCHING AND PSYCHOPHYSICAL SCALING

One of the standard methodological and analytic tools of modern psychophysical scaling is cross-modality matching, brought to the attention of psychophysicists and others largely by the work of S. S. Stevens (1959) and his collaborators (e.g., J. C. Stevens, Mack, & Stevens, 1960)—though the method's roots go back to the 19th century (e.g., Jastrow, 1886; Münsterberg, 1890). From Stevens's perspective, cross-modality matching was important because it could be used to test the internal consistency (and, he seems to have believed, the validity) of his power law, or more generally to test the consistency of any set of magnitude-estimation functions. The argument is well known: If Ψ_a and Ψ_b are the magnitude-estimation representations of perceptual dimensions a and b, respectively, related to stimulus domains D_a and D_b by functions F and G, so that $\Psi_a = F(D_a)$ and $\Psi_b = G(D_b)$, then when subjects match, they set $\Psi_a = \Psi_b$, from which $D_a = F^{-1}[G(D_b)]$. When F and G are power functions, the stimulus settings derived from cross-modality matching should also conform to a power function, with an exponent equal to the ratio of the exponents obtained in magnitude estimation.

Figure 12.1 gives an example: cross-modality matches that J. C. Stevens and I reported in 1965, obtained by asking subjects either to adjust loudness to match brightness (open circles), or to adjust brightness to match loudness (filled circles). In this instance, power functions do provide a reasonably good fit to the

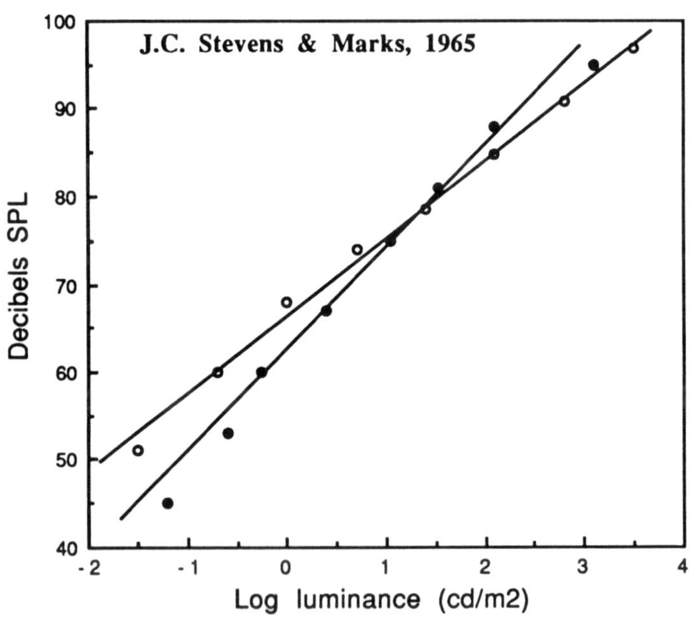

FIG. 12.1. The matching of loudness and brightness, where subjects set sound levels when presented white lights varying in luminance (open circles) or set luminances when presented noises varying in SPL (filled circles). (Data from J. C. Stevens & Marks, 1965).

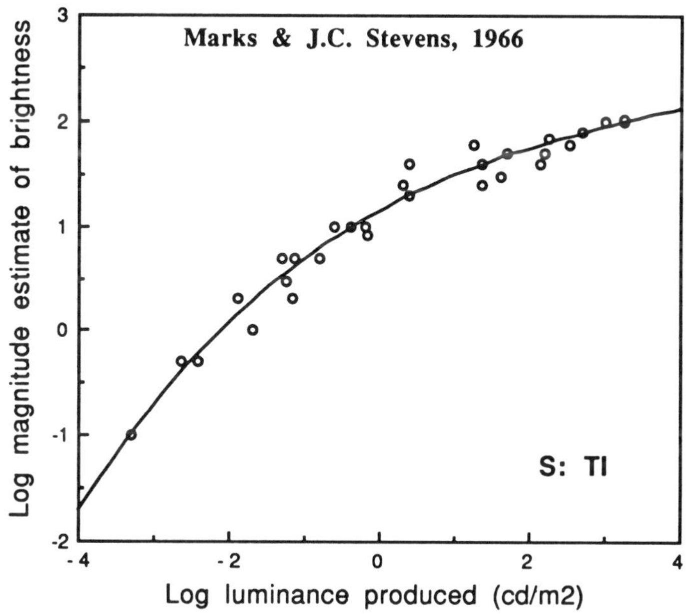

FIG. 12.2. An example of a psychophysical function for brightness of white light, obtained from subject TI. Shown are results obtained on each trial, where the subject both set the luminance level of the light and gave a numerical estimate of its brightness. (Data from Marks & J. C. Stevens, 1966).

resulting matches (straight lines to the plot of SPL versus log luminance), although they reveal the subjects' tendency to give slightly different functions when adjusting loudness to match brightness and when adjusting brightness to match loudness—the ubiquitous effect that Stevens and Greenbaum (1966) called "regression."

In fact, the rather good fit of power functions to matches between brightness and loudness is somewhat surprising, given that magnitude estimates of brightness often deviate substantially from a power function at low luminances. Figure 12.2 gives an example (Marks & Stevens, 1966). In that study, each subject both set the visual stimulus to various luminance levels and judged its brightness, thereby generating many pairings of luminance and brightness judgment, each of which could be plotted separately. Although individual subjects typically provide orderly results with both cross-modality matching and magnitude estimation (e.g., J. C. Stevens & Guirao, 1964), detailed examination of individual scaling functions often shows deviations from power functions (e.g., Luce & Mo, 1965). And in fact, close inspection of Figure 12.1 indicates a slight bowing (downward concavity) in both sets of cross-modality matches, resembling the curvature in the scaling function of Figure 12.2.

A COMMON MENTAL REPRESENTATION FOR PERCEIVED INTENSITY?

What kinds of mental processes mediate cross-modality matching? Krantz (1972) argued that cross-modal matching could rely on a process of relative judgment, the mapping of *ratios* of sensory magnitudes from one domain to another rather than magnitudes per se; such a mapping would suffice to account for the way exponents of cross-modality matching functions correspond to exponents governing the component perceptual modalities (though relational theory ipso facto cannot by itself account for equivalence between absolute levels). A relational model like Krantz's needs to assume very little communality between, say, loudness and brightness, save that subjects implicitly quantify values on both dimensions, and that these values can be related by ratio (or log-interval) representations. In fact, relational theory need not assume that both dimensions represent intensity per se. Still, it is commonly assumed (sometimes implicitly, sometimes explicitly) that cross-modality matching rests on the existence of something more than the possibility of computing common quantitative relations—the "something more" being some sort of superordinate dimension of psychological intensity, instantiated in different senses through such modal dimensions as brightness, loudness, heaviness, and so forth. By this view, while people may be capable of using ratios as their common denominator (so to speak) for making cross-modal comparisons, people may also be able to compare directly the perceptual magnitudes experienced on different modalities. This last notion finds elaboration in Zwislocki's work on "absolute scaling" (e.g., Zwislocki, 1983; Zwislocki & Goodman, 1980).

CROSS-MODAL SIMILARITY

The attribution of a common (and commensurable) dimension of intensity to percepts belonging to different sense modalities implies a kind of similarity among them—a similarity among qualitative dissimilars. Sounds have their defining qualities such as pitch or musical tonality; lights have their colors, tastes their qualities such as saltiness or sourness; feelings have their qualities such as vibration or stickiness or pain. But these qualitative dissimilarities become transparent, in some fashion, through the equivalences of perceived intensity.[1]

[1]Lest this issue seem quintessentially contemporary, I note that Plato repeatedly raised questions about "the One and the Many," and he squarely faced a near cousin to the matter at hand when he

Evidence for the primitive "givenness" of intensity as a common, intermodal dimension comes from studies tracing cross-modality matching back to childhood, even to infancy. Stevens was pleased to show that five-year-old children can give systematic and reliable cross-modality matches (Bond & Stevens, 1969) when appropriately instructed to do so (this is a pertinent qualification, to be considered in the next section); two decades later, Collins and Gescheider (1989) reported cross-modality matches obtained in individual four- to seven-year-olds (with several studies intervening; see, e.g., Teghtsoonian, 1980). In a related vein, Veronica Stone and I extended the method of magnitude matching (J. C. Stevens & Marks, 1980)—where subjects apply a common scale to judge the intensities of stimuli taken from two modalities—to children as young as six to seven years. These children used a graphic (line) scale to rate the loudness of bursts of noise and the taste intensity of salt solutions, all judged within a single session. Young children do this systematically and consistently, and much as adults do (see Figure 12.3). Although the children's ratings of loudness are virtually identical to those of adults (obtained earlier with the same stimulus levels and method; see Marks et al., 1988), the children tended to give relatively greater ratings to taste intensity, suggesting a possibly greater responsiveness to NaCl.

In this regard, an intriguing experiment by Lewkowicz and Turkewitz (1980) claimed evidence for cross-modal equivalence between brightness and loudness in one-month old infants. Using cardiac deceleration as a measure of the "novelty" of a stimulus, Lewkowicz and Turkewitz measured reactions to light after infants habituated to a sound. The smallest change in heart rate (indicating least novelty) came in response to a subsequent light whose brightness "matched" the habituated sound's loudness (where matching was defined by the cross-modality functions of Figure 12.1); this outcome is compatible with Smith's (1985) hypothesis that young children often behave as though they are responding to the overall perceptual magnitude of stimuli, regardless of the sense modality (but see also Smith & Sera, 1992).

Note that the method used by Lewkowicz and Turkewitz (1980) shifts the topic of inquiry across an important theoretical boundary, taking it from the domain of cross-modal matching (of the typical psychophysical sort) to that of cross-modal similarity. Studies of cross-modality matching, which typically concern themselves with the quantitative properties of subjects' judgments—whether with the nature of individual or group matching functions, or with the stochastic properties of individual responses—have generally taken for granted a monotonic ordering between modalities or, more properly, between dimensions on two or more modalities. To wit, the instructions in cross-modality matching

sought to characterize what it is that underlies the Good. In seeking what it is that makes different values comparable and commensurable, in one sense, Plato sought to provide an ontology to decision theory.

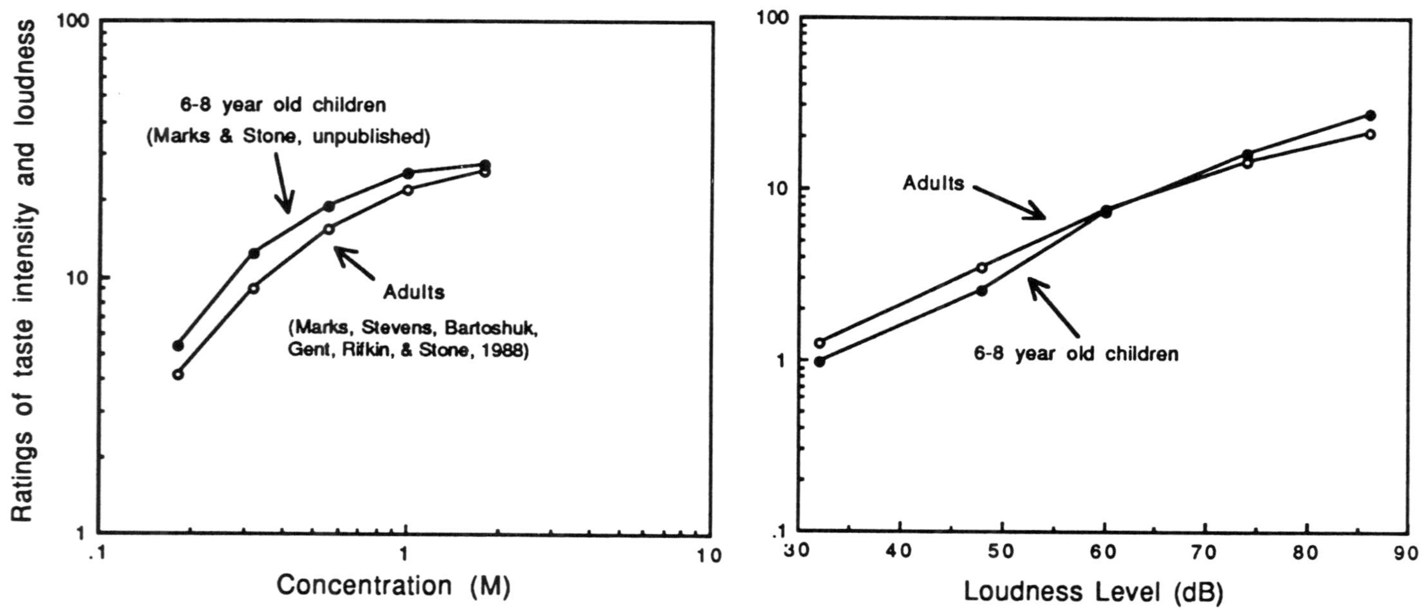

FIG. 12.3. Ratings of taste intensity of sodium chloride (left panel) and loudness of noise (right panel) obtained from six- to eight-year old children (filled circles, data from Marks & Stone, unpublished) and from adults (open circles, data from Marks, Stevens, Bartoshuk, Gent, Rifkin, & Stone, 1988). Salt solutions and sounds alternated from trial to trial, and subjects attempted to judge all of the stimuli on a common graphic rating scale.

may urge subjects to "match loudness to brightness," or "make the sound as much louder or softer as the light is brighter or dimmer." In doing this, such instructions serve to anchor the two dimensions with respect to their polar endpoints (defining soft = dim and loud = bright).

But Lewkowicz and Turkewitz (1980) asked, albeit indirectly, a propaedeutic question, "Are (or in what way are) lights and sounds varying in intensity similar?" Given the measure of cross-modal equivalence—maximal generalization of habituation—there was no a priori reason that increasing brightness should necessarily correspond to increasing loudness. Increasing brightness might just as well correspond to decreasing loudness.

And even if increasing brightness does have a perceptual correlate in increasing loudness, may it not also have an auditory correlate in another auditory dimension, such as increasing pitch? In fact, there is considerable evidence to back the contention that cross-modal similarities go well beyond the realm of intensity, for there turns out to be a constellation of multidimensional relations connecting perception in different sense modalities.

The remainder of this chapter will consider several intermodal similarities, with special emphasis on the visual and auditory. Of particular concern are three central matters: First is the matter of origins: Here I summarize the results of developmental studies concerning cross-modal similarity at different ages, indicating which intersensory relations are most primitive or "primary." Next is the matter of method: Although matching methods have a long history, there is evidence that similarity-scaling methods, especially when evaluated with techniques of multidimensional scaling, can usefully evaluate intermodal relations. And last is the matter of mechanism: Although cross-modal similarities may initially derive from intrinsic perceptual connections, perhaps grounded in common neural coding mechanisms, these similarities need not always rely on perceptual processes per se. In some instances at least, cross-modal similarities may also be mediated by semantic, lexical, or other relatively higher-level processes.

VISUAL-AUDITORY SIMILARITIES IN ONE DIMENSION

Three Intermodal Relations

One can use a relatively simple paradigm, pairwise cross-modality matching, to learn whether a given pair of psychological dimensions have a common polar structure. Note that the traditional cross-modality matching paradigm, as outlined earlier, may instruct subjects to "set loudness equal to brightness," or vice versa, thereby presuming that loud tones are equivalent to bright lights, while soft tones are equivalent to dim lights. A more "basic" version of this task, in a simple form, presents subjects with two tones (one louder than the other) and two lights (one brighter than the other) and ask the subjects to match them up. In other words, the procedure asks, psychophysically, whether soft resembles dim and loud, bright.

Adults do this quite readily. Using this pairwise paradigm, a group of adults without exception matched the louder of two tones with the brighter of two spots of white light (loud = bright, soft = dim) (Marks, Hammeal, & Bornstein, 1987). My colleagues and I defined this pattern as a *normative* loudness-brightness match, indicating by the expression that a majority of adults match this way, in a paradigm that does not restrict (or assume) the mapping of polarities but actually determines it.

But loudness is not the sole simulacrum of brightness. In addition, our adult subjects without exception matched pitch (of a constant-loudness tone) with brightness (of identical spots of white light) (high pitch = bright, low pitch = dim). And finally, we found that most, though not all, adults also agreed about the polar correspondence between the pitch of a tone and the two-dimensional size of drawn circles; in the case of pitch and size, the polarities were inverted (high pitch = small, low pitch = large). So it is evident that vision and hearing reveal, even with simple stimuli (spots of lights, circles, tones), three distinct (and, by definition, normative) dimensional similarities, at least in adults.

Children Versus Adults

What about children? Using the identical pairwise matching procedure, equipment, and stimuli, about 75% of four- to five-year-old children made adultlike (normative) matches between loudness and brightness, and by age 8 nearly 90% did (Marks et al., 1987). This is quite good performance, indicating widespread recognition of loudness-brightness similarity in young children. But at the same time, these results imply that just about one-fourth of four- to six-year-olds *failed* to match soft tones with dim lights, and loud tones with bright lights—though presumably they can do this when they are appropriately instructed (as were the children in Bond & Stevens's, 1969, study in which the polarities were defined through instruction and example). So the recognition of similarity between loudness and brightness, though widespread, is not universal among young children.

On the other hand, it turned out that children are even more consistent than this when asked to assess possible similarity between pitch and brightness: Nearly 90% of four-year-olds and practically 100% of six-year-olds matched normatively. By way of contrast, very young children did not give any evidence at all of consensually recognizing similarity between pitch and size. This similarity was not recognized by a majority of children until after age 9.

Figure 12.4 summarizes all of these results and thereby shows the multiplicity of visual-auditory similarities, even in young children. The figure shows how the percentage of normative pairwise matches varies with age: Clearly, the developmental trajectories for these three kinds of visual-auditory similarity are strikingly different. Perhaps surprisingly, intensity fails to lead the pack. Despite

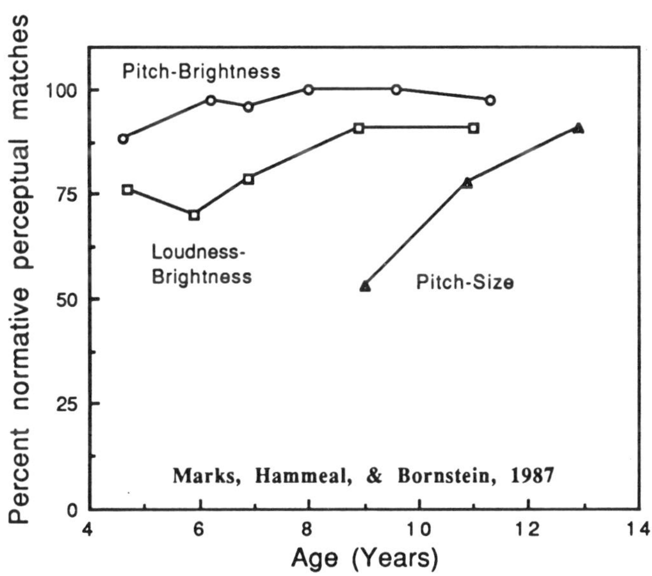

FIG. 12.4. Percentages of children of various ages who matched visual and auditory stimuli in the same way as most adults: For pitch and brightness, this meant matching a low-pitched tone with a dim light and a high-pitched tone with a bright light. For loudness and brightness, it meant matching a soft tone with dim light and a loud tone with bright light. For pitch and size, it meant matching a low-pitched tone with a large circle and a high-pitched tone with a small circle. (Data from Marks, Hammeal, and Bornstein, 1987).

the overwhelming focus that studies of cross-modality matching provide to the dimension of intensity, it is not the similarity of loudness and brightness—in the usual psychophysical view, the auditory and visual instantiations of intensity—that shows the greatest consistency in childhood; the similarity of pitch and brightness does. Nevertheless, these findings show that at least a few cross-modal similarities have early roots, suggesting thereby that these similarities are likely rooted directly in some kind of perceptual equivalence, and are probably not, at least initially, mediated by language.[2] This characterization seems undoubtedly true for the relation between pitch and brightness, for which it is difficult to imagine any source of mediation (indeed, we observed that, by and large, four-year-old children typically do not relate the labels "low pitch" and

[2]The work of Smith and Sera (1992) suggests, however, that the relation between perception and language in structuring dimensions can be considerably more complex than the present discussion may imply. With some dimensions, such as visual lightness (running from dark to light), language and perception may interact so that different polar structures and mappings can emerge at different ages.

"high pitch" systematically to sound frequency; see also Roffler & Butler, 1968). On the other hand, the evidence that recognition of a connection between pitch and size first appears relatively late in childhood, just before adolescence, may well mean that there is no intrinsic communality between these dimensions; their similarity may instead derive from experience—and, in particular, experience with the inverse correlation between mass (a correlate to volume or size) and resonance frequency. Small objects ping, while large ones thud.

Cross-Modal Similarity and Perception

Cross-modal matching, in its various forms, has proved a useful tool for elucidating quantitative properties of perceptual/decisional systems (e.g., Luce & Mo, 1965; Stevens, 1959), and for evaluating various models of perception and judgment (e.g., Baird, Green, & Luce, 1980), as well as for assessing the existence or strength of cross-modal similarities. But if resemblances between attributes of experience in different sense modalities have any substantial role in organizing or guiding perceptual experiences, then we should not be surprised to find these resemblances revealing themselves spontaneously, in various kinds of ongoing perceptual behaviors.

Cross-Modal Similarity and Synesthetic Perception

One example, which I merely mention, is synesthetic perception. A small portion of the population (conservatively estimated, something under 1%) claims to experience the world, on a regular basis, in what seems to most of us a peculiar way, reporting that stimuli usually judged appropriate to one modality (typically, sounds) spontaneously, regularly, and reliably produce some kind of perceptual qualities or images appropriate to another modality (shapes, colors). Although people with synesthetic perception reveal diverse and often idiosyncratic associations—such individuals have been studied off and on for well over a century now—they also show considerable agreement. For example, synesthetic perceivers consistently report that the lightness or brightness of their visual images varies directly with the loudness of the inducing sounds; that the lightness or brightness of the images varies directly with the pitch of the sounds; and that the size of the images varies inversely with the pitch of the sounds (for reviews, see Marks, 1975, 1978). All three synesthetic correspondences mimic patterns of cross-modal similarity seen in nonsynesthetic individuals.

Cross-Modal Similarity and Language

Perceptual experiences are linked, or can be linked, through words. We call soft sounds and dim lights "weak" but loud and bright ones "strong" or "intense." Moreover, words are themselves represented perceptually, available in visual form when written and in auditory form when spoken. In fact, there is a long-standing literature on the topic of phonetic symbolism, the power of speech sounds in and of themselves to convey meanings. Perhaps it should not be surprising to find, in a review of this literature (Marks, 1975, 1978), two of the same triumvirate of dimensional similarities outlined earlier: pitch-brightness and pitch-size. Consider the following paradigm. Subjects are given nonsense trigrams of the form consonant-vowel-consonant, in which the vowel varies while the consonants remain constant; when the vowel is made higher rather than lower in pitch (e.g., [i] or [I] versus [a]), subjects judge the "referents" of these nonsense syllables to be correspondingly smaller and brighter.

Within the realm of natural language, the ways that both children and adults understand cross-modal metaphors provides still further evidence. Edgar Allan Poe's "sound of coming darkness" wanes in intensity, growing ever softer, whereas Conrad Aiken's "music [that] suddenly opened like a luminous book" waxes in loudness. Indeed, by examining how people rate the meanings of words and short phrases, it is possible to derive verbal measures of "cross-modal similarity" that are in some sense analogous to perceptual measures. For instance, when subjects are asked to rate verbal stimuli with respect to a dimension of loudness or pitch, brightness or size, they not only produce orderly judgments in response to expressions that are literally related to the dimension being scaled (e.g., by rating "sunlight" to be brighter than "moonlight"), but they also produce orderly judgments in response to expressions that are metaphorically connected to the dimension (e.g., by rating "sunlight" to be higher in pitch than "moonlight") (Marks, 1982).

Adult subjects systematically rate the noun "sunlight" not only brighter than "moonlight" but also higher pitched, and rate the verb "roars" not only louder but also brighter than "whispers." They judge a "drum note" to be brighter as well as louder than a "piano note," a "sneeze" to be brighter as well as higher pitched than a "cough," and "white" to be higher pitched as well as lighter than "black." Judgments of this sort are consistent and orderly, revealing parallels between loudness and brightness, between pitch and brightness, and between pitch and size.

Figures 12.5 and 12.6 give two complementary examples. Expressions denoting acoustic events that are interpreted to be low or high in pitch are judged to be equally low or high in brightness, as Figure 12.5 makes clear. People know, in some fashion, that prototypical "thunder" and "cough" are low in pitch, whereas "squeak" and "sneeze" are high. The brightnesses of these acoustic events correspond directly. And color words follow the same rule, providing a set of especially neat examples: Colors are readily ascribed to be low or high in lightness or brightness: People know that "black" and "brown" are prototypically dark colors, while "white" and "yellow" are light (dark yellows being called "brown"). And, correspondingly, these color words are rated equally low or high in pitch. Figure 12.6 shows that ratings of lightness and pitch of color words are virtually collinear.

Thus intermodal relations are structured, at least implicitly, within the domain

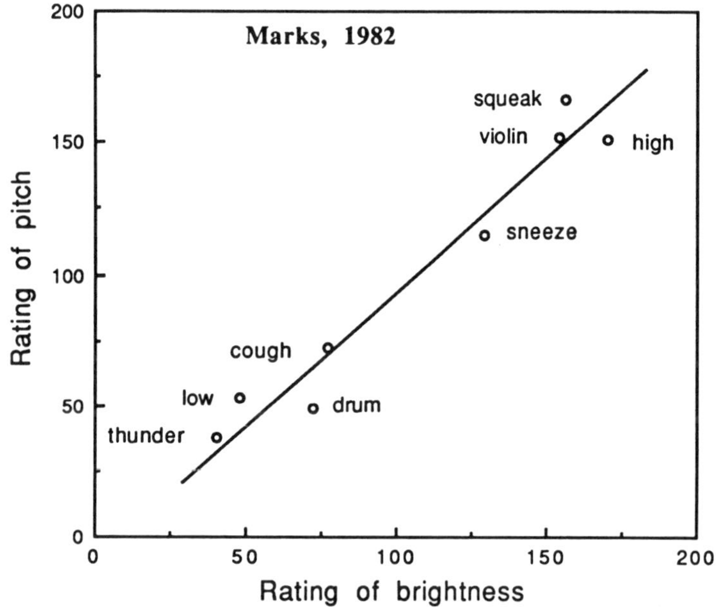

FIG. 12.5. Ratings of pitch given to words describing various sounds or acoustic events, plotted against ratings of brightness of the same words. (Data from Marks, 1982).

of linguistic or semantic knowledge: Despite the many differences between perceptual and linguistic representations of the world (Miller & Johnson-Laird, 1976), there is much that perception and language share. To help determine the degree of overlap, it would be valuable to find cultures in which cross-modal similarities are absent from the linguistic/semantic systems and then to determine whether individuals raised in these cultures nevertheless do recognize the cross-modal similarities when posed with appropriate perceptual tasks. Unfortunately, such cultures may not exist. Within western cultures, cross-modal relations seem terribly well ingrained in language; for instance, Hartshorne (1934) pointed out that the German adjective for visually bright, "hell," originally referred to high-pitched sounds. Such relations may exist in non-Western linguistic cultures too. According to Kainz (1943), words in the African language Ewe that refer to small versus large objects are expressed in high rather than low pitch.

Lastly in this regard, it is notable that children behave much like adults, although the intermodal correspondences are not always as clear, systematic, or reliable in children (Marks et al., 1987).[3] To see this most clearly, one can

[3]Nevertheless, it is notable that children find cross-modal metaphorical expressions to be among the easiest of metaphorical expressions to understand (Winner, Rosentiel, & Gardner, 1976).

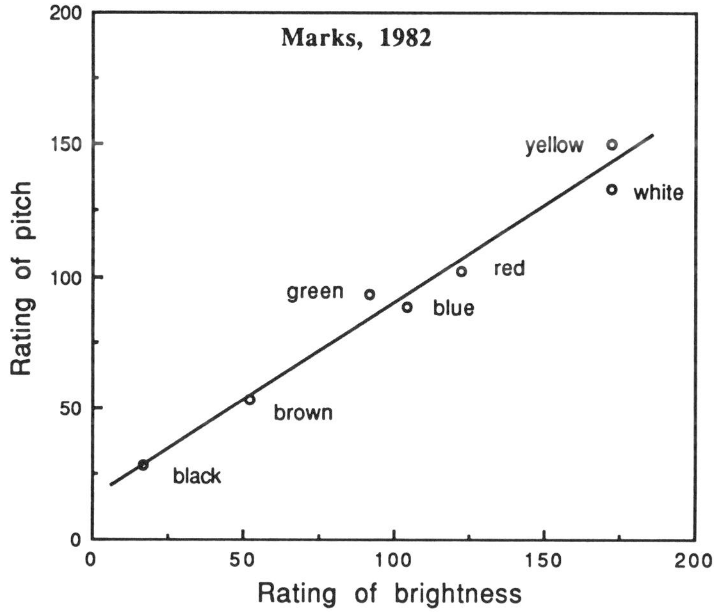

FIG. 12.6. Ratings of pitch given to words naming various colors, plotted against ratings of brightness of the same words. (Data from Marks, 1982).

condense each subject's (child's or adult's) entire set of ratings into a single score, which then defines a verbal cross-modality match or mismatch. Again using the performance of the preponderance of adults to define the norms, these measures of verbal "matches," given in Figure 12.7, are similar in their general form to the perceptual matches of Figure 12.4.

As comparison of Figure 12.7 with Figure 12.4 shows, at all ages, and for each kind of cross-modal similarity, performance gauged from these verbal tasks was inferior to performance gauged from perceptual matching. For example, with pitch-brightness metaphors, only 75% of four-year-olds gave adultlike matches, whereas 90% gave adultlike matches on the corresponding perceptual task. Indeed, with loudness-brightness metaphors, four-year-olds responded essentially randomly (splitting 50/50), whereas a substantial proportion of them (75%) made adultlike perceptual matches. A reasonable inference is that loudness-brightness and pitch-brightness similarities are intrinsically perceptual in nature, but eventually become incorporated into semantic systems.

I shall defer for the moment the issues concerning semantic representation of cross-modal relations. This will be the subject of the last part of this article, which considers the possibility that verbal representations, even if derivative, may be capable of taking on psychological preeminence.

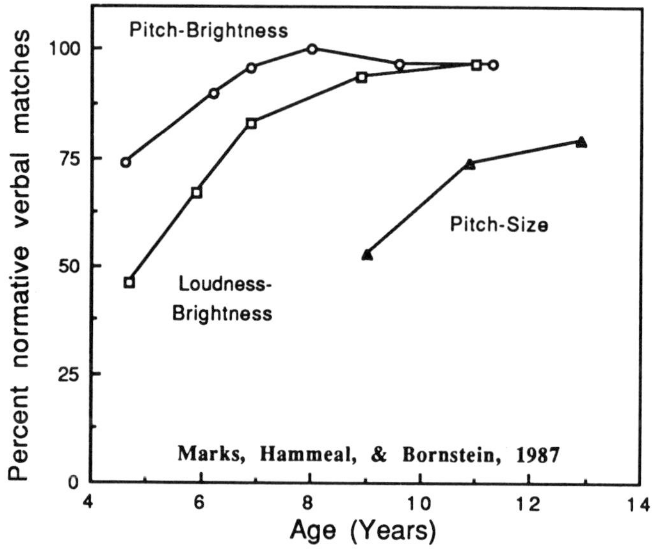

FIG. 12.7. Percentages of children of various ages whose ratings of metaphorical verbal expressions on the dimensions of pitch, loudness, brightness, and size conformed to the pattern of adults, as in Figure 12.4. (Data from Marks, Hammeal, & Bornstein, 1987).

MULTIDIMENSIONAL AND MULTIMODAL REPRESENTATIONS

The evidence just summarized suggests that, at least within the modalities of vision and hearing, cross-modal similarities and relations constitute a coherent structural system. Not only do perceptual experiences vary multidimensionally within a given sense modality, but similarities across different senses may in some fashion tap into these sense-specific, multidimensional structures. To say this is, of course, to transcend the evidence adduced thus far, and to suggest that cross-modal relations may themselves capitalize on the multidimensional perceptual structures evident in the individual modalities.

If this is so—if *inter*modal similarity comprises a natural extension of multidimensional, *intra*modal similarity—then it should be possible to capture these multimodal structures through various procedures of multidimensional comparison and scaling. One of the first to recognize this possibility was Wicker (1968), who sought to learn whether similarity judgments among tones and colors alike could be represented within a single perceptual space. Because Wicker presented his visual stimuli (Munsell colors) against a background of medium lightness, the main visual dimension turned out to be lightness contrast, which turned out in his study to have a strong cross-modal correlate in the loudness of the tones. On the other hand, when Webb (1976) had subjects rate similarities of colors and musical chords, the spatial representation lacked a common intermodal dimension, revealing a main dimension of visual lightness and an auditory dimension of "compactness" of the chords. Webb's findings are perhaps the consequence of using compound (bimodal) stimuli; that is, on each trial the subjects compared one color-chord combination to another, a procedure that may have encouraged the subjects to compare color to color and chord to chord.

From the various findings considered so far, we may infer that the visual dimension of brightness has (at least) two correlates in hearing: loudness and pitch. How might loudness and pitch combine their similarities? If both increasing loudness and increasing pitch correspond, psychologically, to increasing brightness, then if we keep brightness constant, it should be possible to offset changes in pitch with corresponding changes in loudness. Let the equation $\Psi_A(L, P) = \Psi_V(B)$ characterize the mapping of visual-auditory similarity, so that a tone of loudness L and pitch P "matches" a light of brightness B; that is, $L = F1(B)$ and $P = F2(B)$ represent monotonic increasing orderings between loudness and brightness and between pitch and brightness, respectively. If L increases, then in order to maintain a match (that is, for B to remain constant), P must decrease ($\Psi_A[L + \Delta L1, P - \Delta P1] = \Psi_V[B]$); if L decreases, P must increase ($\Psi_A[L - \Delta L2, P + \Delta P2] = \Psi_V[B]$). The trade-off between L and P depends on the still-unknown rule governing their integration (e.g., linear sum, vector sum), as well as the functions relating loudness ($F1$) and pitch ($F2$) to brightness.

Cross-Modal Comparison

Several psychophysical means offer themselves by which we could examine multidimensional and cross-modal trade-offs of this sort. A relatively straightforward method is cross-modal comparison: On each trial, give the subject two lights differing in brightness, together with a single tone; the task is to decide which light is more similar to the tone. By letting the frequency and SPL of the tone vary from trial to trial, while the two luminances remain the same, we produce psychometric functions relating the probability of responding "more similar to the brighter light" to both parameters of the auditory signal. Points of subjective equality (50% points on the psychometric functions) provide the cross-modality "matches," that is, the set of acoustic signals, varying in frequency and SPL, that are equally similar to the two lights. This set of values forms a kind of equivalence function or indifference curve, making it possible to assess the trade-off between frequency and SPL (and if the appropriate psychophysical functions are known, the trade-off between loudness and pitch).

A study of loudness, pitch, and brightness similarity determined indifference functions in 16 subjects (Marks, 1989). Figure 12.8 gives representative examples, each of the four panels showing data derived from a single subject. Each of

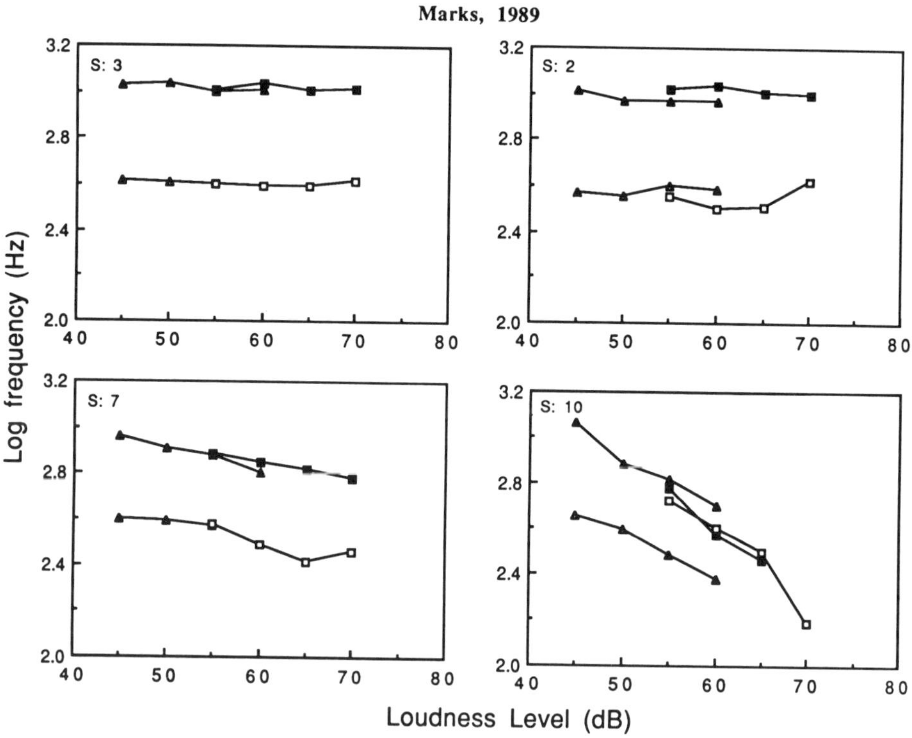

FIG. 12.8. How pitch and loudness combine in cross-modal similarity between sound and light: Each line shows how frequency and SPL trade off so that a tone appears equally similar to each of two lights of fixed luminance. Different lines within a panel represent results obtained in different contextual conditions, and each panel gives results from an individual subject. (Data from Marks, 1989).

the four lines within each panel represents a different condition of stimulus context (not a matter of primary concern here). What are of interest are the slopes of these lines. By plotting log frequency on one axis against dB loudness level on the other (the use of loudness level serves to equate signals for loudness at different frequencies), the slope of each contour indicates the magnitude of the trade-off between loudness and pitch. A horizontal line would indicate that the derived cross-modality matches were independent of loudness, but depended solely on pitch; a vertical line would indicate the reverse, that the derived matches were independent of pitch, but depended solely on loudness.

In Figure 12.8's examples, subjects 2 and 3 showed that only pitch mattered, not loudness; subject 7 showed that pitch mattered mostly, loudness much less so; and subject 10 showed that loudness mattered still more. For most of the 16 subjects tested, pitch counted more than loudness. This generalization should be tempered by the possibility that the results were influenced by the relative ranges of pitch and loudness used (though I did seek to make the perceptual ranges roughly equivalent). In any case, the frequency-loudness level trade-offs in nearly half of the subjects had flat or near-zero slopes, implying that the dimension of pitch predominated in cross-modal similarity; by way of contrast, only one subject gave anything like a set of vertical lines indicating that loudness predominated. In a nutshell, to most of these subjects, pitch proved stronger or more salient than loudness as an auditory analogue to brightness.

Multidimensional Scaling

Another approach to articulating the structure of multidimensional, intermodal relations comes through multidimensional similarity scaling. Multidimensional scaling assumes that similarity relations among stimuli within a given modality can be represented by a metric space; stimuli are positioned at points within the space so that the distance between any pair of points represents the degree of difference or dissimilarity between the stimuli (e.g., Schiffman, Reynolds, & Young, 1981). Given this representational scheme, the extension to multimodal similarity is straightforward: Corresponding dimensions in different modalities imply common projections or axes within a unitary, or uniform, multidimensional space. Thus, Wicker (1968) was able to embed color surfaces and tones within a single space, in which a visual dimension of lightness contrast corresponded to an auditory dimension of loudness, while lightness corresponded to pitch.

From the range of findings already reviewed, it is clear that people judge increases in the brightness of a light as if they were similar to increases in both the loudness and the pitch of a sound. Within the framework of multidimensional spatial models, we may surmise that visual and auditory experiences share dimensions in a bimodal perceptual space, such that the axis through the space that represents visual brightness projects onto the axes representing both loudness and pitch. To test this supposition, a study asked subjects to rate the degree of dissimilarity between all possible pairs of signals comprising several white lights, varying in luminance, and several pure tones, varying in frequency and SPL (Marks, 1989). (White light and pure tones were used because color per se [hue] seemed not to matter in Wicker's, 1968, study; nevertheless, it would be important ultimately to examine possible cross-modal correspondences between qualitative dimensions of perceptual experience.)

Multidimensional scaling yielded the two-dimensional spatial representation shown in Figure 12.9. One dimension of the space reflects loudness and the other pitch, with brightness of the lights projecting onto both auditory dimensions. Tones of uniform pitch fall at roughly the same locus on dimension 2 but increase with greater loudness on dimension 1; tones of uniform loudness fall roughly at

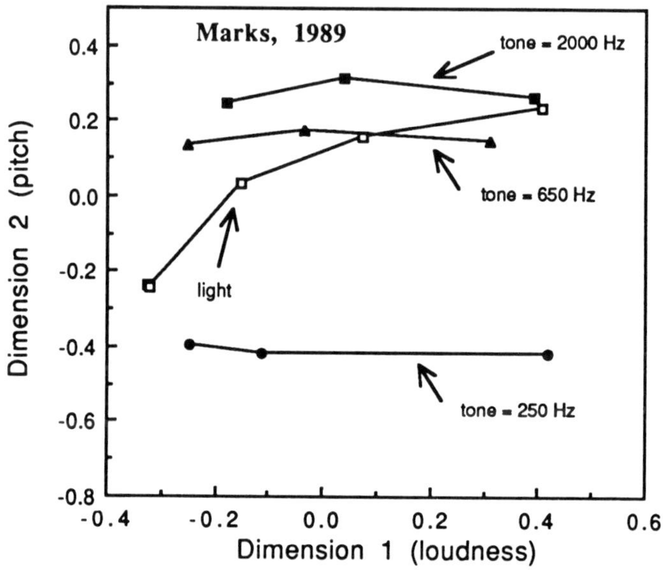

FIG. 12.9. Multidimensional scaling solution obtained by applying the individual-differences method SINDSCAL to ratings of dissimilarity between tones varying in frequency and SPL and white lights varying in luminance. (Data from Marks, 1989).

the same locus on dimension 1 but increase with greater pitch on dimension 2; meanwhile, lights cut a curvilinear trajectory through both dimensions, with increases in brightness corresponding to increases in both loudness and pitch. Oddly enough, no substantial dimension turned up that corresponded to the difference between the two modalities themselves.

Figure 12.9 is based on a scaling solution obtained with canonical decomposition by SINDSCAL, a symmetrical version of the individual-differences method INDSCAL (Carroll & Chang, 1970). The techniques of individual-differences scaling provide several advantages over other techniques and models of multidimensional scaling, including the capacity to identify the orientation of the primary psychological axes in the similarity space; the scaling routine accomplishes this by assuming that individuals vary in the extent to which they rely on or weight the different dimensions in Euclidean space, but that everyone structures the dimensions in a similar fashion. Consequently, the scaling solution that underlies Figure 12.9 includes weighting coefficients for each subject on the two dimensions, and these coefficients in turn characterize the subject's relative reliance on loudness and pitch. Because these same subjects had participated in the previous study using cross-modal comparison, examples of which appeared in Figure 12.8, it was possible to compare performance, subject by subject, on the two tasks. The visual-auditory comparisons obtained in the first study made it possible to derive, for each subject in each condition, a coefficient α that is related to the slope of the pitch-loudness trade-off. The size of α reflects the subject's relative reliance on loudness and pitch. Comparing values of α to the difference, obtained in the second study, between individual weighting coefficients from multidimensional scaling, we find a rank-order correlation coefficient of 0.86. Clearly, the tendency in each subject for similarity to depend on pitch or loudness transcended the particular psychophysical task.

Do Pitch and Loudness Combine?

Let me make an historical note. During the early decades of this century, with cognitive psychology still in the final thralls of Titchenerian structuralism, investigators sought to determine and characterize psychophysically the full complement of dimensions of perceptual experience. Besides loudness and pitch (and duration), even the simplest of auditory stimuli—a pure tone—was said to display two additional attributes, called *volume* and *brightness* (e.g., Boring & Stevens, 1936; Rich, 1919; Stevens, 1934). While both volume and brightness turn out to depend on both frequency and SPL (as do both loudness and pitch), they do so in different ways. Brightness increases with greater frequency, holding SPL constant, and increases with greater SPL, holding frequency constant, as if auditory brightness represented some combination of loudness and pitch. Later on, it became clear that brightness was the same as yet another putative auditory dimension, *density* (Boring & Stevens, 1936; Guirao & Stevens, 1964).

Given this, it is tempting to infer that visual-auditory similarity is governed by the following principle: Colors and tones are most similar when, ceteris paribus, they are equal in brightness, where auditory brightness, or density, is a joint function of loudness and pitch. Although this principle provides a reasonably good account of empirical evidence on cross-modal similarity, it does not speak directly to the ontological status of auditory brightness/density. Several other findings do, however, and they suggest that the dimensions of pitch and loudness have a psychological primacy that brightness/density and volume lack. For one, the vast majority of multidimensional scalings of tones (e.g., Melara, Marks, & Lesko, 1992; Schneider & Bissett, 1981) reveal pitch and loudness as primary axes of auditory space, although one study, using bands of noise, did suggest that brightness/density and volume may be primary (Chipman & Carey, 1975). And for another, measures of response times to classify tones show clear evidence that loudness and pitch are processed more efficiently, and presumably with priority, whereas brightness/density and volume are presumably derivative (Grau & Kemler Nelson, 1988; Melara & Marks, 1990).

CROSS-MODAL SIMILARITY AND PERCEPTUAL INFORMATION PROCESSING

Does the existence of cross-modal similarities bear any consequences for the ways that people process perceptual information? In particular, does the proficiency in processing information presented in one modality—in, say, classifying or identifying different signals—depend on the concomitant presence of matching or mismatching signals in another? Melara and O'Brien (1987) reported one kind of intermodal interaction: When subjects tried, as rapidly as possible, to classify two tones differing in frequency, performance was impaired by the presentation of a visual stimulus that varied from trial to trial in its spatial location (compared to a baseline condition in which the visual location did not change). The interaction was more or less symmetrical, in that performance was also impaired when irrelevant tones varied in frequency and the subjects' task was to classify visual position. Later, Melara (1989) reported analogous findings when the visual stimuli varied in lightness (white versus black) and tones again varied in sound frequency.

These interactions constitute failures to attend selectively to information in different modalities (see Garner, 1974). A common feature of such interactions is the presence of cross-modal congruence (Bernstein & Edelstein, 1971; Marks, 1987; Melara, 1989; Melara & O'Brien, 1987). Speed and accuracy in classifying two stimuli are not only depressed in toto by concomitant fluctuations in stimuli on another modality; in addition, responses to a particular test stimulus can depend on whether the irrelevant stimulus "matches" or "mismatches," according to the rules of cross-modal similarity. That is to say, congruence characterizes the pattern of results obtained with a set of four stimuli, comprising all possible pairs taken across dimensions on two modalities. For instance, a subject may be asked to make one response if the test light is dim, another if the test light is bright. Simultaneous with the light is a tone, which can be either low or high in pitch, the trial-by-trial choice of frequency being independent of the luminance of the visual stimulus, and thus uninformative of the correct response. Nevertheless, the contingent relation between irrelevant and relevant stimuli on each trial produces a small but systematic effect: On trials where the light is dim, responses are faster *and* more accurate when the tone is low in frequency rather than high; but on trials where the light is bright, responding is more proficient when the tone is high in frequency rather than low (e.g., Marks, 1987). The changes in response time and errors do not represent a speed-accuracy trade-off, but instead a positive speed-accuracy correlation indicative of changes in efficiency of processing.

Again, the cross-modal interactions are reasonably symmetrical, in that an irrelevant visual stimulus varying in brightness affects, to about the same degree, speed and accuracy in responding to tones varying in pitch (Marks, 1987). Moreover, these congruence effects are widespread, being evident also in interactions between pitch and lightness (analogous to pitch and brightness, except that the visual stimuli vary along the dimension that runs from dark to light: Marks, 1987; Melara, 1989), in interactions between loudness and brightness (where responding is more efficient when loudness and brightness match than when they mismatch: Marks, 1987), and in interactions between pitch and spatial location (responding more efficient when low or high pitch is conjoined with low or high spatial location: Bernstein & Edelstein, 1971).

WHAT MECHANISMS MEDIATE CROSS-MODAL CONGRUENCE?

Although the intermodal congruence observed in tasks requiring speeded classification of stimuli parallels cross-modal similarities, some of which, at least, seem to be quintessentially or at least originally perceptual in nature and not derived from language, it is nevertheless possible that what I have called congruence interactions do reflect higher-order processing, perhaps at a semantic level. Surely it is unlikely that intermodal effects take place very "early" in perceptual processing; and their appearance in Stroop-like tasks suggests a critical role for relatively "late" processes, because the Stroop effect—the slowing in naming a color word when printed in a competing color—is commonly ascribed to response competition (e.g., Treisman, 1969).

As a matter of fact, three pieces of evidence that I will describe all point to the possible role of semantic or other high-level processes in cross-modal congruence. One piece comes from Walker and Smith (1984). They reported that sound frequency could influence the speed in identifying words whose meanings were metaphorically related to low and high pitch.

Second, I have recently repeated some of the experiments testing pitch-lightness congruence, but with a small change in the stimuli used: Instead of presenting patches of color (black or white), as visual stimuli, I presented subjects a word ("black" or "white," the word being constant throughout any given experiment); this word could take on either of the two colors to be classified (black or white).

If subjects either explicitly or implicitly encode every stimulus semantically, then presenting the same, constant word throughout an experiment should encourage the subjects to label that word (encode it semantically) in the same way on every trial; that is, presenting a constant color word should prevent the subjects from differentially labeling the stimuli according to their color. Consequently, if congruence interactions depend on matches or mismatches in semantic codes, then presenting a constant word, by blocking the differential labeling, should reduce congruence between color and pitch. And it did. On average, the degree of congruence diminished substantially, with the interaction in response time decreasing from 13.6 ms (obtained with color patches) to 8.1 ms (obtained with constant color words), and with the interaction in errors decreasing from 4% (errors) to −0.3% (words). At the very least, the findings suggest that stimuli that are directly encoded semantically can modify cross-modal interactions; further, the findings may mean that interactions between such perceptual attributes as low versus high pitch and low versus high brightness are typically mediated by implicit semantic responses, which can be blocked when a competing word is presented.

Finally, a more complex experimental design manipulated all three of the relevant stimulus dimensions: the perceptual dimension of visual lightness, the semantic dimension of color word, and the auditory dimension of pitch. On each trial, the subject saw a visual stimulus, consisting of the word "black" or "white," and this word was presented in either white or black color against the gray background; at the same time, the subject heard a tone that could be low or high in pitch. As in earlier studies, the frequency of the tone was uninformative, uncorrelated with the visual stimulus, and thus irrelevant to the subject's task, which was to integrate the information on the visual and semantic dimensions, word and color, by making one response if the word and color agreed (i.e., if the word "black" appeared in black, or if the word "white" in white) and making another response if word and color disagreed ("black" in white color or "white" in black).

Both speed and accuracy depended jointly on all three dimensions, as shown in Figure 12.10, which plots mean errors against mean response times for all eight possible stimuli. Notice that response times and errors obtained with visually congruent stimuli ("black"/black and "white"/white) lie in the upper left, while those obtained with visually incongruent stimuli ("black"/white and "white"/black) lie in the lower right. Visually congruent stimuli gave shorter response times but poorer accuracy. Perhaps the difference between responses to visually congruent and incongruent stimuli represents a speed-accuracy trade-off, and thus reflects an asymmetry between the criteria used to make the two responses. That is, the subjects may have set a "laxer" criterion to respond "word and color agree" than to respond "word and color disagree": As a consequence, the former response would be made more quickly on average than the latter, but at the cost of greater errors.

Be this as it may, the congruent stimuli are not particularly diagnostic of the underlying semantic and perceptual processes linking the visual and auditory modalities. Because the semantic and perceptual attributes are positively correlated in this subset of stimuli, responses to visually congruent stimuli do not help us to distinguish semantic from sensory/perceptual interactions—though they do nicely reveal an overall pitch-lightness congruence; for example, performance was impaired (response times and errors were greater) when the low-pitched tone, rather than the high-pitched tone, accompanied the word "white" displayed in a white color, compared to performance when the low-pitched tone accompanied the word "black" displayed in black.

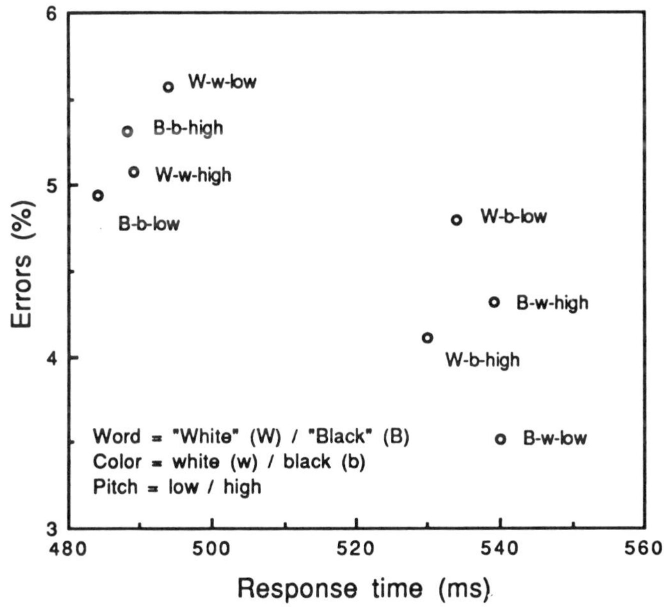

FIG. 12.10. Relation between errors and response times, obtained in a divided-attention task: Subjects had to decide whether a color word ("white" or "black") and the color in which it appeared (white or black) were the same or different. Simultaneously the subjects heard an irrelevant tone (low or high frequency) that, though irrelevant to the decision, nevertheless influenced the accuracy and speed of response.

By way of contrast, the visually incongruent stimuli are diagnostic, because ipso facto they pit a code that is primarily visual against one that is primarily semantic. As Figure 12.10 shows, the measures of response time and errors together suggest that the main interaction, or at least the net interaction, took place between pitch and word, not between pitch and color. Though small, the interactions in response times are reliable (though interactions in errors are not). Performance was better when the high-pitched tone accompanied the word "white" shown in black color (the tone being congruent with the word but incongruent with the color) when compared to performance when the low-pitched tone accompanied the same visual stimulus, "white" shown in black (where now the tone is incongruent with the word but congruent with the color). Similarly, performance was better (at least with regard to accuracy) when the low-pitched tone, compared to the high-pitched tone, accompanied the word "black" shown in white; again, congruence between tone and word led to superior performance, or, equivalently, incongruence led to inferior performance. Of course, these findings must be considered in light of possible inequalities in the magnitude of the psychological differences between pairs of colors, words, and pitches (that is, words, colors, and pitches may not be "scaled" equivalently). Even so, the preponderance of evidence suggests that semantic processes can intervene in tasks of stimulus classification. Perhaps intermodal interactions rely on very abstract, high-level amodal attributes or "semes," which both words and perceptual stimulus attributes can activate.

I raise this last possibility because of the substantial problem posed by the very notion that cross-modal congruence relies on semantic processes, whether this congruence is measured by response times or by some other psychophysical means: Prototypical congruence effects connect auditory pitch (low-high continuum) with visual brightness (dim-bright continuum) and with auditory lightness (dark-light continuum). Yet neither of these intermodal relations relies in any obvious fashion on *common* semantic elements or labels. Unlike interactions in Stroop-type tasks, where colors and color words share verbal labels, cross-modal interactions often involve attributes that do not. "Low" is neither "dim" nor "dark," and "high" is neither "bright" nor "light." What the "silver needle note of a fife" (to use the poet Joseph Auslander's expression) shares with light, bright colors (like that of silver) is an equivalence at a level of meaning that perhaps transcends language per se, but may be one into which linguistic mechanisms, as well as perceptual mechanisms, feed. Consequently, I shall entertain the hypothesis that intermodal similarity structures exist concurrently at several psychological levels—in perception; in language; and maybe even in higher-level, postlinguistic, or supralinguistic processes.

It remains to elucidate precisely what are the relations among these structures and their underlying mechanisms. Some very recent and still preliminary findings suggest that the structural organization of perceptual meanings can differ in important ways from the organization of perceptions: Unlike the similarity judgments of tones and lights described earlier (Figure 12.9), which provided distinctive dimensions of loudness and pitch, similarity judgments of words referring to acoustic and optic events imply the existence of a single, unitary dimension of brightness/density. So the mental structures of perceptions and meanings may differ.

Last of all, many of these findings pose a mild conundrum, given my claim that certain intersensory similarities, such as those between loudness and brightness and between pitch and brightness, likely have their origins in perception itself, and infiltrate semantic systems second-hand. Perhaps with experience, and linguistic development, we come to make use of multiple cross-modal mechanisms, which various psychophysical tasks may tap differentially. (The notion that tasks are central, with the mental apparatus a kind of opportunistic chameleon, has particular appeal.) Distinguishing the multiple processes and mechanisms is itself not an easy task. For instance, both cross-modality matching and cross-modal similarity have strong relativistic components. Indeed, Krantz (1972) argued that cross-modal intensity matches are based on communality of relations in different modalities, not matches of magnitudes per se. And while relativity seems especially consistent with linguistic labeling, it may also characterize perceptual coding. In any case, the structural relations and functional interactions between and among perceptual dimensions of different modalities continue to provide a rich domain both for modeling the quantitative characteristics and for evaluating the qualitative properties of stimulus processing.

ACKNOWLEDGMENTS

Research was supported by NIH Grant DC00271.

REFERENCES

Baird, J. C., Green, D. M., & Luce, R. D. (1980). Variability and sequential effects in cross-modality matching of area and loudness. *Journal of Experimental Psychology: Human Perception and Performance, 6,* 277–289.

Bernstein, I. H., & Edelstein, B. A. (1971). Effects of some variations in auditory input upon visual choice reaction time. *Journal of Experimental Psychology, 87,* 241–247.

Bond, B., & Stevens, S. S. (1969). Cross-modality matching of brightness to loudness by 5-year-olds. *Perception & Psychophysics, 6,* 337–339.

Boring, E. G., & Stevens, S. S. (1936). The nature of tonal brightness. *Proceedings of the National Academy of Sciences, 22,* 514–521.

Carroll, J. D., & Chang, J. J. (1970). Analysis of individual differences in multidimensional scaling via an *N*-way generalization of Eckart-Young decomposition. *Psychometrika, 35,* 283–319.

Chipman, S. F., & Carey, S. (1975). Anatomy of a stimulus domain: The relation between multidimensional and unidimensional scaling of noise bands. *Perception & Psychophysics, 17,* 417–424.

Collins, A. A., & Gescheider, G. A. (1989). The measurement of loudness in individual children and adults by absolute magnitude estimation and cross-modality matching. *Journal of the Acoustical Society of America, 85,* 2012–2021.

Garner, W. R. (1974). Attention: The processing of multiple sources of information. In E. C. Carterette and M. P. Friedman (Eds.), *Handbook of perception: Vol. II. Psychophysical judgment and measurement* (pp. 23–59). New York: Academic Press.

Grau, J. W., & Kemler Nelson, D. G. (1988). The distinction between integral and separable dimensions: Evidence for the integrality of pitch and loudness. *Journal of Experimental Psychology: General, 117,* 347–370.

Guirao, M., & Stevens, S. S. (1964). Measurement of auditory density. *Journal of the Acoustical Society of America, 36,* 1176–1182.

Hartshorne, C. (1934). *The philosophy and psychology of sensation.* Chicago: University of Chicago Press.

Jastrow, J. (1886). The perception of space by disparate senses. *Mind, 11,* 539–554.

Kainz, F. (1943). *Psychologie der Sprache.* Stuttgart: Enke.

Krantz, D. H. (1972). A theory of magnitude estimation and cross-modality matching. *Journal of Mathematical Psychology, 9,* 168–199.

Lewkowicz, D. J., & Turkewitz, G. (1980). Cross-modal equivalence in early infancy: Auditory-visual intensity matching. *Developmental Psychology, 16,* 597–607.

Luce, R. D., & Mo, S. S. (1965). Magnitude estimation of heaviness and loudness by individual subjects: A test of a probabilistic response theory. *British Journal of Mathematical and Statistical Psychology, 18,* 159–174.

Marks, L. E. (1975). On colored-hearing synesthesia: Cross-modal translations of sensory dimensions. *Psychological Bulletin, 82,* 303–331.

Marks, L. E. (1978). *The unity of the senses: Interrelations among the modalities.* New York: Academic Press.

Marks, L. E. (1982). Bright sneezes and dark coughs, loud sunlight and soft moonlight. *Journal of Experimental Psychology: Human Perception and Performance, 8,* 177–193.

Marks, L. E. (1987). On cross-modal similarity: Auditory-visual interactions in speeded discrimination. *Journal of Experimental Psychology: Human Perception and Performance, 13,* 384–394.

Marks, L. E. (1989). On cross-modal similarity: The perceptual structure of pitch, loudness, and brightness. *Journal of Experimental Psychology: Human Perception and Performance, 15,* 586–602.

Marks, L. E., Hammeal, R. J., & Bornstein, M. H. (1987). Perceiving similarity and comprehending metaphor. *Monographs of the Society for Research in Child Development, 52* (Serial No. 215).

Marks, L. E., & Stevens, J. C. (1966). Individual brightness functions. *Perception & Psychophysics, 1,* 17–24.

Marks, L. E., Stevens, J. C., Bartoshuk, L. M., Gent, J. F., Rifkin, B., & Stone, V. K. (1988). Magnitude-matching: The measurement of taste and smell. *Chemical Senses, 13,* 63–87.

Melara, R. D. (1989). Dimensional interaction between color and pitch. *Journal of Experimental Psychology: Human Perception and Performance, 15,* 69–79.

Melara, R. D., & Marks, L. E. (1990). Perceptual primacy of dimensions: Support for a model of dimensional interaction. *Journal of Experimental Psychology: Human Perception and Performance, 16,* 398–414.

Melara, R. D., Marks, L. E., & Lesko, K. (1992). Optional processes in similarity judgments. *Perception & Psychophysics, 51,* 123–133.

Melara, R. D., & O'Brien, T. P. (1987). Interaction between synesthetically corresponding dimensions. *Journal of Experimental Psychology: General, 116,* 323–336.

Miller, G. A., & Johnson-Laird, P. N. (1976). *Language and perception.* Cambridge, MA: Harvard University Press.

Münsterberg, H. (1890). *Beiträge zur experimentellen Psychologie. III.* Freiburg: Mohr.

Rich, G. J. (1919). A study of tonal attributes. *American Journal of Psychology, 30,* 121–164.

Roffler, S. K., & Butler, R. A. (1968). Localization of tonal stimuli in the vertical plane. *Journal of the Acoustical Society of America, 43,* 1260–1266.

Schiffman, S. S., Reynolds, M. L., & Young, F. W. (1981). *Introduction to multidimensional scaling: Theory, methods, and applications.* New York: Academic Press.

Schneider, B. A., & Bissett, R. J. (1981). The dimensions of tonal experience: A nonmetric multidimensional scaling approach. *Perception & Psychophysics, 30,* 39–48.

Smith, L. B. (1985). Young children's attention to global magnitude. Evidence from classification tasks. *Journal of Experimental Child Psychology, 39,* 471–492.

Smith, L. B., & Sera, M. D. (1992). A developmental analysis of the polar structure of dimensions. *Cognitive Psychology, 24,* 99–142.

Stevens, J. C., & Guirao, M. (1964). Individual loudness functions. *Journal of the Acoustical Society of America, 36,* 2210–2213.

Stevens, J. C., Mack, J. D., & Stevens, S. S. (1960). Growth of sensation on seven continua as measured by force of handgrip. *Journal of Experimental Psychology, 69,* 60–67.

Stevens, J. C., & Marks, L. E. (1965). Cross-modality matching of brightness and loudness. *Proceedings of the National Academy of Sciences, 54,* 407–411.

Stevens, J. C., & Marks, L. E. (1980). Cross-modality matching functions generated by magnitude estimation. *Perception & Psychophysics, 27,* 379–389.

Stevens, S. S. (1934). The attributes of tones. *Proceedings of the National Academy of Sciences, 20,* 457–459.

Stevens, S. S. (1959). Cross-modality validation of subjective scales for loudness, vibration, and electric shock. *Journal of Experimental Psychology, 57,* 201–209.

Stevens, S. S., & Greenbaum, H. B. (1966). Regression effect in psychophysical judgment. *Perception & Psychophysics, 1,* 439–446.

Teghtsoonian, M. (1980). Children's scales of length and loudness: A developmental application of cross-modal matching. *Journal of Experimental Child Psychology, 30,* 290–307.

Treisman, A. M. (1969). Strategies and models of selective attention. *Psychological Review, 76,* 282–289.

Walker, P., & Smith, S. (1984). Stroop interference based on the synaesthetic qualities of auditory pitch. *Perception, 13,* 75–81.

Webb, N. (1976). Impressions in the dark: A multidimensional scaling approach to multi-modal experience. *Perceptual and Motor Skills, 42,* 1031–1036.

Wicker, F. W. (1968). Mapping the intersensory regions of perceptual space. *American Journal of Psychology, 81,* 178–188.

Winner, E., Rosentiel, A. K., & Gardner, H. (1976). The development of metaphoric understanding. *Developmental Psychology, 12,* 289–297.

Zwislocki, J. J. (1983). Group and individual relations between sensation magnitudes and their numerical estimates. *Perception & Psychophysics, 33,* 460–468.

Zwislocki, J. J., & Goodman, D. A. (1980). Absolute scaling of sensory magnitudes: A validation. *Perception & Psychophysics, 28,* 28–38.

13 Judgment Windows in Psychophysical Scaling

John C. Baird
Dartmouth College

ABSTRACT

A model is proposed to describe the decision process engaged in by subjects making judgments on a trial-by-trial basis under the procedural constraints imposed by the experimenter in magnitude estimation tasks. The chief assumption is that subjects attempt to maintain rank order among successive responses that match rank order among stimulus intensities. The implications of the model were explored by comparing the output of a computer program written to simulate magnitude estimation with empirical results produced by a single subject who judged the loudness of a 1000-Hz tone (Green, Luce, & Duncan, 1977). The model mimicked the empirical data very well in terms of a variety of dependent measures. However, the assumption that subjects keep rank order among their responses on successive trials was not supported. As it stands, the theoretical approach is promising but incomplete.

INTRODUCTION

According to most standard accounts, direct scaling methods reveal the operation of fundamental sensory mechanisms (Stevens, 1975). The functioning of such mechanisms should be impervious to a wide array of procedural choices concerning how such judgments are collected. If the human being is considered a measuring instrument comparable to those in the physical sciences, then this instrument should be stable when applied to a variety of situations. Unfortunately, it has been known for some time that procedural variables such as the range of stimulus intensities, the available response options, and the position of a standard in the stimulus series all affect the outcome of such experiments (Poulton, 1989).

I doubt if context effects will ever be understood by continuing the search for sensory scales through the use of traditional scaling methods. This enterprise has deteriorated into a battle over the validity of psychological measurement, the identification of the true psychophysical law, and the relevance of linking psychophysics and neurophysiology. Unfortunately, there are no means available within psychophysics to decide these issues one way or the other. Perhaps it is time to explore alternative approaches that are not driven by the desire to establish sensory scales.

I am concerned with the decision process revealed by subjects making judgments on a trial-by-trial basis under the procedural constraints imposed by the experimenter through such variables as instructions, stimulus spacing, number of trials, and the use of a judgment standard. In considering previous approaches to psychophysics, the present effort is closest in spirit to the position stated over 30 years ago by Frank Restle (1961). In discussing the possibility of discovering a single psychophysical law, he states:

> It seems that the results of a psychophysical scaling experiment depend heavily on the psychophysical method used, and therefore the results may usefully be interpreted not as measures of "perceived magnitude," which lead to an input-output relationship, but rather as *judgments* which lead to an understanding of the process of judging stimuli. (Restle, 1961, pp. 207–208)

Throughout this chapter I will be treating psychophysical outcomes as judgments, and as such, dependent in large part on psychological factors that may be independent of the sensory processes involved.

THE JUDGMENT CONCEPT

The model under consideration rests on five assumptions: (1) the mapping between the mean of an underlying evidence variable and discrete stimulus intensities, (2) the evidence distributions, (3) the trial-by-trial constraints on these distributions that guide judgment, (4) the decision rules employed by subjects in making a judgment within the constraints, and (5) transformation of the judgment into an overt response.

Throughout this discussion it is important to bear in mind two general points. First, the evidence distribution is hypothetical and is not to be confused with the "observed" response distribution, based on actual performance in the experiment. Second, except for the postulation of overlapping evidence distributions arising from different stimulus intensities, the model is not Thurstonian in character. In particular, it is not the case here that responses are based on a comparison of the magnitude of two items of evidence, one taken from the evidence distribution associated with the present stimulus, the other taken from the evidence distribution associated with the previous stimulus. A different decision rule is invoked here that will be described below.

Stimulus-Response Mapping

The stimulus set in direct scaling experiments consists of intensities that are clearly distinguishable one from the other in a paired comparison test. In order to make this aspect even more prominent, I assume the sensory variance in such an experiment is zero or inconsequential. It is assumed further that the means of the evidence distributions are in the same rank order (on an equal-interval scale) as the stimulus intensities. Although rank order is maintained among stimuli, this does not necessarily mean that rank order is maintained among the responses, because of the presence of additional contextual factors that can influence both underlying judgments and overt responses. In the special application of the model presented here, it is assumed that subjects do not violate rank order in assigning responses to successive stimuli (However, it is possible for the same response to be given to successive stimuli which are of different intensity).

Although it is not a factor in the present application, different spacing of the stimulus intensities would not change the mapping into the means along the evidence axis. That is, unequal spacing between stimuli is not reflected in the judgment domain.

Evidence Distributions

The hypothetical evidence variable is normally distributed with standard deviation the same for each stimulus intensity. The psychological or physiological underpinnings of these distributions need not be specified further in order to incorporate them into a theoretical model, but it may help to think of them in the same light as one would the "noise" and "signal-plus-noise" distributions in the theory of signal detectability (Green & Swets, 1966). Since the spacing between the means is always the same (equal interval), the overlap of adjacent distributions is solely a function of the standard deviation.

Judgment Constraints

The chief constraint on the subject is the response given on the previous trial, though the model is generalizable to incorporate the impact of earlier responses. The influence of the previous response depends on where its corresponding judgment magnitude falls in respect to the evidence distribution associated with the current stimulus. In this respect there are two subconstraints.

First, a symmetric "decision band" is centered on the mean of the evidence distribution associated with the current stimulus. For the applications described here, this band is bounded by ± 1.5 standard deviations. If the response (as represented by the underlying judgment) given on the previous trial falls within the band, then it influences the current judgment; if the response (judgment) on the previous trial falls outside this band, it has no influence on the current judgment (response). This assumption is in the spirit of the "attention band" model proposed earlier to explain sequence effects (Luce, Baird, Green, & Smith, 1980).

Second, a "judgment window" exists within the decision band. The position of the judgment window depends on the psychophysical task, but it is closely tied to the location of the judgment selected on the previous trial and on the mean of the evidence distribution associated with the current stimulus. The actual judgment on the current trial is selected according to the probability density of the doubly truncated normal distribution defined by the judgment window. The truncation points are the lower (L) and upper (U) boundaries of the judgment window, which cannot fall outside of the lower and upper boundaries of the decision band. That is, the x variable (evidence) has a probability density function with mean ξ and standard deviation σ (Johnson & Kotz, 1970):

$$\frac{1}{\sqrt{2\pi}\sigma} e^{(x-\xi)^2/2\sigma^2} \left[\frac{1}{\sqrt{2\pi}\sigma} \int_L^U e^{(t-\xi)^2/2\sigma^2} \, dt \right]^{-1}.$$

Figure 13.1 illustrates what an hypothetical situation might look like on a single trial, showing the two normal distributions contingent upon presentation of two successive stimulus intensities, the previous response as represented in the judgment domain (J_{N-1}), and the judgment window (ω) along the x-axis with the appropriate probability density represented by the cross-hatching. In this model, ω is a free parameter. In the actual implementation of the model, each of these values is defined in standard deviation units, as opposed to the arbitrary units shown in the figure. On any given trial the judgment is selected according to the probability density in the judgment window.

Decision Rules

The key concept that distinguishes this approach from a simple Thurstonian model concerns the decision process on a given trial. This process is assumed to be different for different experimental tasks, as suggested by Restle (1961), but only one type of task will be examined here—that in which a subject gives values along a response continuum to match stimulus intensities (magnitude estimation, magnitude production, cross-modality matching). I refer to these tasks generically as "magnitude matching." As a specific example, in a magnitude matching experiment 10 subjects may give two judgments of each of seven stimuli. The

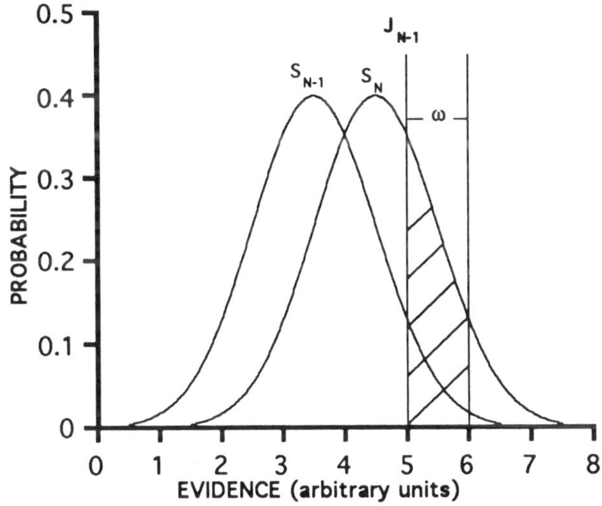

FIG. 13.1. Theoretical situation on a single trial of magnitude matching.

model holds that these subjects are attempting to rank order their response magnitudes in such a way as to match the rank order among the perceived stimulus intensities. Hence, the judgment window is located on a trial-by-trial basis to satisfy this constraint (as shown by example in Figure 13.1).

The decision rule works as follows. Let N = trial number. On the first trial, the judgment J_1 is selected according to the probability density of the decision band. For all subsequent trials, if J_{N-1} falls outside the decision band (not shown in the figure) for S_N, then J_N is selected from within this band according to its probability density. If J_{N-1} falls within the band for S_N, then J_N is selected from within the judgment window. Let ω be the width of the judgment window in standard deviation units, and let its lower (L) and upper (U) bounds be defined to keep rank order. Then

(i) If $S_N > S_{N-1}$, then $L = J_{N-1}$ and $U = J_{N-1} + \omega$.
(ii) If $S_N < S_{N-1}$, then $L = J_{N-1} - \omega$ and $U = J_{n-1}$.
(iii) If $S_N = S_{N-1}$, then $J_N = J_{N-1}$.

If by this procedure U is greater than the upper bound of the decision band, then U is made equal to the upper bound of the decision band; similarly, if L is less than the lower bound of the decision band, then L is made equal to the lower bound. No judgment can ever fall outside of the decision band appropriate for the stimulus on trial N.

Output Transformation

All of this decision making on the part of the subject is being carried out in the hypothetical judgment domain, which is not the same as the response domain. In order to produce an observable response, the judgment selected from the evidence distribution undergoes a transformation at the output stage. A single type of transformation is not appropriate for all experimental tasks. An exponential function is assumed to hold between the judgment and the response domains for the continuous-response case of magnitude production and cross-modality matching (cf. Birnbaum, 1982). For the computer simulation reported here, the following equation was used to transform judgments into responses:

$$R = .03(2)^{J_N}$$

For magnitude estimation, the judgments were mapped into the preferred number domain (Baird & Noma, 1975), details of which are described below.

COMPUTER SIMULATIONS

At present, analytic solutions to most aspects of the model are not available, so a computer program was written to simulate subjects' behavior in a variety of tasks. The main independent variables are the standard deviation, number of trials over which a response influences succeeding judgments, width of the decision band, and window size. The dependent measures are the mean and standard deviation of the responses actually emitted by the hypothetical subjects, as well as other statistical measures employed for such scaling applications (Luce et al., 1980). Since the logarithms of the stimulus intensities are always mapped into the same interval scale (defined by the evidence variable), the slope of the psychophysical function is arbitrary. In particular, the exponent of the power function varies with the range of stimulus intensities employed in the simulation.

The simulation was run under several experimental situations with 500 subjects making one judgment of each of seven logarithmically spaced stimuli. Different random orders of presentation were used for each subject, and data analyses were performed on a group level. One of the chief distinctions among tasks is between the continuous case of magnitude production and cross-modality matching and the discrete case of magnitude estimation. In general, it is easier to follow the rationale behind the continuous case, and such results will be presented first. However, the discrete case is the one most often studied in the laboratory, and when comparing the output of the model to empirical data, I will only consider magnitude estimation.

The output of the model mimicks the empirical data very well for many different conditions, and is especially successful in simulating a variety of context effects, such as the effect of stimulus range, position of the standard, and sequential dependencies. The major emphasis here is on sequence effects observed when individual subjects render a large number of judgments of the same set of stimulus intensities.

Simulation Results: Continuous Case

The logarithm of the geometric mean of 500 estimates is plotted as a function of the logarithm of the stimulus intensity in Figure 13.2 (top) for two standard deviations of the evidence distributions (.5 and 2). The width of the judgment window (ω in Figure 13.1) was set at one-half the standard deviation. The

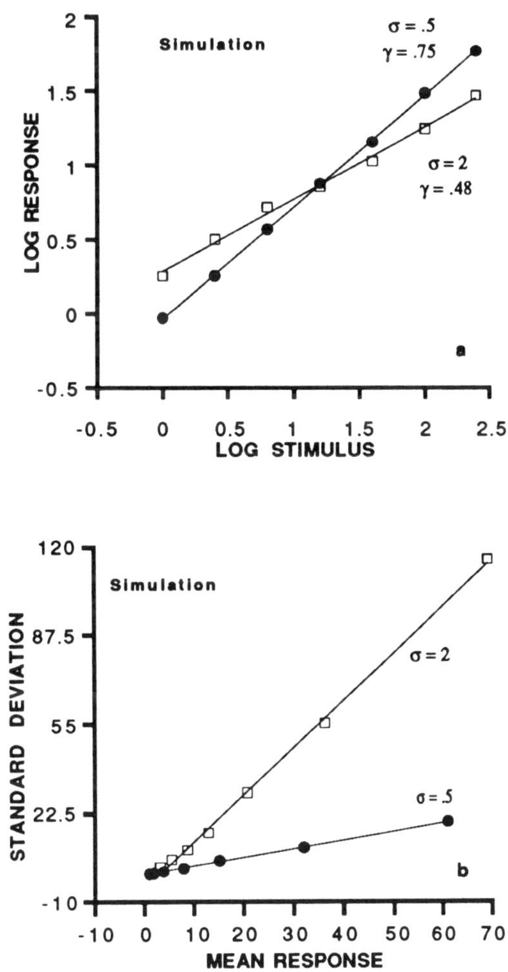

FIG. 13.2. Stevens functions (top) and Ekman functions (bottom) based on computer simulation with two standard deviations of the evidence distributions.

Stevens functions (Baird & Noma, 1978) are linear with slopes (exponents) of .75 for the small standard deviation and .48 for the large standard deviation. Figure 13.2 (bottom) presents the standard deviation of the responses as a function of the arithmetic mean. As expected, the Ekman functions (Olsson, Harder, & Baird, 1993) are linear with a steeper slope (more variability) for the evidence distribution possessing greater variability.

The slope of the Stevens function decreases with increasing standard deviation because the judgment window is bounded by the previous response located at the lower boundary of the window for the situation where $S_N > S_{N-1}$, and at the upper boundary when $S_N < S_{N-1}$. Since the pressure on responses for the extreme stimulus intensities (very weak or very strong) comes from the responses given previously to stimuli in the middle region of the series, this particular arrangement tends to pull down judgments for the more intense stimuli and pull up those for the weak stimuli. This can be thought of as an assimilation effect, whose degree is a function of the overlap among the evidence distributions. With greater overlap, the chances of having the previous judgment (response) fall within the decision band is greater; therefore, the greater the standard deviation, the more the assimilation, and the lower the exponent.

Since the model assumes no variance in the sensory representation, the source of this variability must be introduced elsewhere. One possibility is that stimulus attributes differ in respect to how well subjects are able to remember previous stimuli in a scaling experiment. According to this interpretation, Stevens exponents depend on differential memory effects rather than on differential transduction properties of the sensory systems.

Simulation Results: Group Effects

For magnitude estimation, the response alternatives are discrete, and, hence, the theoretical model must be altered to accommodate this fact. To do this, the simulation assigned a set of response values to the means of the evidence variable that were based on experimentation showing that subjects typically use numerical responses that are multiples of 5, 10, and 100 (Baird, Lewis, & Romer, 1970: Baird & Noma, 1975). In particular, the means of the seven evidence distributions (in the judgment domain) were mapped into the integer responses 15, 20, 25, 50, 75, 100, and 150. In addition, the distributions were truncated at the means of the smallest and largest stimuli, based again on the observation that subjects often self-select a minimum and maximum response that serve as actual limiting boundaries. The consequence of this variation in the model is that the only responses that can occur are these seven integer values. Details of the results (though not the overall patterns) would of course change if a different response set were employed.

A second variation in the model was that instead of applying a judgment window over a continuous interval of the evidence scale, it was necessary to deal with discrete alternatives. This was accomplished by using the following decision rule to establish the boundaries for a judgment, where all judgment values are integers:

(i) If $S_N > S_{N-1}$ then $L = J_{N-1}$ and $U = J_{N-1} + 1$.
(ii) If $S_N < S_{N-1}$ then $L = J_{N-1} - 1$ and $U = J_{N-1}$.

The relative probabilities of the two response options were then standardized according to their respective probability density within the truncated normal distribution (with truncation points L and U). This is merely a procedure to discretize the available judgment options. Therefore, in this simulation it is possible for the same response to be given to successive stimuli of different intensity, but it is impossible for rank order to be violated.

An example of a sequence effect in magnitude estimation is given in Figure 13.3 (top), based on a study conducted at Stockholm University (Baird, Berglund, Berglund, & Lindberg, 1991). Here, 12 subjects employed magnitude estimation (with a standard called "100" at 68 dB) to judge the loudness of tones (3 repititions of each of 12 intensities). A stimulus was not allowed to follow itself in the series. Each point on the graph indicates the momentary exponent based on the two responses given to adjacent stimulus intensities (ascending and descending order). A similar, though less robust, pattern emerges when pairs of successive stimuli are taken which are further apart in intensity.

The momentary exponent depends on the absolute stimulus intensities involved and on the relative intensities of successive stimuli. On ascending trials for weak stimulus intensities, the exponent is close to 0, whereas for the same pair of intensities, the exponent is between .4 and .5 for descending trials. The effect is not as pronounced at the upper end of the stimulus continuum. Here, the ascending trials produce higher exponents than the descending trials. Briefly, these findings show that when the second of two successive stimuli is close to the ends of the stimulus series, then the exponent tends to be higher than the reverse situation in which the second stimulus is located closer to the center of the series.

In conducting a simulation for these conditons, I sought to capture the pattern in the data rather than the details, by employing the rank-order rule described above. The order of stimulus presentation and location within the stimulus series has a marked effect on the empirical exponent, as is clear in the simulation data shown in Figure 13.3 (bottom).

The reason for these latter effects is tied closely with the truncation at the smallest and largest response magnitude. To understand why this is so, consider the response to the weakest stimulus intensity. The evidence distribution for this stimulus is truncated at the mean, and therefore, only half the normal distribution (the half greater than the mean) is available. Thus, it is almost always true that when the weakest stimulus is presented it falls within the decision band of

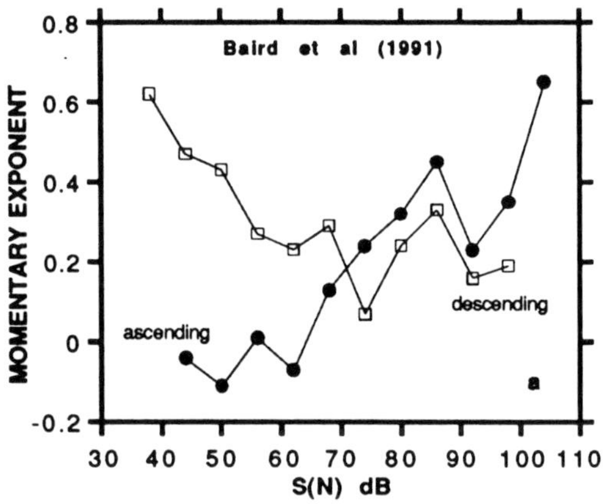

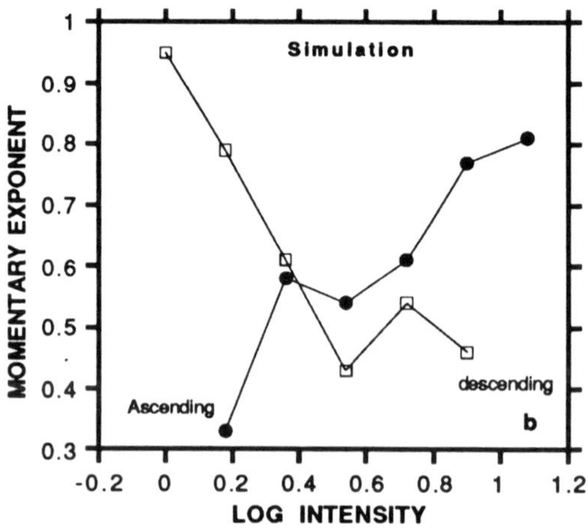

FIG. 13.3. Momentary exponent for adjacent pairs of stimuli as a function of log intensity on trial *N* for experimental (top) and simulation (bottom) data.

stronger, nearby stimuli and, hence, tends to pull the response for these stimuli downward. This creates a low momentary exponent, based on the responses to this single pair of stimuli. In the reverse situation, when the weakest stimulus follows one slightly stronger, the previous response does not fall within the decision band of the weakest stimulus as often because its evidence distribution is broader and less affected by the truncation at the bottom of the continuum. On those occasions when the previous response falls outside the decision band of the weakest stimulus, no assimilation takes place. This has the effect of producing a smaller ratio between the response on trial N to the weaker stimulus and the response on trial $N - 1$ to the stronger stimulus. Hence, the momentary exponent is relatively large. The opposite effect occurs at the upper end of the stimulus series.

Simulation Results: Individual Subjects

Further insight into the cause of sequence effects can be obtained by modeling the trial-by-trial behavior of individual subjects who give multiple estimates of each stimulus intensity. Green et al. (1977)[1] had each of five subjects give magnitude estimates (without a standard) of the loudness of a 1000-Hz tone of 500-msec duration which varied in intensity from 40 to 80 dB in 2-dB steps. Subjects reported their estimates by typing integers into a specially designed response box.

Unlike the situation in the Baird et al. (1991) study, the same stimulus could follow itself in the series. Three subjects gave 50 judgments of each intensity; two other subjects gave 100 judgments. Not all subjects behaved in the same fashion, and therefore, it is necessary to treat each data set separately. In particular, one subject (LT) performed the task with greater reliability than the others, and committed the fewest violations of rank order among responses to successive stimuli. The model described above to handle group sequence effects can be applied directly to the behavior of this subject, whereas the model must be modified to accommodate the behavior of the other four subjects. Here I only consider the data of LT (see Baird, 1990, for a somewhat different analysis of data averaged across all five subjects). The only modification of the model necessary here is when $S_N = S_{N-1}$, for which it is assumed that $R_N = R_{N-1}$.

No attempt was made to actually fit the empirical data; rather, the goal was to capture the general character of the results by assuming that the subject was behaving according to the judgment model.

Psychophysical Functions

The Stevens function is shown in Figure 13.4 (top) for the empirical data and in Figure 13.4 (bottom) for the simulation (based on 500 estimates of seven stimuli, randomized as a single block of 3500). As indicated on the graph, the simulation used a standard deviation of 1.5, and the evidence distributions were truncated at the means assigned to the weakest ($lz = 0$) and strongest ($uz = 0$) intensities. A straight line fits the data points very well in both instances, indicating the adequacy of the power law. The exponent is approximately .3. There is evidence of a slight sinuosity in both functions, a result that may be due to the subject's tendency to use a relatively small number of "preferred" responses (Baird & Noma, 1975; Teghtsoonian, Teghtsoonian, & Baird, 1994).

Response Variability

The variability of the responses for both LT and for the simulation is depicted in Figure 13.5, where the standard deviation is plotted against the arithmetic

[1]I am grateful to the authors for providing me with these data.

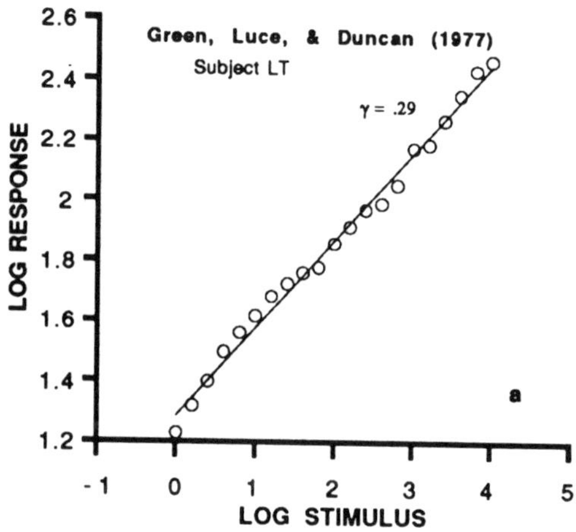

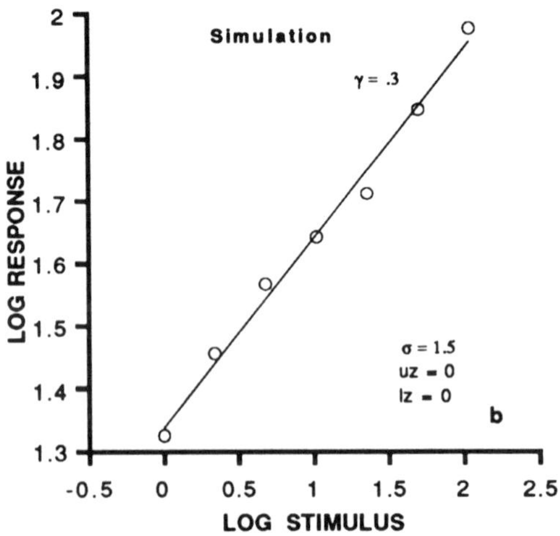

FIG. 13.4. Stevens functions for a single subject: top (empirical data), bottom (simulation).

mean. In both instances (Figures 13.5 (top) and 13.5 (bottom)) the trend of the data is linear over most of the range, but is definitely nonlinear overall. A quadratic equation was fit to the data points merely to highlight this overall nonlinearity. The departure from linearity in the simulation (as compared to the continuous case shown in Figure 13.2) arises because of the truncation at the upper end of the response continuum. This suggests that a similar truncation was present in the empirical data, and detailed examination of the data show that LT used a small set of responses when estimating the loudness of the maximum stimulus (80 dB).

The quadratic form is quite common in studies of both magnitude and cate-

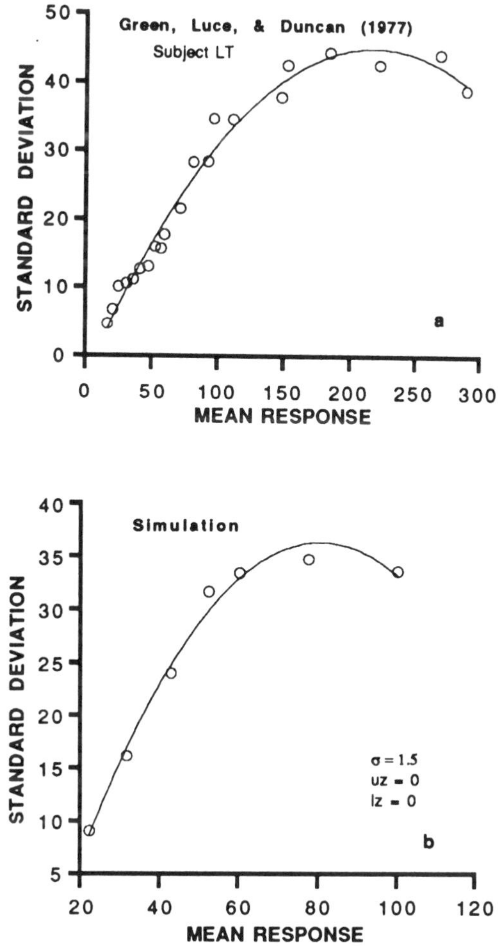

FIG. 13.5. Ekman functions for a single subject: top (empirical data), bottom (simulation).

gory estimation, as amply documented by Montgomery (1975). The reason for this is obvious for category estimation, since the truncation of the response scale at the minimum and maximum categories is an inevitable consequence of the experimental procedure. A similar truncation can occur in magnitude estimation; presumably due to the inclination of some subjects to set self-imposed boundaries on their response scale.

Triangle Pattern

Figure 13.6 shows the correlation between responses on trial N and trial $N - 1$ as a function of stimulus separation in dB. The correlation is highly positive

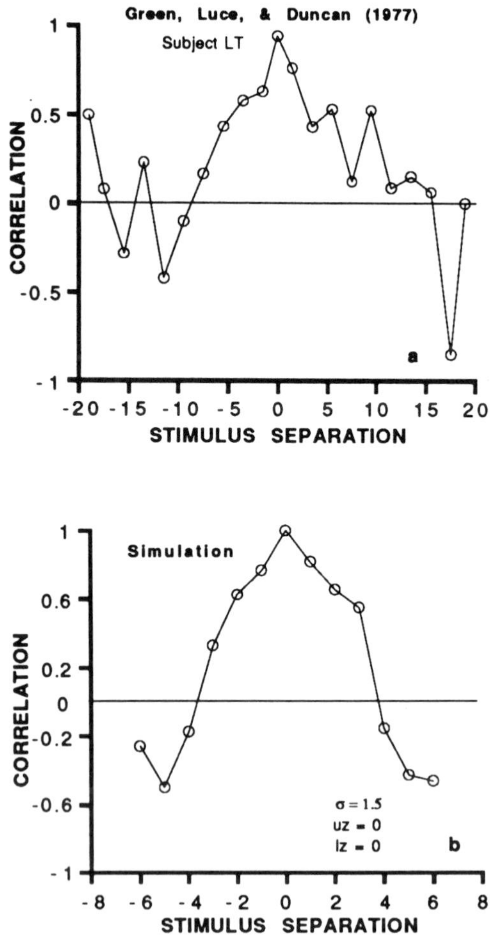

FIG. 13.6. Triangle pattern of correlations for a single subject: top (empirical data), bottom (simulation).

when stimuli on successive trials are similar in intensity, whereas the correlation approaches zero as the separation between successive intensities increases and is negative for several of the extreme separations. (To obtain a nonspurious triangle pattern, calculations must be performed for each stimulus pair in the series and then averaged over pairs having the same separation; Green et al., 1977). The pattern produced by the simulation (Figure 13.6, bottom) is similar in character to that seen in the empirical data (Figure 13.6, top), including negative correlations at the extreme separations.

The positive correlations occur because of the overlap among the evidence distributions together with the fact that a response is selected from a judgment window bounded by the previous response on one side and plus or minus one integer step (on the interval scale) on the other side. This leads to assimilation of R_N to R_{N-1}, and the degree of assimilation, as reflected in the magnitude of the correlation, depends on the amount of overlap. When two successive stimuli are identical, $R_N = R_{N-1}$, so the correlation is 1 for this condition.

The negative correlations at the extremes are produced in the simulation as follows. First, it is necessary to have considerable variability in the evidence distributions, such that even the extreme intensities yield responses that occasionally fall within the decision band of the evidence distributions at the opposite end of the judgment continuum. As a specific example, consider the situation when the weakest intensity in the series follows the strongest. Whenever the response (judgment) on trial $N - 1$ (strong) falls within the decision band of the evidence distribution on trial N (weak), it is likely that R_{N-1} will be at the upper edge of the decision band (a relatively large z-score in respect to the mean of the distribution for S_N). Hence, the response to the weakest stimulus (N) will be drawn strongly upward. On the other hand, when R_{N-1} does not fall in the decision band of S_N (which will usually happen), R_N will be selected from the entire evidence distribution and, therefore, will on average be closer to the mean. In sum, a very small value of R_{N-1} leads to a relatively large value of R_N, and all other values of R_{N-1} lead to relatively lower values of R_N. This produces a negative correlation. The slight upturn of the simulation function for separations -5 and -6 is not always obtained in simulations with different random orders of stimulus presentation.

Rank Order

One of the major implications of the present model is that subjects should maintain rank order between responses given to successive stimuli. Figure 13.7 presents such data for subject LT where the percentage of trials on which rank order was maintained is plotted against the intensity of the stimulus on trial N. Clearly this subject does keep rank order on a very high percentage of trials. When looked at from the standpoint of the dB separation on successive trials (not shown here), it is seen that most violations of rank order occur for the smallest stimulus separations, though even in these instances, the percent correct is always greater than 50%. This is not true for the other four subjects in the Green et al. (1977) study, and, as mentioned earlier, a modification of the present model is required to handle these data (Baird, 1990).

IMPLICATIONS

The judgment model has implications for experimentation, statistical analysis, and analytic description. Concerning scaling experiments, the model emphasizes the decision making strategies of the subject on a trial-by-trial basis. Any vari-

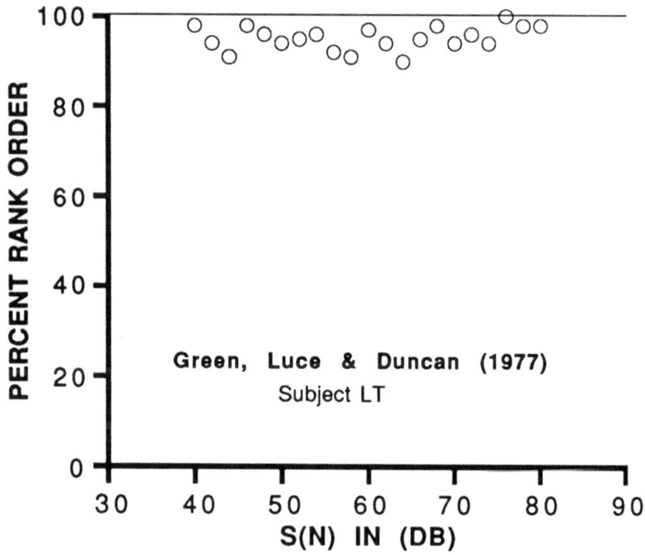

FIG. 13.7. Percent rank order maintained on successive trials as a function of the intensity of the stimulus on trial *N*.

able that affects this decision process can be presumed to affect the standard psychophysical parameters, such as the exponent of the Stevens function, or the shape of the Ekman function. In addition, any experimental variable that affects the ability of subjects to recall stimuli over trials will have an impact on judgments. The most obvious variable in this regard is the time between trials. By increasing the time between stimulus presentation one probably increases the memory variance and, hence, reduces the exponent of the Stevens function (Graf, Baird, & Glesman, 1974). Another effective variable might be one that impacts on the subject's ability to maintain rank order among successive responses appropriate for representing successive stimulus intensities. Such variables can be expected to alter the exponent.

On a statistical level, the approach stresses the variance of the evidence (judgment) distributions and, consequently, directs attention to variability measures. In particular, the shape of the correlation triangle will depend on the underlying variability. The greater the overlap among the evidence distributions, the greater we can expect the correlation to be between responses given on successive trials.

A second statistical implication concerns the truncation of the evidence distributions. Here too, one could look for truncation in the response distributions, though this is not as straightforward as it might appear, because the observed result (e.g., skewness) will depend both on the mapping between the underlying normal distributions at the level of the evidence variable and on the spacing between the response alternatives.

Finally, the model suggests that the percent of rank-order violations be routinely checked in scaling studies. In the past, it seems not to have been appreciated that a change in exponent could be due either to the fact that subjects have difficulty keeping rank order or to a genuine difference in the perceived magnitude of individual stimuli.

It would be advantageous to have an analytic description of the model's predictions, but it is by no means obvious how to procede in this respect. The mean and standard deviation of doubly truncated normal distributions is known (Johnson & Kotz, 1970), and this helps. However, the sticky problem does not seem to lie with the statistics but with the enumeration of the possible chains of sequential stimulus presentations and each of their judgment consequences.

REFERENCES

Baird, J. C. (1990). Modeling sequence effects among many magnitude estimates. In F. Müller (Ed.), *Proceedings of the Sixth Annual Meeting of the International Society for Psychophysics.* Würzburg, Germany.

Baird, J. C., Berglund, B., Berglund, U., & Lindberg, S. (1991). Stimulus sequence and the exponent of the power function for loudness. *Perceptual & Motor Skills, 73,* 3–17.

Baird, J. C., Lewis, C., & Romer, D. (1970). Relative frequencies of numerical responses in ratio estimation. *Perception & Psychophysics, 8,* 358–362.

Baird, J. C., & Noma, E. (1975). Psychophysical study of numbers: I. Generation of numerical responses. *Psychological Research, 37,* 281–297.

Baird, J. C., & Noma, E. (1978). *Fundamentals of scaling and psychophysics.* New York: Wiley.

Birnbaum, M. H. (1982). Controversies in psychological measurement. In B. Wegener (Ed.), *Social attitudes and psychophysical measurement.* Hillsdale, NJ: Lawrence Erlbaum.

Graf, V., Baird, J. C., & Glesman, G. (1974). An empirical test of two psychophysical models. *Acta Psychologica, 38,* 59–72.

Green, D. M., Luce, R. D., & Duncan, J. E. (1977). Variability and sequential effects in magnitude production and estimation of auditory intensity. *Perception & Psychophysics, 22,* 450–456.

Green, D. M., & Swets, J. A. (1966). *Signal detection theory and psychophysics.* New York: Wiley.

Johnson, N. L., & Kotz, S. (1970). *Continuous univariate distributions: 1. Distributions in statistics.* New York: Wiley.

Luce, R. D., Baird, J. C., Green, D. M., & Smith, A. F. (1980). Two classes of models for magnitude estimation. *Journal of Mathematical Psychology, 22,* 121–148.

Montgomery, H. (1975). Direct estimation: Effect of methodological factors on scale type. *Scandinavian Journal of Psychology, 16,* 19–29.

Olsson, M. J., Harder, K., & Baird, J. C. (1993). What Ekman really said. *The Behavioral and Brain Sciences, 16,* 157–158.

Poulton, E. C. (1989). *Bias in quantifying judgments.* Hillsdale, NJ: Erlbaum.

Restle, F. (1961). *Psychology of judgment and choice: A theoretical essay.* New York: Wiley.

Stevens, S. S. (1975). *Psychophysics: Introduction to its perceptual, neural and social prospects.* New York: Wiley.

Teghtsoonian, R., Teghtsoonian, M., & Baird, J. C. (1994). *On the nature and meaning of sinuousity in magnitude estimation functions.* Manuscript submitted for publication.

14 The Psychophysical Functions for Time Perception: Interpreting Their Parameters

Hannes Eisler
Stockholm University, Sweden

ABSTRACT

Given measurements of reproduced durations, plus a model for time perception, the psychophysical function for time was a power function with a break, or discontinuity. Accordingly it consists of two segments, and features five parameters: the exponent β, the ratio of the unit of the upper segment over the lower segment α, the subjective zeros $\Phi_{0\ell}$ and Φ_{0u} for the lower and the upper segments, respectively, and the position of the break, Φ_b. Upper and lower boundaries for the subjective zeros in terms of the other parameters, and the shortest and longest standard durations were derived under the assumption that the segments do not overlap. Differences in β and α could account for differences in reproduced durations observed with (1) low and high sound intensities (β); (2) men and women (α); (3) young and old (α); and (4) African immigrants and native Swedes (β), for which tentative explanations were given. The power function for duration derived for H. M., an individual with loss of long-term memory, was shown to deviate from functions for healthy subjects (segments overlap and Φ_{0u} is extremely large and negative).

This chapter deals with time perception—to be more specific, with subjective duration—from a quantitative point of view. It blends empirical results with theory that can both be deduced from the results and derived mathematically. There are three parts: (1) a brief summary of earlier work, necessary for an understanding of the following presentation, (2) mathematically derived restrictions of the parameters as a consequence of the empirical finding of a discontinuity or break in the psychophysical function, and (3) a description of the effect of stimulus and group differences in terms of differing parameter values.

THE PARALLEL-CLOCK MODEL

Type of Experiments

The method used throughout is reproduction of durations. Compared to numerical scaling, reproduction data not only show less scatter (Block, 1989, 1990; Zakay, 1990, 1993) but also make the construction of a subjective scale possible without the subjects using numerals, thereby avoiding possible number biases.

A trial in a reproduction experiment consists of presenting a standard duration, e.g., of a sound. After a short pause, experienced as an interruption, the sound resumes and is terminated by the subject when she judges that the sound after the interruption has lasted as long as the standard.

The Psychophysical Function

H. Eisler (1974, 1975, 1989) and H. Eisler and Eisler (1991) showed that a linear relation between set ratios and standards in a ratio-setting task, as is typically the case, entails a power function as psychophysical function, for both standards and reproductions, with the same exponent β. The unit κ and the subjective zero Φ_0 may differ between standard and variable. The equation of the psychophysical function is

$$\Psi = \kappa(\Phi - \Phi_0)^{\beta}, \qquad (14.1)$$

where Ψ denotes subjective duration and Φ physical (clock) time.

The Parallel-Clock Model

The parallel-clock model differs from the common conception of ratio setting or reproduction in that it requires no memory in which to store the standard duration. Instead, the model posits two sensory registers, which continuously accumulate subjective time units. The first register starts accumulating at the onset of the standard duration and stops when the subject terminates the variable duration (the reproduction). Thus, it accumulates the subjective counterpart of the *total* duration, standard + reproduction. The second sensory register accumulates the subjective duration of the reproduction. At the point in time when the difference between the contents of the two registers equals the contents of the second register, the subject experiences the two successive durations as equal and terminates the reproduction. Figure 14.1 makes this clear.

A consequence of the parallel-clock model is that what for the subject is reproduction—that is, equal setting—is, from the point of view of the researcher, halving; this makes it possible to construct a scale of subjective duration:

$$\Psi_r = \tfrac{1}{2}\Psi_t, \qquad (14.2)$$

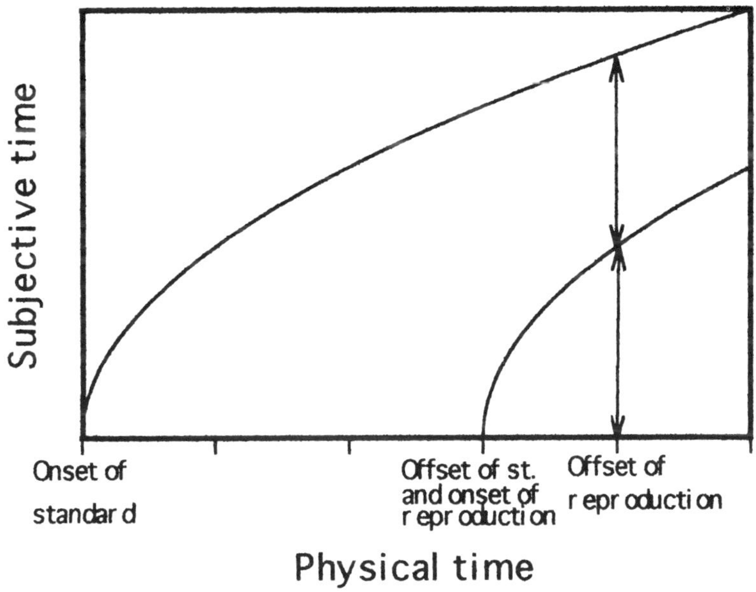

FIG. 14.1. Duration reproduction according to the parallel-clock model. Subjective versus total physical duration (left curve) and versus reproduction duration (right curve). When the difference between these two subjective durations (upper arrow) equals the subjective reproduction duration (lower arrow), the subject reports equality between standard and reproduction by shutting off the sound. (From *Psychophysics in Action* (p. 13) by G. Ljunggren and S. Dornic, 1989, Berlin: Springer. Copyright 1989 by Springer. Reprinted by permission.)

where subscript r refers to reproduced and subscript t to total subjective time. Applying Equation 14.1 yields

$$\kappa(\Phi_r - \Phi_0)^\beta = (\alpha/2)(\Phi_t - \Phi_0)^\beta, \quad (14.3)$$

where $\Phi_t = \Phi_s + \Phi_r$; Φ_s is the standard duration. Solving Equation 14.3 for Φ_r shows that Φ_r is a linear function of Φ_t. The κ's cancel, and since Φ_s is known and Φ_r empirically obtained, parameters β and Φ_0 can be determined; see, e.g., Eisler (1975).

The model is also applicable to ratios other than 1 (i.e., reproduction; see Eisler, 1975, 1976), but for these ratios the advantage of avoiding numbers is lost.

The Break in the Psychophysical Function

Figure 14.2 shows the outcome of a duration reproduction experiment for one subject. Consider first the lower panel, which shows the psychophysical function

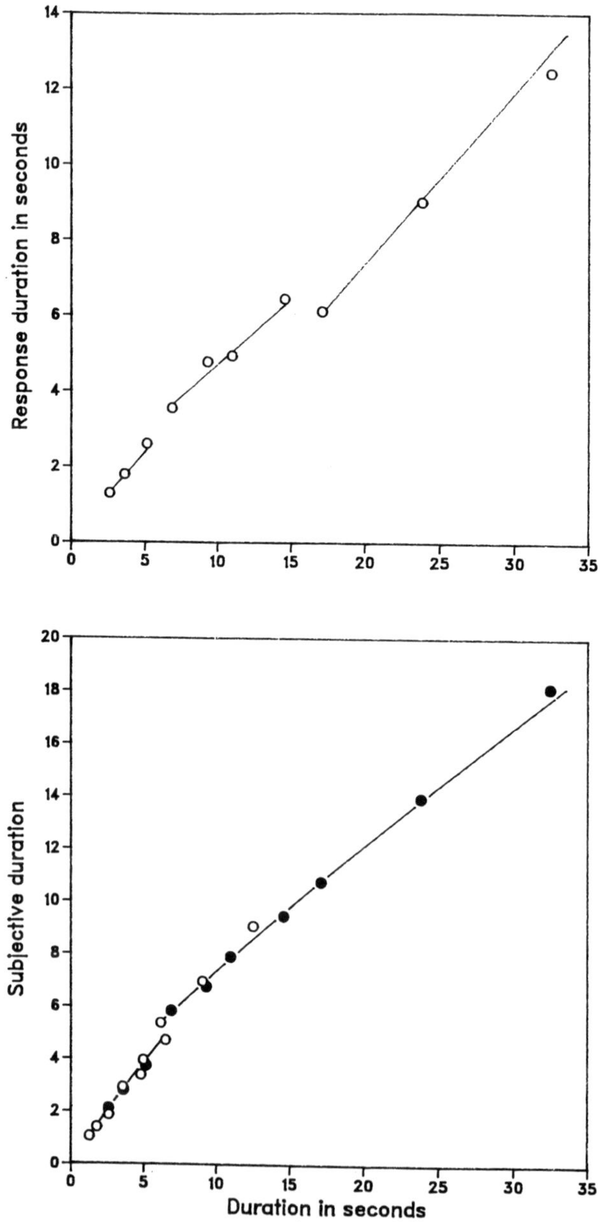

FIG. 14.2. (Upper panel) Reproduced versus total duration; (lower panel) the psychophysical function (subjective versus physical duration) for one subject. ● denote total durations and ○ reproductions in the lower panel. Note that the break at about 6 s (lower panel) corresponds to a break at the same duration at both the abscissa and the ordinate in the upper panel.

for duration with the break at about 6 s, dividing the function into two segments. The experimental points are linked pairwise: filled symbols correspond to total durations and empty symbols to reproductions. Remember that reproductions are subjectively half of total durations. For example, the subjective duration for the longest total duration is about 18 subjective duration units (the uppermost filled symbol), and for the corresponding reproduction about 9 units (the uppermost empty symbol). The data split into three groups: (1) data for both total duration and reproduction lie on the upper segment (the uppermost three pairs of points), (2) data for both total duration and reproduction lie on the lower segment (the lowermost three pairs of points), and (3) the data for total duration lie on the upper and those for reproduction lie on the lower segment (the middle four pairs of points). This partition into three groups is shown in the raw data plot, where reproductions are plotted against total durations as three straight lines with clear breaks or discontinuities between them (see upper panel of Figure 14.2). Note that the two breaks (upper panel) agree with the break of 6 s of the psychophysical function (lower panel); the lower break in the upper panel occurs at an abscissa of 6 s (total duration), and the upper break occurs at an ordinate of 6 s (reproduction). The two outer lines must be parallel, since the exponent β is the same throughout. The deviations in the slope of the middle line express a difference in the units of the two segments of the psychophysical function; see Equation 14.4b.

Assuming the break to be at Φ_b, we have to replace Equation 14.3 by the following three equations, with different subjective zeros, $\Phi_{0\ell}$ and Φ_{0u}, for the lower and the upper segment, respectively, and in which α denotes the ratio of the unit or weight of the upper segment over the lower:

$$(\Phi_r - \Phi_{0\ell})^\beta = {}^1\!/\!{}_2(\Phi_t - \Phi_{0\ell})^\beta, \qquad \Phi_t < \Phi_b, \qquad (14.4a)$$

$$(\Phi_r - \Phi_{0\ell})^\beta = (\alpha/2)(\Phi_t - \Phi_{0u})^\beta, \qquad \Phi_r < \Phi_b < \Phi_t, \qquad (14.4b)$$

$$(\Phi_r - \Phi_{0u})^\beta = {}^1\!/\!{}_2(\Phi_t - \Phi_{0u})^\beta, \qquad \Phi_r > \Phi_b. \qquad (14.4c)$$

There is no overlap between the two segments, a fact that gives rise to the restrictions described in the next section. Other examples of breaks can be found in H. Eisler (1990). An exceptional subject (brain-damaged individual), showing overlap, will be considered later. It should also be pointed out that certain raw data sets show only two straight lines. In these cases, the experimental range is probably too small, so Equation 14.4a or 14.4c is not applicable.

RESTRICTIONS ON THE PARAMETERS OF THE PSYCHOPHYSICAL FUNCTION

The restrictions to be described emerge from the condition that the two segments of the psychophysical function for duration do not overlap. The function is supposed to be strictly one to one, though, exceptionally, an empirical data set may show a small overlap (H. Eisler, 1990). The restrictions to be derived concern the subjective zeros, $\Phi_{0\ell}$ and Φ_{0u}, assuming the values of the other parameters are known.

Restrictions on $\Phi_{0\ell}$

Replacing Φ_t by $\Phi_s + \Phi_r$ in Equation 14.4a and solving for Φ_r, we obtain

$$\Phi_r = \frac{(1/2)^{1/\beta}}{1 - (1/2)^{1/\beta}} \Phi_S + \Phi_{0\ell} > 0. \tag{14.5}$$

To simplify the expression, we set $(1/2)^{1/\beta} = k < 1$, and $k/(1 - k) = K > 0$. Then $\Phi_{0\ell} > -K\Phi_s$, and, denoting the shortest standard duration used by Φ_1,

$$\Phi_{0\ell} > -K\Phi_1. \tag{14.6}$$

It may be noted that, in contrast to the general psychophysical function (Equation 14.1), where Φ_0 has to be less than Φ, $\Phi_{0\ell}$ has no upper limit, since, as can be seen in Equation 14.5, a large value of $\Phi_{0\ell}$ entails a still larger value of Φ_r.

Taking the break into account, one more inequality can be obtained. Replacing Φ_r by $\Phi_t - \Phi_s$, Equation 14.4a yields

$$\Phi_t = \frac{\Phi_s}{1 - k} + \Phi_{0\ell} < \Phi_b. \tag{14.7}$$

It is not possible to know the largest value of Φ_s that does not violate the condition that $\Phi_t < \Phi_b$. However, we are on the safe side when, again replacing Φ_s by Φ_1, we combine Equations 14.6 and 14.7, giving

$$-K\Phi_1 < \Phi_{0\ell} < \Phi_b - \frac{\Phi_1}{1 - k}. \tag{14.8}$$

The limits on $\Phi_{0\ell}$ are rather wide. For a numerical example with comparatively narrow limits, see, e.g., subject F2, at 55 dB SL, in Eisler and Eisler (1992, Table 3). In this case, $\Phi_1 = 1.3$ s, $\beta = 0.59$, and $\Phi_b = 3.2$ s. $\Phi_{0\ell}$ equals 1.07 s, which is in accordance with Equation 14.8: $0.58 < 1.07 < 1.3$.

Restrictions on Φ_{0u}

Solving Equation 14.4c for Φ_r yields

$$\Phi_r = K\Phi_s + \Phi_{0u} > \Phi_b; \tag{14.9}$$

that is,

$$\Phi_{0u} > \Phi_b - K\Phi_s. \tag{14.10}$$

Again, to play safe, let Φ_s equal the longest empirical duration used, denoted by Φ_m, and thus

$$\Phi_{0u} > \Phi_b - K\Phi_m. \tag{14.11}$$

In order to simplify the use of Equation 14.4b, set $(\alpha/2)^{1/\beta} = v < 1$, and $v/(1 - v) = V > 0$. Solving for Φ_r yields

$$\Phi_r = V\Phi_s - V\Phi_{0u} + \frac{\Phi_{0\ell}}{1 - v} < \Phi_b, \tag{14.12}$$

whence

$$\Phi_{0u} > \Phi_s - \frac{\Phi_b}{V} + \frac{\Phi_{0\ell}}{v}, \tag{14.13}$$

and setting $\Phi_s = \Phi_1$

$$\Phi_{0u} > \Phi_1 - \frac{\Phi_b}{V} + \frac{\Phi_{0\ell}}{v}. \tag{14.14}$$

Neither of the right-hand sides of inequalities 14.11 and 14.14 is consistently greater than the other; thus both equations are kept.

The fact that Φ_r has to be positive can also be used in Equation 14.4b (cf. Equation 14.12):

$$\Phi_r = V\Phi_s - V\Phi_{0u} + \frac{\Phi_{0\ell}}{1 - v} > 0, \tag{14.15}$$

leading to

$$\Phi_{0u} < \Phi_s + \Phi_{0\ell}/v, \tag{14.16}$$

and, setting $\Phi_s = \Phi_1$,

$$\Phi_{0u} < \Phi_1 + \Phi_{0\ell}/v, \tag{14.17}$$

Finally, rewriting Equation 14.4b as

$$\Phi_t - \Phi_s - \Phi_{0\ell} = v(\Phi_t - \Phi_{0u}), \tag{14.18}$$

i.e.,

$$\Phi_t = \frac{\Phi_s + \Phi_{0\ell}}{1 - v} - V\Phi_{0u} > \Phi_b, \tag{14.19}$$

yields

$$\Phi_{0u} < \frac{\Phi_s}{v} + \frac{\Phi_{0\ell}}{v} - \frac{\Phi_b}{V}. \tag{14.20}$$

For $\Phi_s = \Phi_1$,

$$\Phi_{0u} < \frac{\Phi_1}{v} + \frac{\Phi_{0\ell}}{v} - \frac{\Phi_b}{V}. \tag{14.21}$$

It is easy to show that the right-hand side of Inequality 14.17 is larger than the right-hand side of Inequality 14.21, and thus only Inequality 14.21 is retained.

In order to get rid of $\Phi_{0\ell}$ the restrictions given in Equation 14.8 are introduced into Inequalities 14.14 and 14.21, and at last the following three restrictions on Φ_{0u} are obtained:

$$\Phi_{0u} > \Phi_b - K\Phi_m, \tag{14.22}$$

$$\Phi_{0u} > \left(1 - \frac{K}{v}\right)\Phi_1 - \frac{\Phi_b}{V}, \tag{14.23}$$

$$\Phi_{0u} < \left(1 - \frac{1}{v(1 - k)}\right)\Phi_1 + \frac{\Phi_b}{v} \tag{14.24}$$

Considering the same subject as before (F2 in Eisler & Eisler, 1992), adding the lacking parameter values ($\alpha = 1.20$, $\Phi_m = 20$ s), and knowing that $\Phi_{0u} = 2.06$, we get $-5.74\ (-4.49) < 2.06 < 4.44$.

The Relation Between Φ_{0u} and α

Two experiments (A. D. Eisler & Eisler, 1994; H. Eisler & Eisler, 1992) reported correlations of about .8 between the parameters Φ_{0u} and α. An increasing function was to be expected, since solving Equation 14.4b for Φ_{0u} yields

$$\Phi_{0u} = \Phi_t - (2/\alpha)^{1/\beta}(\Phi_r - \Phi_{0\ell}). \tag{14.25}$$

A plot of values obtained from these two experiments, with 96 data points in all, showed surprisingly little scatter (see Figure 14.3). This finding may mean either that there is an empirical relation between these two parameters, implying that one can be predicted from the other for each subject individually, or that the finding is a statistical artifact, holding only for a family of functions.[1] To decide which interpretation is correct, a simulation of 100 psychophysical functions was carried out, based on the SDs from one subject. The plot of the 100 pairs of Φ_{0u} and α agreed rather closely with Figure 14.3, and thus the hypothesis of a possible empirical connection between these parameters had to be abandoned.

Conclusion

For any one data set, the values of the subjective zeros $\Phi_{0\ell}$ and Φ_{0u} are restricted if the parameters β and α and the position of the break, Φ_b, are fixed. These

[1]Compare correlation obtained between intercept and slope of a family of straight lines fitted individually for data from, say, a number of subjects. Such a correlation is a statistical artifact.

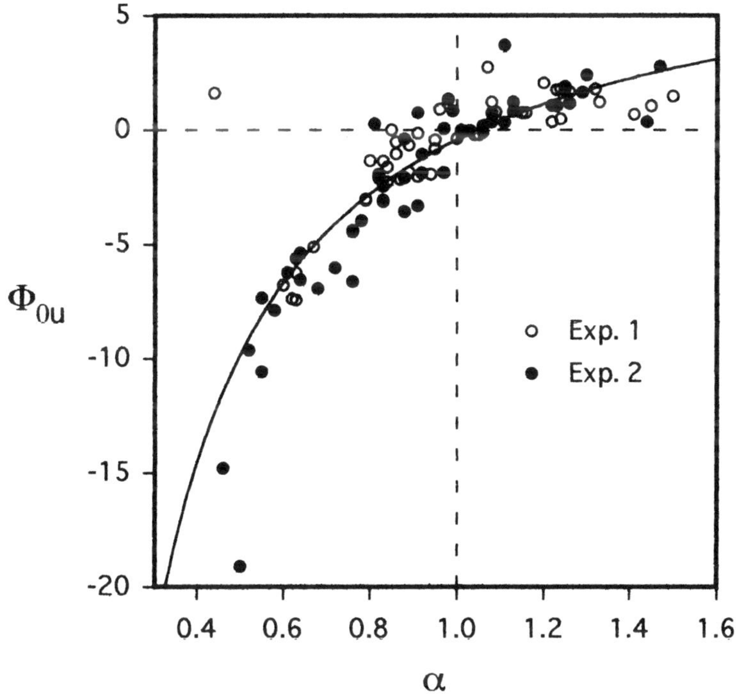

FIG. 14.3. Φ_{0u} plotted versus α for 96 data sets, showing an apparent hyperbolic function. Note that the curve passes closely through the coordinates $\alpha = 1$, $\Phi_{0u} = 0$, corresponding to no break.

restrictions pertain to a psychophysical power function with a break, as derived from the parallel-clock model. It would be possible, but more complicated, to fix the values of the subjective zeros and seek restrictions in the other parameters. The latter procedure was approximated, when the relation between Φ_{0u} and α was investigated over 96 data sets, revealing an approximately hyperbolic function, which apparently resulted from stochastic variation within the mathematical constraints in the model.

SIGNIFICANCE OF THE PARAMETER VALUES

Two studies already mentioned (A. D. Eisler & Eisler, 1994; H. Eisler & Eisler, 1992) examined effects on time perception of the following variables: (1) sound intensity, (2) gender, (3) age, and (4) Type A–Type B behavior. Furthermore, in a study by A. D. Eisler (1992, in press) time perception was compared between Africans living permanently in Sweden and native Swedes. This study included seven female and seven male subjects in each group, matched in age.

Statistically significant differences appeared in all raw data. These differences could clearly be attributed to differences in the values of the two parameters β (the exponent of the power function), and α (the ratio between the units of the upper over the lower segment of the function).

Higher sound intensity entailed smaller values of β. Perhaps most interesting was the finding that gender affected α, with male subjects showing a lower value. This gender difference was found in all three experiments; also African males had a lower value of α than African females. Age also affected α, younger subjects of both sexes having lower values. Further, African subjects reproduced standard durations shorter than did native Swedes of the same sex, a result that could be attributed to lower values of β. Finally, there was no difference in the raw data or between parameter values in Type A and Type B subjects.

Several interpretations might be offered for these results. The variations in parameter values probably express effects of psychological or biological (neurophysiological) variables, or both.

Vroon (1975) and Vroon, Timmers, and Tempelaars (1977) claimed that timekeeping ability is better in the left brain than right brain. If in fact time perception is somewhat lateralized, such lateralization could result in gender differences and hence in turn could explain the obtained differences in α.

As regards the age differences, besides a biological explanation, one can hypothesize that older people have gathered more experiences, and thus have developed a richer and larger network of associations, compared to younger people. Older people may therefore be exposed to more associations per unit clock time, entailing a greater value of α (A. D. Eisler, 1993).

Concerning the lower value of β found at higher sound intensities, one can speculate that this is a consequence of a higher frequency of neural firing, perhaps through heightened arousal. A more psychological interpretation would state that the change in β reflects a greater disturbance caused by noise.

Finally, a biological difference might underlie the difference between African and Swedish subjects' values of β. Alternatively, the difference might be cultural; or perhaps assimilation to a new culture, combined with loss of own culture ("detribalization"), creates acculturative stress, which in turn could influence time perception, perhaps physiologically.

To wind up, I present duration reproductions obtained from the brain-damaged individual, H. M., collected by Richards (1973) and processed according to the parallel-clock model. Surgery was performed on H. M. in 1953 in order to relieve incapacitating nonfocal epileptic seizures. The anterior hippocampus, the hippocampal gyrus, uncus, and amygdala were removed bilaterally, resulting in almost complete loss of long-term memory (Scoville & Milner, 1957). Figure 14.4 shows a conspicuous abnormality: a clear overlap between the two segments of the psychophysical function, and a value of Φ_{0u} of about -50 s. The latter implies that, for long durations (the upper segment), H. M. reproduced durations as if they had begun 50 s before they actually started. But even in this case at least two interpretations are possible: The findings might reflect a direct

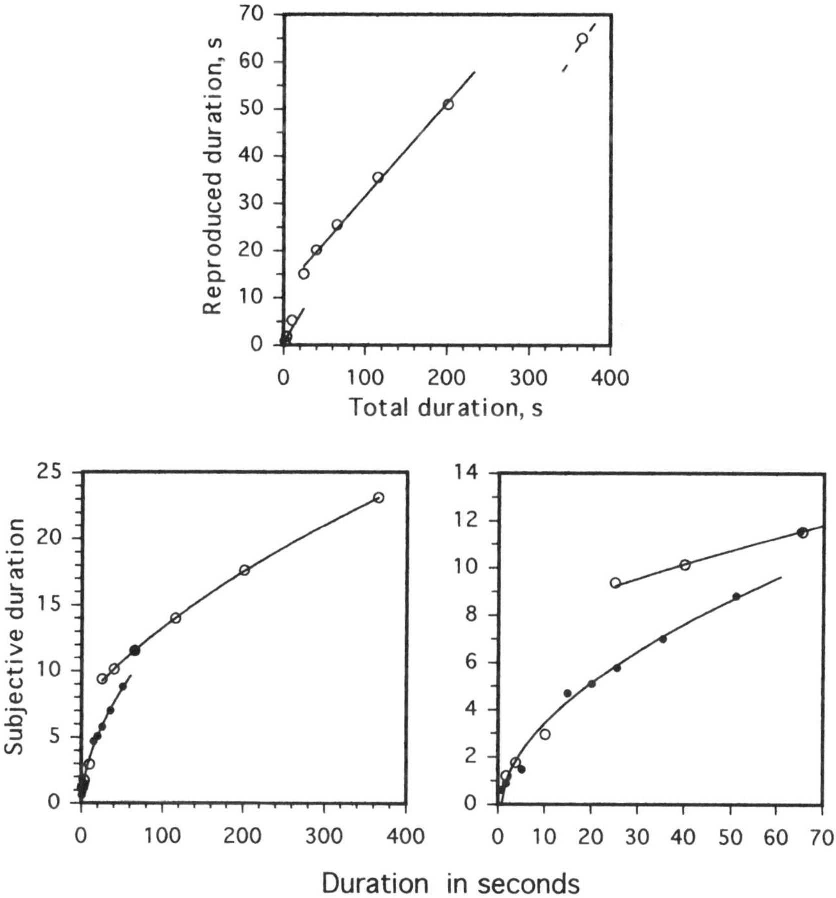

FIG. 14.4. Data for H. M. (from Richards, 1973), processed according to the parallel-clock model. (Upper panel) Reproductions plotted versus total duration. (Lower panels) The psychophysical function, where ○ denotes total and ● reproduced duration. The right-hand panel is a blown-up part of the left-hand panel for shorter durations, demonstrating the overlap of the two segments.

result of the injury, or a kind of coping, based on knowledge of an impaired memory.

Concluding Remark

I have described the evidence that experimental as well as group differences in the perception or reproduction of durations can be attributed to differing parameter values of the psychophysical power function. All of the interpretations of these values that I have offered, whether psychological or biological, are speculative. Nevertheless, the experimental findings, analyzed through the parallel-clock model, may help eventually to bridge the gap between psychology proper and neurophysiology.

ACKNOWLEDGMENTS

This investigation was supported by the Swedish Council for Research in the Humanities and Social Sciences.

REFERENCES

Block, R. A. (1989). Experiencing and remembering time: Affordances, context, and cognition. In I. Levin & D. Zakay (Eds.), *Time and human cognition: A life-span perspective* (pp. 333–363). Amsterdam: North-Holland.

Block, R. A. (1990). Models of psychological time. In R. A. Block (Ed.), *Cognitive models of psychological time* (pp. 1–35). Hillsdale, NJ: Erlbaum.

Eisler, A. D. (1992). Time perception: Reproduction of duration by two cultural groups. In S. Iwawaki, Y. Kashima, & K. Leung (Eds.), *Innovations in cross-cultural psychology* (pp. 304–310). Amsterdam: Swets & Zeitlinger.

Eisler, A. D. (1993). *Time perception: Theoretical considerations and empirical studies of the influence of gender, age, and culture on subjective duration.* Stockholm: Akademitryck.

Eisler, A. D. (in press). Cross-cultural differences in time perception: Comparison of African immigrants and native Swedes. In G. Neely (Ed.), *Perception and psychophysics in theory and application.* Stockholm: HSFR.

Eisler, A. D., & Eisler, H. (1994). Subjective time scaling: Influence of age, gender, and type A and type B behavior. *Chronobiologia, 21,* 185–200.

Eisler, H. (1974). The derivation of Stevens' psychophysical power law. In H. R. Moskowitz, B. Scharf, & J. C. Stevens (Eds.), *Sensation and measurement* (pp. 61–64). Dordrecht, Holland: Reidel.

Eisler, H. (1975). Subjective duration and psychophysics. *Psychological Review, 82,* 429–450.

Eisler, H. (1976). Experiments on subjective duration 1868–1975: A collection of power function exponents. *Psychological Bulletin, 83,* 1154–1171.

Eisler, H. (1989). Data-equivalent models in psychophysics: Examples and reflections. In G. Ljunggren & S. Dornic (Eds.), *Psychophysics in action* (pp. 11–24). Berlin, FRG: Springer.

Eisler, H. (1990). Breaks in the psychophysical function for duration. In H.-G. Geissler, M. H. Müller, & W. Prinz (Eds.), *Psychophysical explorations of mental structures* (pp. 242–252). Göttingen, Germany: Hogrefe & Huber.

Eisler, H., & Eisler, A. D. (1991). A mathematical model for time perception with experimentally obtained subjective time scales for humans and rats. *Chronobiologia, 18,* 79–88.

Eisler, H., & Eisler, A. D. (1992). Time perception: Effects of sex and sound intensity on scales of subjective duration. *Scandinavian Journal of Psychology, 33,* 339–358.

Richards, W. A. (1973). Time reproductions by H. M. *Acta Psychologica, 37,* 279–282.

Scoville, W. B., & Milner, B. (1957). Loss of recent memory after bilateral hippocampal lesions. *Journal of Neurology, Neurosurgery, & Psychiatry, 20,* 11–21.

Vroon, P. A. (1975). *On the hemispheric representation of time* (Tech. Rep. No. 10). Utrecht, Netherlands: University of Utrecht, Psychological Laboratory.

Vroon, P. A., Timmers, H., & Tempelaars, S. (1977). On the hemispheric representation of time. In S. Dornic (Ed.), *Attention and performance. VI* (pp. 231–245). Hillsdale, NJ: Lawrence Erlbaum.

Zakay, D. (1990). The evasive art of subjective time measurement: Some methodological dilemmas. In R. A. Block (Ed.), *Cognitive models of psychological time* (pp. 59–84). Hillsdale, NJ: Lawrence Erlbaum.

Zakay, D. (1993). Time estimation methods—do they influence prospective duration estimates? *Perception, 22,* 91–101.

15 Scaling Semantic Domains

A. Kimball Romney
William H. Batchelder
Tim Brazill
University of California, Irvine

ABSTRACT

The aim of this chapter is to present a method for scaling judged similarity judgments, among items in a semantic domain, from many subjects into a single representation. We present a single spatial representation that contains scaled information on where each of 125 subjects locates 21 animals. Data were collected from subjects using two formats, namely triadic comparisons and paired comparison ratings. The method serves three main purposes. First, it enables one to comprehend and examine very large data sets that would otherwise not be accessible in a single coherent view. Second, it allows one to describe and test comparisons among individuals and subgroups. Third, it provides an optimally aggregated representation that can be used to predict cognitive behaviors that relate to cognitive structure.

The aim of this chapter is to present a new approach to the scaling of homogeneous semantic domains. The method is illustrated on empirical data from the semantic domain of names of animals. The proposed scaling enables us to compare variations in both individual and group representations in a detailed and precise way that complements current practices. "A homogeneous semantic domain consists of a set of words (exemplars) that are all members of a superordinate category, such as fish, furniture, or vehicles" (Romney, Brewer, & Batchelder, 1993, p. 28). All of the words are on the same level of contrast. Thus, for example, if the semantic domain were "animals," it would include words such as horse, cow, cat, dog, etc., but not superordinate terms like "animal" nor subordinate terms like names of breeds of dogs or cats, etc.

Various lines of evidence suggest that homogeneous semantic domains are important psychologically. Neuropsychological studies have found that aphasic patients sometimes have selective impairment of specific semantic categories such as fish, vegetables, or animals. This suggests that semantic domains may be localized functional units in the brain (see, for example, Goodglass, Wingfield, Hyde, & Theurkauf, 1986; Warrington & Shallice, 1984; Hart, Berndt, & Caramazza, 1985; McCarthy & Warrington, 1988; Sartori & Job, 1988; Silveri & Gainotti, 1988; Warrington & McCarthy, 1987). Semantic domains also appear in Chan, Butters, Paulsen, Salmon, Swenson, and Maloney, 1993, and Chan, Butters, Salmon, and McGuire, 1993, who are investigating the semantic structure of animals in studies of Alzheimer's and Huntingtons's diseases. Their results suggest a breakdown of semantic structure as well as deterioration of its accessibility.

Improvements in measuring an individual's cognitive representation of a semantic domain is important since such spatial representations are known to relate to a number of cognitive functions. For example, distances in such models have been shown to predict categorical judgment time (Caramazza, Hersh, & Torgerson, 1976; Rips, Shoben, & Smith, 1973; Shoben, 1976), completion of analogies (Rips et al., 1973; Rumelhart & Abrahamson, 1973), the strength of semantic clustering in memory (Romney et al., 1993), and reaction time to solve triadic comparison problems (Hutchison & Lockhead, 1977; Romney, 1989). The fact that a number of cognitive functions can be predicted so well from these representations illustrates their usefulness in cognitive science and the potential for their wider application. As Nosofsky (1992) says, "The beauty of deriving a similarity-scaling representation by modeling performance in a given task is that the derived representation can then be used to predict performance in independent tasks involving the same objects and stimulus conditions" (p. 26).

The semantic domain of animals was chosen because it has appeared as one of the most frequently used domains in psychological studies. Several of the studies mentioned in the previous paragraphs used the domain of animals (Caramazza et al., 1976; Chan, Butters, Paulsen, Salmon, Swenson, & Maloney, 1993; Chan, Butters, Salmon, & McGuire, 1993; Hutchison & Lockhead, 1977; Rips et al., 1973; Romney, 1989; Rumelhart & Abrahamson, 1973; Shoben, 1976). Many other occurrences of the use of the domain of animals can be found in the literature (e.g., Baker & Young, 1975; Cunningham, 1978; Friendly, 1979; Henley, 1969; Howard & Howard, 1977; Rips, 1975; Sattath & Tversky, 1977; Shepard, 1974; Smith, Shoben, & Rips, 1974).

The motivation for scaling a semantic domain is to provide spatial representations that serve three purposes. First, to provide an optimally aggregated representation to predict cognitive behaviors like those mentioned above. Second, to describe and test comparisons among individuals and groups. Such comparisons might focus on different data collection techniques, differences between experimental and control group (for example, Alzheimer's versus normals), or differences among individuals. Third, to enable one to comprehend and examine very large data sets that would otherwise not be accessible in a single coherent view.

The scaling of semantic domains obtained from data aggregated across subjects using nonmetric multidimensional scaling has been common since the classic study of Henley (1969) on the domain of animals. The most common model for the study of individual variations in semantic domains up to now has been INDSCAL (Carroll & Chang, 1970). The INDSCAL model provides an aggregate representation with fixed dimensions common to all subjects. Thus, though each subject has the same configuration of stimulus points as every other subject, they do vary in the extent to which they weight each dimension. In the method that we propose, no restrictions are imposed on the configuration for a given subject although each subject may be compared to a summary aggregated configuration, representing a common culture.

The scaling of semantic domains occupies an important role not only in cognitive theory but also in anthropological theory, since the central concept of culture can be defined by the shared elements of the individual semantic structures. Thus the scaled representations of semantic domains at the individual level may be thought of as individual cognitive maps, while at the aggregate group or societal level they may be thought of as representing shared cultural patterns.

DATA COLLECTION

Judged similarity data from a total of 125 university student subjects on the semantic domain of 21 animals were collected. Subjects were assigned to one of four different experimental groups. Two of the groups completed a triadic comparison task and two groups completed a 20-point rating scale task. The four groups are characterized as follows:

Group 1 (Triad 1). The same (each student received the same triads) lambda-one balanced incomplete block design with 70 triads (Weller & Romney, 1988), with order within and between triads individually randomized, was administered to 33 students. The task was to pick, for each triad, the animal most different from the other two.

Group 2 (Triad 2). A different (each student received a different set of triads) lambda-one balanced incomplete block design with 70 triads, individually randomized, was administered to 39 students. The task was to pick the animal most different from the other two.

Group 3 (Pairs 1). Each of 26 students rated the dissimilarity of each of the 210 pairs of animals on a scale of 1 to 20, with 20 as most dissimilar. Order within each animal pair was constant, but pairs were individually randomized.

Group 4 (Pairs 2). Design administered to 27 students identical to Group 3 except order within each animal pair was reversed.

Both triadic comparison groups received the following set of written instructions:

> Thank you for participating in this study. On the next page, you will find a set of three words on each line. For each set, please circle the word which is MOST DIFFERENT in meaning from the other two. For example, for the set
>
> HOUSE WOMAN BUILDING
>
> you would circle WOMAN, since it is the word most different in meaning. Here is another example:
>
> DOG CAT ROCK
>
> In this case, you would circle ROCK.
>
> Please give an answer for EVERY set of three, even if you are not sure of the answer. DO NOT SKIP ANY sets: if you don't know the answer, just guess. Thank you.

Both paired comparison groups were given the following set of written instructions, which were adapted from those used by Krantz and Tversky (1975) who also used a 20-point rating scale in their study of perceived dissimilarities among rectangles:

> JUDGMENTS OF DISSIMILARITY TASK
>
> In this experiment we will show you pairs of animals and we'll ask you to circle a number on a scale from 1 to 20, according to the degree of dissimilarity between the animals.
>
> For example: if the animals are almost identical, that is, the dissimilarity between them is very small, circle a low number. If the animals are very different from one another, circle a high number. In the same fashion, for all intermediate levels of dissimilarity between the animals, circle an intermediate number.
>
> We are interested in your subjective impression of degree of dissimilarity. Different people are likely to have different impressions. Hence, there are no correct or incorrect answers. Simply look at the animals for a short time, and circle the number which appears to correspond to the degree of dissimilarity between the animals. THANK YOU.

The animals included in the domain were the 21 most frequently named in a free recall task originally administered by Henley (1969) and were as follows: antelope, beaver, camel, cat, chimpanzee, chipmunk, cow, deer, dog, elephant, giraffe, goat, gorilla, horse, lion, monkey, rabbit, rat, sheep, tiger, zebra. For a more detailed discussion of both the triadic comparison and paired comparison rating scale methods see Weller and Romney (1988). Questionnaires were constructed with the software package ANTHROPAC (Borgatti, 1992).

The choice of tasks given to the different groups allows us to make various comparisons among groups to illustrate the possible utility of the methods employed. The primary comparison is between the tasks involving triadic comparisons and the paired comparison rating scales. Since the paired comparison task involves more items, 210 compared to 70, and more bits of information per item (defined as the log to the base 2 of the alternatives per question), approximately 4.32 compared to 1.58, we would expect to find the paired comparison task giving "better" results. The idea of computing and comparing the amount of information per item goes back to Coombs (1964, pp. 34ff).

Our method requires scaling the data in a way that represents all subjects in the same space. This allows comparisons among all possible subsets of subjects on the basis of comparable scales. Only then can we define what is meant by saying that one method is "better" than another.

ANALYSIS

In order to represent all subjects in the same space we use a form of multiple correspondence analysis (Gifi, 1990; Weller & Romney, 1990) in which data from all subjects are stacked into a single data matrix. Table 15.1 shows the first 21 and last 21 rows of such a data matrix that contains 2625 rows and 21 columns in total. Each 21 × 21 matrix representing a subject's data is symmetric and contains duplicate copies (one below the diagonal and one above the diagonal) of the coded data from the triadic and paired comparison rating tasks. The rows and columns of each matrix represent the 21 animals in alphabetic order. Since correspondence analysis assumes similarity data, the paired comparison rating task matrices have been reversed in direction, so 20 represents the most similar (rather than dissimilar) response (Weller & Romney, 1990, p. 70ff). Since correspondence analysis requires similarity data a "large" number is required on the diagonal (each item is maximally similar to itself). Accordingly, one has been added to the diagonals for subjects who performed the triadic comparisons and 20 to the diagonals for subjects who performed paired comparison ratings (Weller & Romney, 1990, p. 71).

The final data matrix, **A,** contains 2625 rows (125 subjects times 21 animals) and 21 columns (the 21 animals). We analyzed this data matrix with correspondence analysis. All calculations were carried out using ANTHROPAC 4.0 (Borgatti, 1992), and the figures were produced using SYGRAPH (Wilkinson, 1989). The four major steps that we followed are summarized in the following paragraphs.

First, the raw data matrix, **A,** with cell entries, a_{ij}, is normalized by computing a new matrix **H,** with the *ij*th cell entry given by

$$h_{ij} = \frac{a_{ij}}{\sqrt{a_{i.}a_{.j}}},$$

where i goes from 1 to 2625, j goes from 1 to 21, a_{ij} is the original cell frequency, $a_{i.}$ is the total count for row i, and $a_{.j}$ is the total count for column j. In matrix notation this may be written as

$$\mathbf{H} = \mathbf{R}^{-1/2}\mathbf{A}\mathbf{C}^{-1/2},$$

TABLE 15.1
Format for Stacking 125 Individual 21 × 21 Matrices to Produce a Single Data Matrix 2625 Rows by 21 Columns for Analysis

1	0	1	0	0	0	1	1	0	1	1	0	1	1	0	1	0	0	0	0	1
0	1	0	0	0	1	0	1	1	0	0	0	0	0	0	0	1	1	0	0	0
1	0	1	0	1	0	0	0	0	1	1	1	0	0	0	0	0	0	0	0	1
0	0	0	1	0	1	0	0	1	0	0	0	0	0	0	1	0	1	0	1	0
0	0	1	0	1	0	0	0	0	0	0	1	0	0	0	1	0	0	0	0	1
0	1	0	1	0	1	0	0	1	0	0	0	0	0	0	0	1	1	1	0	0
1	0	0	0	0	0	1	1	0	1	1	1	1	1	1	0	0	0	1	0	1
1	1	0	0	0	0	1	1	0	1	1	0	0	1	1	0	0	0	1	1	1
0	1	0	1	0	1	0	0	1	0	0	1	0	0	0	0	0	1	0	1	0
1	0	1	0	0	0	1	1	0	1	1	0	1	1	0	0	0	0	1	0	1
1	0	1	0	0	0	1	1	0	1	1	0	0	1	0	0	0	0	0	0	1
0	0	1	0	1	0	1	0	1	0	0	1	0	0	1	0	0	0	1	1	0
1	0	0	0	0	0	1	0	0	1	0	0	1	0	1	1	0	0	0	0	0
1	0	0	0	0	0	1	1	0	1	1	0	0	1	0	1	0	0	0	0	1
0	0	0	0	0	0	1	1	0	0	0	1	1	0	1	0	0	0	1	1	0
1	0	0	1	1	0	0	0	0	0	0	0	1	1	0	1	0	0	1	0	0
0	1	0	0	0	1	0	0	0	0	0	0	0	0	0	0	1	1	0	1	0
0	1	0	1	0	1	0	0	1	0	0	0	0	0	0	0	1	1	0	0	0
0	0	0	0	0	1	1	1	0	1	0	1	0	0	1	1	0	0	1	0	1
0	0	0	1	0	0	0	1	1	0	0	1	0	0	1	0	1	0	0	1	0
1	0	1	0	1	0	1	1	0	1	1	0	0	1	0	0	0	0	1	0	1

.
.

.

[123 other matrices with a total of 2583 rows]

.

.

.

20	6	18	10	7	5	13	16	7	10	16	16	10	16	16	7	6	2	14	13	15
6	20	5	15	8	18	4	8	15	5	4	6	12	8	7	12	16	15	12	5	6
18	5	20	6	9	5	12	13	8	16	17	16	7	16	14	10	8	4	12	16	18
10	15	6	20	14	15	6	13	16	5	5	12	14	7	13	10	17	16	11	15	6
7	8	9	14	20	11	9	11	14	8	13	9	14	7	10	19	5	8	12	5	7
5	18	5	15	11	20	6	8	11	6	6	7	8	5	5	14	14	9	6	2	7
13	4	12	6	9	6	20	14	13	15	13	16	10	18	13	11	5	7	17	13	13
16	8	13	13	11	8	14	20	13	12	16	15	10	16	16	9	11	3	15	16	15
7	15	8	16	14	11	13	13	20	7	6	15	15	12	11	13	13	16	15	11	10
10	5	16	5	8	6	15	12	7	20	13	6	12	15	16	9	7	5	9	8	14
16	4	17	5	13	6	13	16	6	13	20	13	9	13	13	7	4	6	8	11	15
16	6	16	12	9	7	16	15	15	6	13	20	7	14	16	10	10	9	17	13	16
10	12	7	14	14	8	10	10	15	12	9	7	20	6	8	19	11	6	5	12	7
16	8	16	7	7	5	18	16	12	15	13	14	6	20	15	9	6	6	16	14	19
16	7	14	13	10	5	13	16	11	16	13	16	8	15	20	3	9	1	17	18	16
7	12	10	10	19	14	11	9	13	9	7	10	19	9	3	20	5	7	7	7	11
6	16	8	17	5	14	5	11	13	7	4	10	11	6	9	5	20	17	11	4	5
2	15	4	16	8	9	7	3	16	5	6	9	6	6	1	7	17	20	9	5	8
14	12	12	11	12	6	17	15	15	9	8	17	5	16	17	7	11	9	20	14	16
13	5	16	15	5	2	13	16	11	8	11	13	12	14	18	7	4	5	14	20	16
15	6	18	6	7	7	13	15	10	14	15	16	7	19	16	11	5	8	16	16	20

where $\mathbf{R}^{-1/2}$ and $\mathbf{C}^{-1/2}$ are diagonal matrices whose entries consist of the reciprocals of the square roots of the row marginal totals and column marginal totals, respectively.

Second, the normalized matrix is analyzed by singular value decomposition (SVD), into its triple product, $\mathbf{UDV}^{\mathrm{T}}$, where $\mathbf{U}$ contains row scores, $\mathbf{V}^{\mathrm{T}}$ contains column scores, and $\mathbf{D}$ is a diagonal matrix of singular values.

Third, the vectors of the $\mathbf{U}$ and $\mathbf{V}^{\mathrm{T}}$ matrices are used to compute maximally discriminating scores (i.e., optimal scores, canonical scores, variates) for the rows and columns of $\mathbf{A}$. The rescaling formulas for the optimal scores are

$$\mathbf{X}_i = \bar{\mathbf{U}}_i\sqrt{a_{..}/a_{i.}} \qquad \text{and} \qquad \mathbf{Y}_j = \mathbf{V}_j\sqrt{a_{..}/a_{.j}}.$$

Fourth, a typical final step in correspondence analysis would weight the $\mathbf{X}_i$ and $\mathbf{Y}_j$ scores by the square root of the singular values. This allows comparisons between the $\mathbf{X}_i$ and $\mathbf{Y}_j$ configurations in a joint weighted Euclidean space. In the present case this would lead to comparing individual "pictures" of very different sizes due to differences in response variances among the various subjects. We are interpreting differences in scale (size) as artifacts of response variance and, therefore, without any further substantive interpretation. Since we want to compare among subjects we need to correct for these differences among subjects.

Correspondence analysis, itself, does not prescribe a standard transformation to bring scores within each subject to scale. Consequently, we devised a method of transforming row scores within each subject that would remedy the problem of scale on the assumption that each subject has the same relative distances in their representation of animals. Our solution is to correct for these differences in scale by standardizing the $\mathbf{X}_i$ scores for each subject to zero mean and variance equal to the square root of the singular values. For a related application of this procedure see Kumbasar, Romney, and Batchelder (1994) in a study of biases in social perception. Below we give comparisons between standardized and unstandardized scores in order to illustrate the effects of this crucial step.

RESULTS

The correspondence analysis of the $\mathbf{A}$ matrix shown in Table 15.1 results in 2625 row optimal scores, $\mathbf{X}_i$, in n dimensions and 21 column optimal scores, $\mathbf{Y}_j$, in n dimensions (we saved five dimensions with singular values of .451, 392, .313, .249, and .218, respectively, accounting for 43.5% of the variance). Illustrative plots are presented only in two dimensions. There are 125 row optimal scores for each animal, where each score represents one individual's placement of a given animal. Figure 15.1 shows a two-dimensional plot of all 2625 standardized row scores. In this form the plot is unreadable. However, by judicious choice of subsets of points to plot we can summarize and contrast any desired aspects of the data.

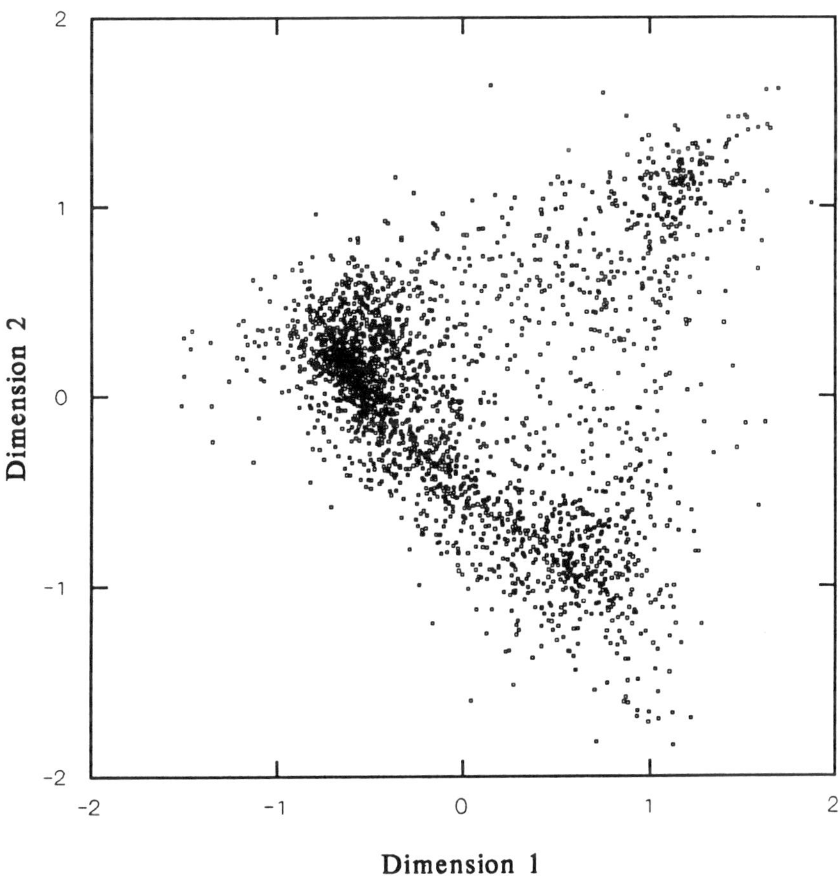

FIG. 15.1. Plot of first two dimensions of 2625 standardized row scores.

For example, we can plot the points for "antelope" for the Triads 1 group and compare them to the plot for "antelope" for the Pairs 1 group, as shown in Figures 15.2 and 15.3. In these plots we have added the 90% confidence ellipses about the means of the points calculated on the assumption of bivariate normal distributions. Note that the apparent spread of points in Figure 15.2 is greater than in Figure 15.3. This suggests that the data from the paired comparison ratings task have less variability than the triadic comparisons task. We will give a more detailed quantitative analysis of this difference below.

An aggregate global view of how all subjects placed each animal can be represented with 90% confidence ellipses on mean scores as shown in Figure 15.4. Each ellipse represents a summary of all 125 subjects' row scores. We

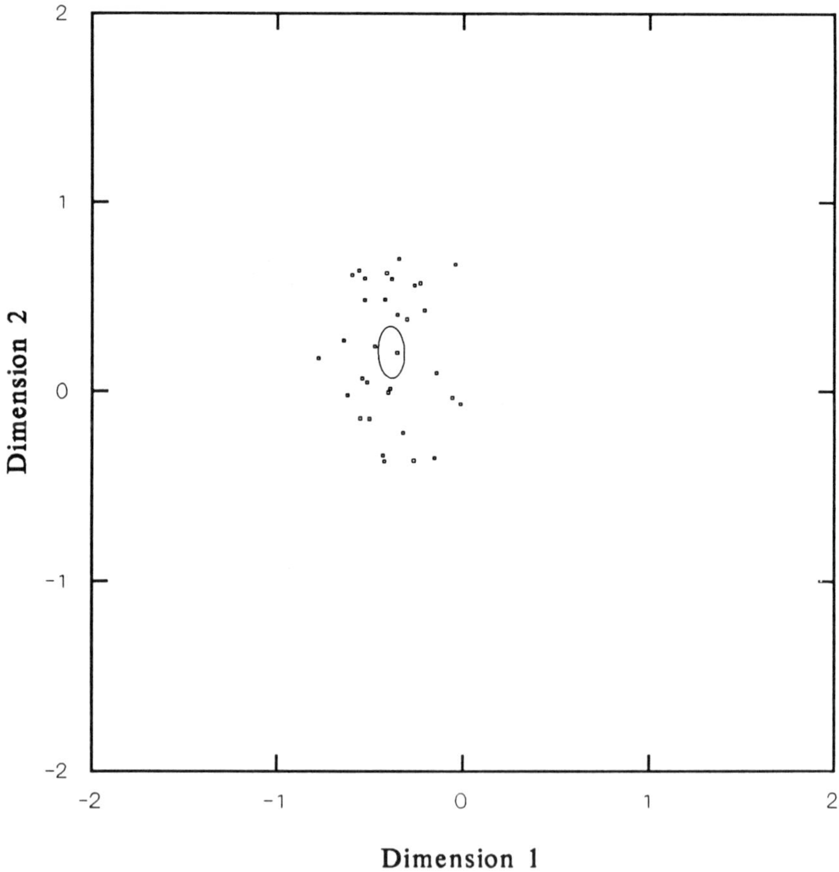

FIG. 15.2. Plot showing placement of antelope in two dimensions with 90% confidence ellipses of means of Triads 1.

present labels for the animals for this comprehensive view (subsequent plots will not show labels since they interfere with the clarity of the figures).

We could use similar confidence ellipse plots for each of the four experimental groups to get a visual idea of relative amounts of variability. Since individuals that received the same task are somewhat comparable in response variance we present the plots of Triads 1 and Pairs 1 groups as examples. The plots are shown in Figures 15.5 and 15.6. The confidence ellipses are noticeably smaller for the Pairs 1 group, reflecting the fact that the paired comparison rating task provides more information than does the triadic comparison task. The size of each ellipse gives us an idea of the "resolving power" of the method.

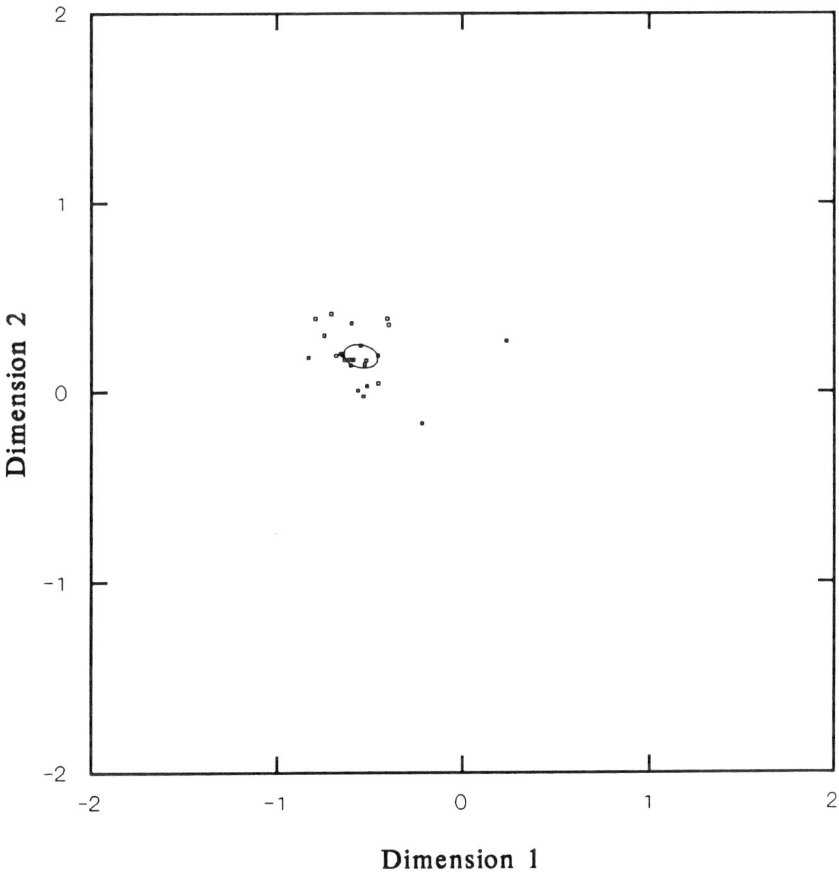

FIG. 15.3. Plot showing placement of antelope in two dimensions with 90% confidence ellipses of means of Pairs 1.

A more formal quantitative method for comparing the resolving power of one method compared to another, as in Figures 15.5 and 15.6, would require some index to measure how much of the variance for any comparison group is accounted for by the animal involved. One appropriate way to do this is to perform an analysis of variance, using "animal" as the grouping, on the row optimal scores, **X,** for each of the dimensions. From this we can calculate a proportion reduction in error (PRE) index that represents how well the subjects of a given group agree with each other in the placement of the animals. A sample calculation of the PRE index is shown in Table 15.2. The PRE is simply the sum of squares explained by the category of animal divided by the total sum of squares.

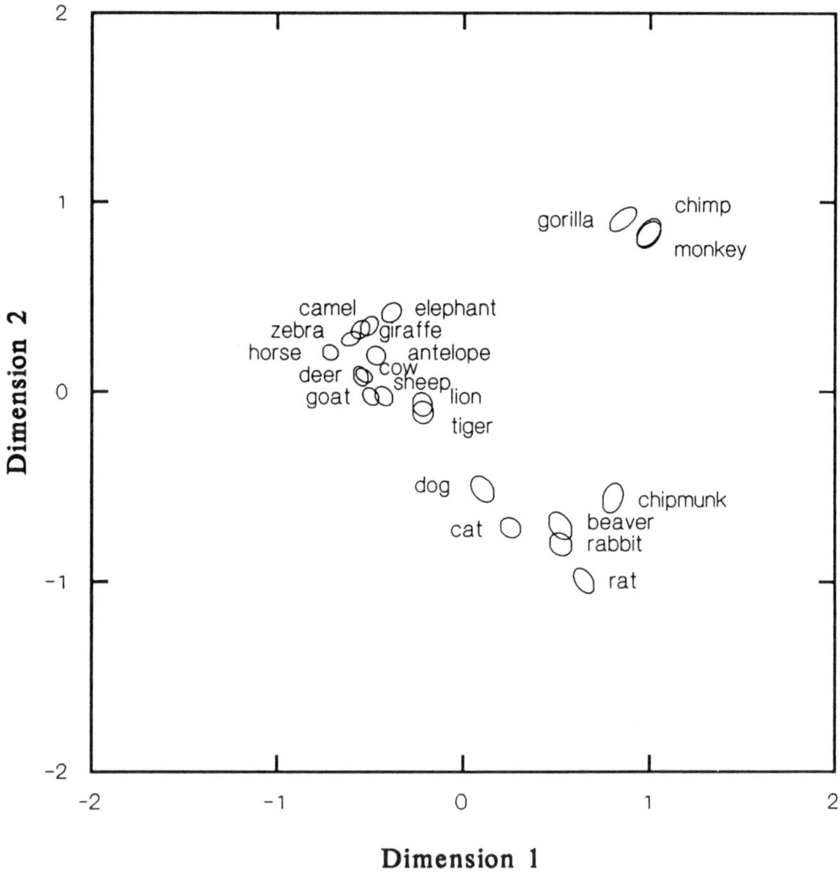

FIG. 15.4. Plot showing 90% confidence ellipses of means for all 125 subjects with animals labels.

For example, an examination of Figures 15.5 and 15.6, would lead us to expect that the Pairs 1 group of Figure 15.6 should have a higher PRE index than the Triads 1 group of Figure 15.5.

We can examine several dimensions with this PRE index in order to obtain an idea of how well each group agrees in the discrimination among the animals. Table 15.3 shows the PRE measures, both unstandardized and standardized, for the first five dimensions for each of the groups. The following characteristics of the results are immediately apparent. First, for every comparison, the paired comparison rating groups have higher PRE than do the groups that completed the triadic comparisons. We would argue that this indicates that the paired comparison rating task does provide more information and that the relative positions of

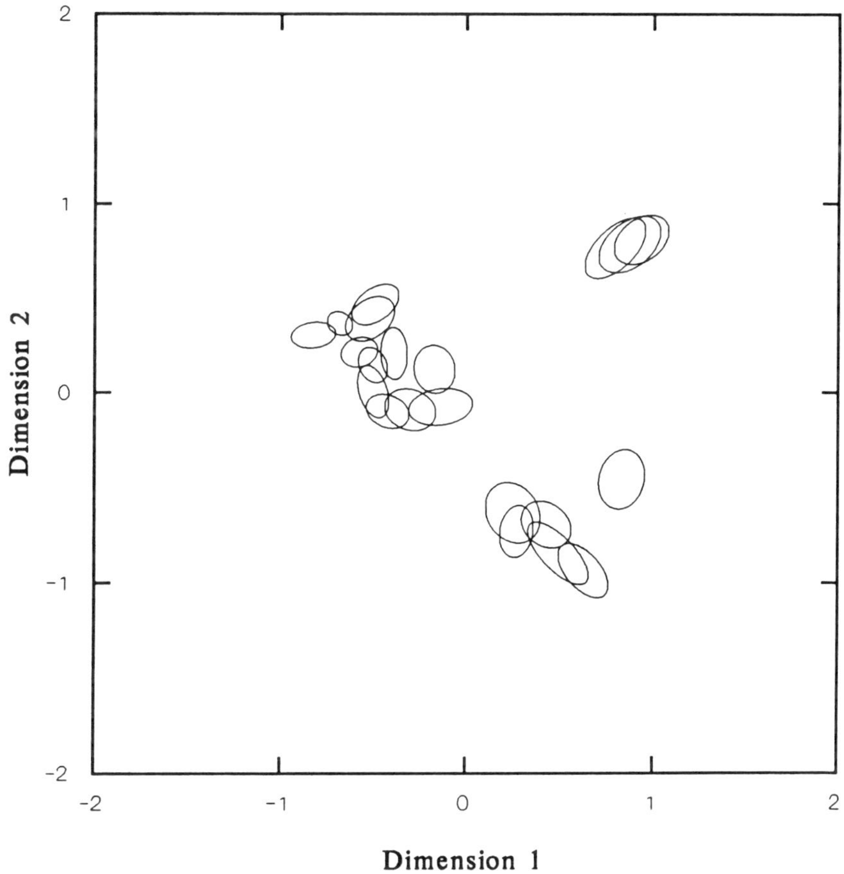

FIG. 15.5. Plot showing 90% confidence ellipses of means for Triads 1 subjects (N = 33).

the animals in the scaled semantic space has higher resolving power for the paired comparison ratings than in the case of the triadic comparisons. Thus, the subjects agree more with one another, which implies a more reliable and precise measurement.

Second, within the paired comparison rating groups the PRE measures computed for the standardized scores are higher than the PRE measures computed for the unstandardized scores. Since the ratings have more variability among subjects than the triadic comparisons, we would expect the standardization to have more effect in the ratings task than the triadic comparisons task. This expectation is consistent with the observed data.

Third, the Triad 1 group has higher PRE measures than the Triad 2 group on

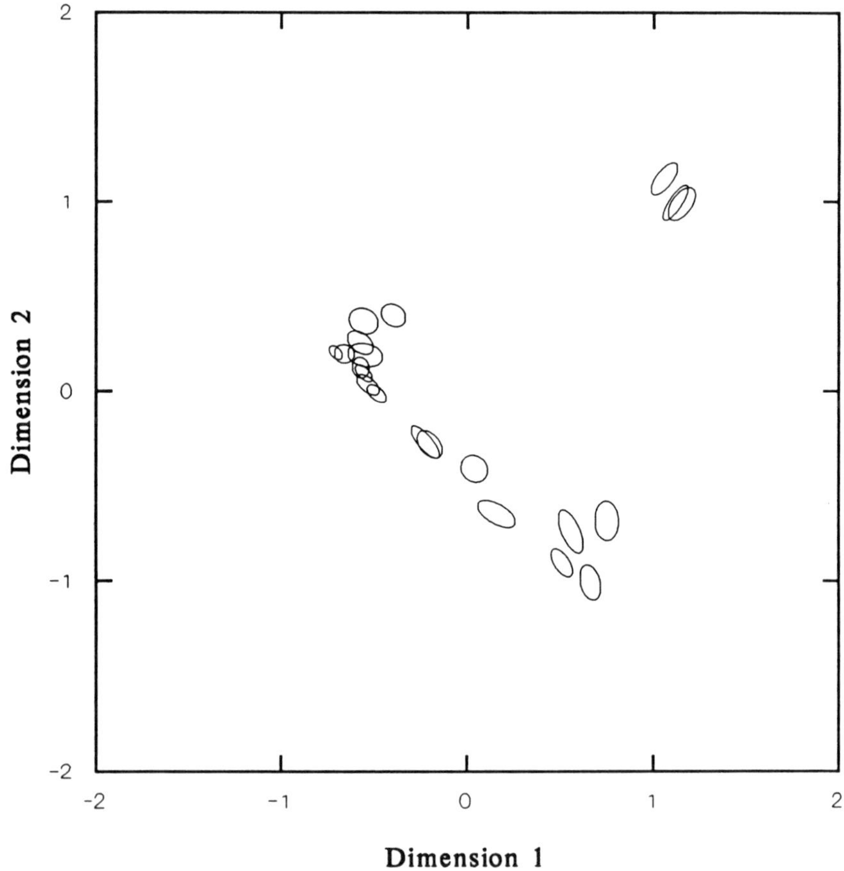

FIG. 15.6. Plot showing 90% confidence ellipses of means for Pairs 1 subjects (N = 26).

TABLE 15.2
Sample Calculation Showing How Proportion Reduction in Error Is Calculated from Analysis of Variance for Unstandardized Data on Dimension 1 for Experimental Group 1 (Triads)

Source	*df*	*Sum-of-Sqs.*	*Mean-Sq.*	*F Ratio*	*P*
Animal	20	613.190	30.660	122.753	0.000
Error	672	167.842	0.250		
		781.032			

$$PRE = \frac{Explained\ SS}{Total\ SS} = \frac{613.190}{613.190 + 167.842} = .785$$

TABLE 15.3
Showing Proportion Reduction in Error Based on a One-Way (by Animal) ANOVA in Four Experimental Groups for First Five Dimensions in Standardized (boldface) and Unstandardized Form

	Experimental Group			
Dimension	*Group 1 Triad 1*	*Group 2 Triad 2*	*Group 3 Pairs 1*	*Group 4 Pairs 2*
1	.785	.739	.875	.853
	.764	**.722**	**.950**	**.926**
2	.706	.624	.859	.809
	.738	**.674**	**.932**	**.889**
3	.615	.436	.754	.743
	.563	**.459**	**.880**	**.851**
4	.542	.261	.689	.717
	.532	**.274**	**.807**	**.810**
5	.374	.206	.627	.693
	.334	**.160**	**.758**	**.798**

every comparison. Again this result is expected in that the subjects in the Triad 1 group all received exactly the same triads (in random order) while in Triads 2 the subjects each received a different set of triads. The error variance should be greater for the group where subjects were responding to different sets of triads.

A very different perspective of the data can be obtained by examining each individual in comparison to an aggregate representation obtained by taking the mean of the scores for each animal across all subjects. This aggregate representation falls, by definition, at the midpoint of the ellipses shown in Figure 15.4. This representation is our best estimate of the shared cultural definition obtained, in effect, by aggregating across all the individual cognitive maps of the 125 subjects.

We examined the similarity of each of the 125 subjects to the aggregate representation. This was done by plotting, for each subject, the aggregate representation and the subject in the same space and connecting the two pictures with a vector for each animal. In general, we found that individuals were closer to the aggregate for the Pairs groups than for the Triads groups. To give an idea of the possibilities we have arbitrarily picked the first two individuals of the Triads 1 and the Pairs 1 groups to illustrate the information contained in such plots. Figures 15.7 to 15.10 compare each of the four individuals with the aggregate scores. The square symbols represent the mean scores for the animals in each plot, and they occur in the same position in each plot (they would fall in the middle of the ellipses in Figure 15.4). The unmarked end of the line represents where each animal is placed by the subject.

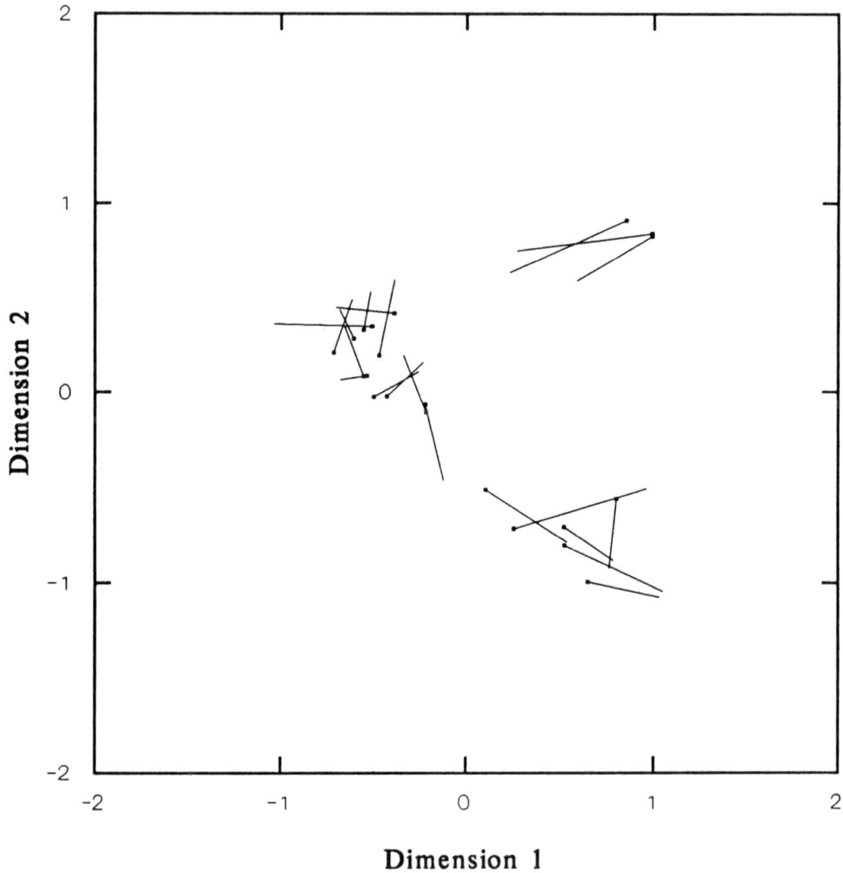

FIG. 15.7. Standardized plot showing Individual 1 of Triads 1 versus aggregated picture.

In effect, each of these figures represents a comparison between the "semantic structure" of the individual and the shared "cultural pattern" as inferred from the group as a whole. It is apparent that the two Pairs 1 individuals are much closer to the aggregate representation than are the Triads 1 individuals. The question arises as to the interpretation of the observed differences represented by the lines in the plots: do they represent idiosyncratic individual differences, systematic differences due to measurement method (rating versus triadic comparison), random effects due to sampling and measurement error, or some combination of the above? A precise answer to this question will require further research, although the methods do provide useful tools for such research. Our own guess at this time is that the observed differences in the plots reflect primarily random effects due to

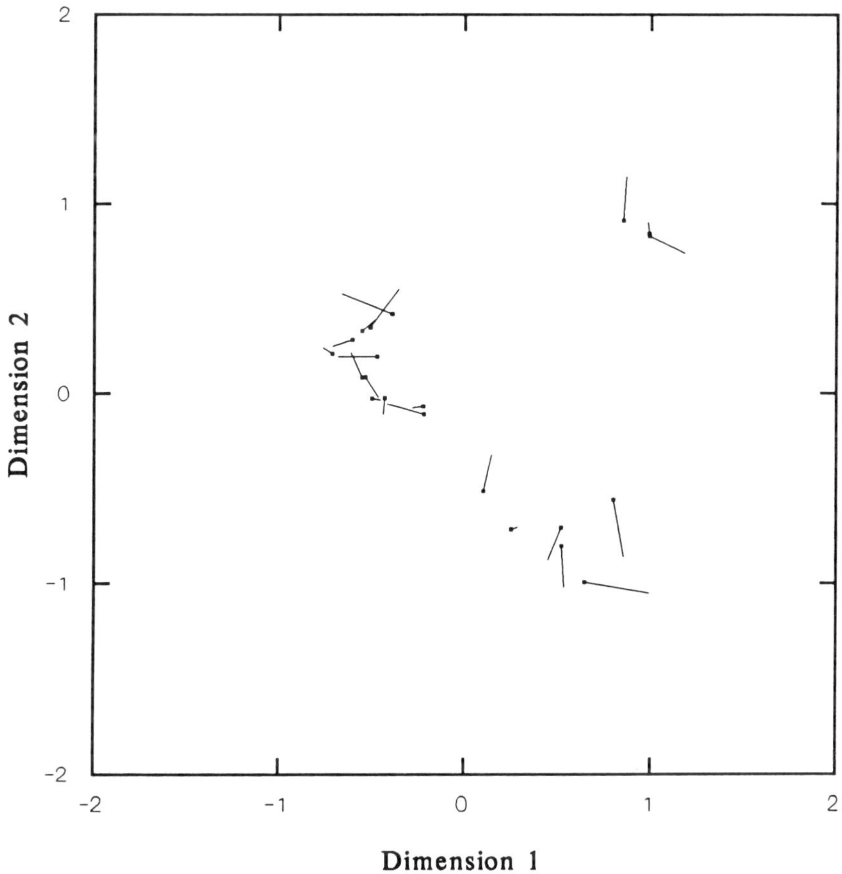

FIG. 15.8. Standardized plot showing Individual 1 of Pairs 1 versus aggregated picture.

sampling and measurement error. The greater variation in the Triads 1 individuals is related to the fact, as mentioned earlier, that the task simply provides a smaller amount of information and hence larger errors in estimating the location of the points in the scaling.

A DETAILED CASE STUDY OF TWO SELECTED SUBJECTS

Insight into various aspects of how individual variation in variance affects outcomes in the scaling can be gained by a fine-grained analysis of two cases

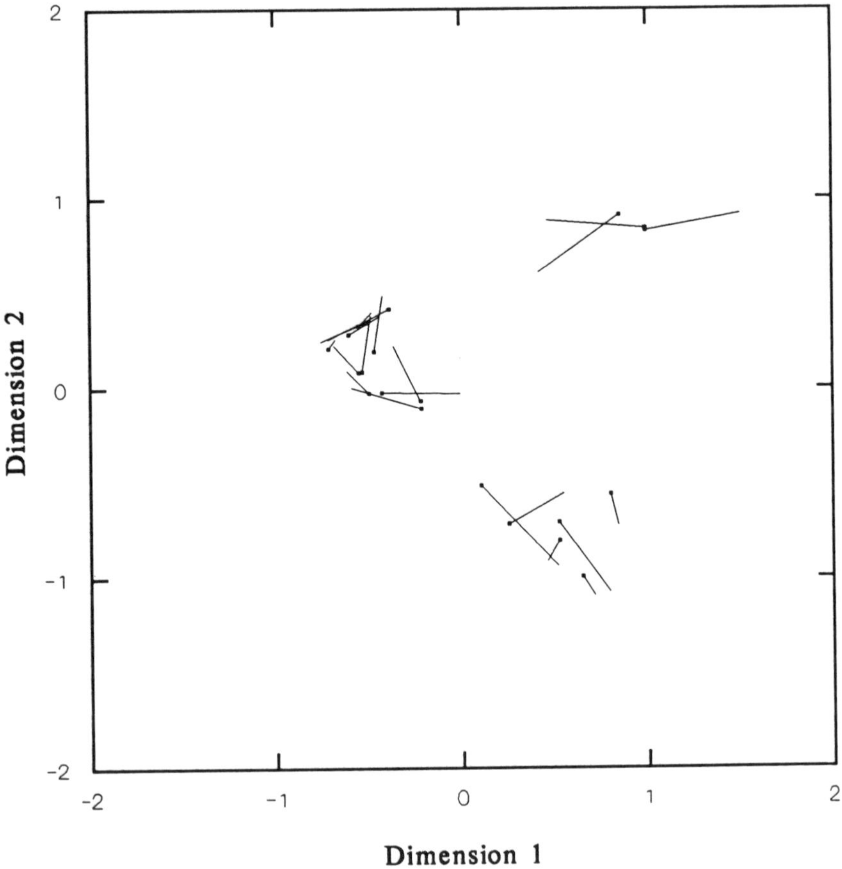

FIG. 15.9. Standardized plot showing Individual 2 of Triads 1 versus aggregated picture.

selected to illustrate a radical difference between two extreme subjects in the paired comparison rating task. We purposefully selected one subject who tended to use extreme judgments (over 90% were either 1 or 20) and another who tended to use all values rather evenly distributed over the allowed range. Table 15.4 presents the raw data for the two subjects and it may be seen that they differ radically in their use of the 20-point rating scale. The first data matrix shows Individual 3 from the Pairs 1 group and the second data matrix shows Individual 7 from the Pairs 1 group.

The first point to mention is that these two pictures differ dramatically in size before the scores are standardized. Figures 15.11 and 15.12 show the two subjects compared to the aggregate picture with the scores in unstandardized form

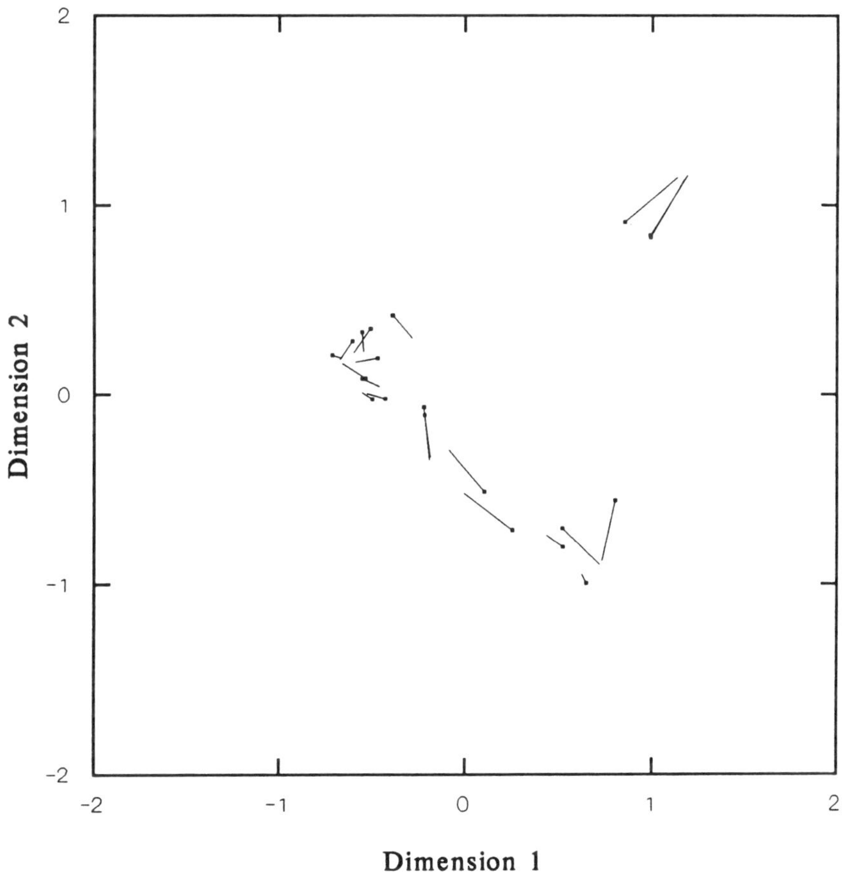

FIG. 15.10. Standardized plot showing Individual 2 of Pairs 1 versus aggregated picture.

(column scores were used for the unstandardized aggregate pictures). Note that most of the lines for Individual 3 point away from the aggregate points. This results in an overall representation for Individual 3 that is much larger than the aggregate representation. This contrasts sharply with the picture for Individual 7 in which most of the lines point toward the center from the aggregate points. In general, in multiple correspondence analysis, the overall size before standardization of a scaled representation of an individual data matrix is a function of its variance.

This size difference, of course, vanishes when each individual's score is standardized. Figures 15.13 and 15.14 illustrate this standardization for Individuals 3 and 7. Visually, both subjects appear to have very reasonable approxima-

TABLE 15.4
Examples of Raw Rating Scale Data from Two Separate Students
Illustrating Differences in Variance

Subject 3																				
20	1	1	1	3	1	1	20	1	1	1	2	1	1	1	1	1	1	2	1	2
1	20	1	1	1	1	1	1	1	1	1	1	1	1	1	1	1	1	1	1	1
1	1	20	1	1	1	1	1	1	1	1	1	1	2	1	1	1	1	1	1	2
1	1	1	20	1	1	1	1	2	1	1	1	1	1	20	1	1	1	1	20	1
3	1	1	1	20	1	1	2	1	2	1	1	20	1	1	20	1	1	1	5	1
1	1	1	1	1	20	1	1	1	1	1	1	1	1	1	2	1	1	1	1	1
1	1	1	1	1	1	20	1	1	1	1	20	1	1	1	1	1	1	11	1	1
20	1	1	1	2	1	1	20	1	1	1	1	1	1	1	1	1	1	1	1	1
1	1	1	2	1	1	1	1	20	1	1	1	1	1	2	1	1	1	1	1	1
1	1	1	1	2	1	1	1	1	20	1	1	1	2	2	1	1	1	1	2	1
1	1	1	1	1	1	1	1	1	1	20	1	1	1	1	1	1	1	1	1	5
2	1	1	1	1	1	20	1	1	1	1	20	1	1	1	1	1	1	20	1	1
1	1	1	1	20	1	1	1	1	1	1	1	20	2	1	20	1	1	1	1	1
1	1	2	1	1	1	1	1	1	2	1	1	2	20	1	1	1	1	1	1	20
1	1	1	20	1	1	1	1	2	2	1	1	1	1	20	1	1	1	1	17	1
1	1	1	1	20	2	1	1	1	1	1	1	20	1	1	20	1	1	1	1	1
1	1	1	1	1	1	1	1	1	1	1	1	1	1	1	1	20	1	1	1	1
1	1	1	1	1	1	1	1	1	1	1	1	1	1	1	1	1	20	1	1	1
2	1	1	1	1	1	11	1	1	1	1	20	1	1	1	1	1	1	20	1	1
1	1	1	20	5	1	1	1	1	2	1	1	1	1	17	1	1	1	1	20	1
2	1	2	1	1	1	1	1	1	1	5	1	1	20	1	1	1	1	1	1	20

Subject 7

20	6	18	10	7	5	13	16	7	10	16	16	10	16	16	7	6	2	14	13	15
6	20	5	15	8	18	4	8	15	5	4	6	12	8	7	12	18	15	12	5	6
18	5	20	6	9	5	12	13	8	16	17	16	7	16	14	10	8	4	12	16	18
10	15	6	20	14	15	6	13	16	5	5	12	14	7	13	10	17	16	11	15	6
7	8	9	14	20	11	9	11	14	8	13	9	14	7	10	19	5	8	12	5	7
5	18	5	15	11	20	6	8	11	6	6	7	8	5	5	14	14	9	6	2	7
13	4	12	6	9	6	20	14	13	15	13	16	10	18	13	11	5	7	17	13	13
16	8	13	13	11	8	14	20	13	12	16	15	10	16	16	9	11	3	15	16	15
7	15	8	16	14	11	13	13	20	7	6	15	15	12	11	13	13	16	15	11	10
10	5	16	5	8	6	15	12	7	20	16	6	12	15	16	9	7	5	9	8	14
16	4	17	5	13	6	13	16	6	13	20	13	9	13	13	7	4	6	8	11	15
16	6	16	12	9	7	16	15	15	6	13	20	7	14	16	10	10	9	17	13	16
10	12	7	14	14	8	10	10	15	12	9	7	20	6	8	19	11	6	5	12	7
16	8	16	7	7	5	18	16	12	15	13	14	6	20	15	9	6	6	16	14	19
16	7	14	13	10	5	13	16	11	16	13	16	8	15	20	3	9	1	17	18	16
7	12	10	10	19	14	11	9	13	9	7	10	19	9	3	20	5	7	7	7	11
6	16	8	17	5	14	5	11	13	7	4	10	11	6	9	5	20	17	11	4	5
2	15	4	16	8	9	7	3	16	5	6	9	6	6	1	7	17	20	9	5	8
14	12	12	11	12	6	17	15	15	9	8	17	5	16	17	7	11	9	20	14	16
13	5	16	15	5	2	13	16	11	8	11	13	12	14	18	7	4	5	14	20	16
15	6	18	6	7	7	13	15	10	14	15	16	7	19	16	11	5	8	16	16	20

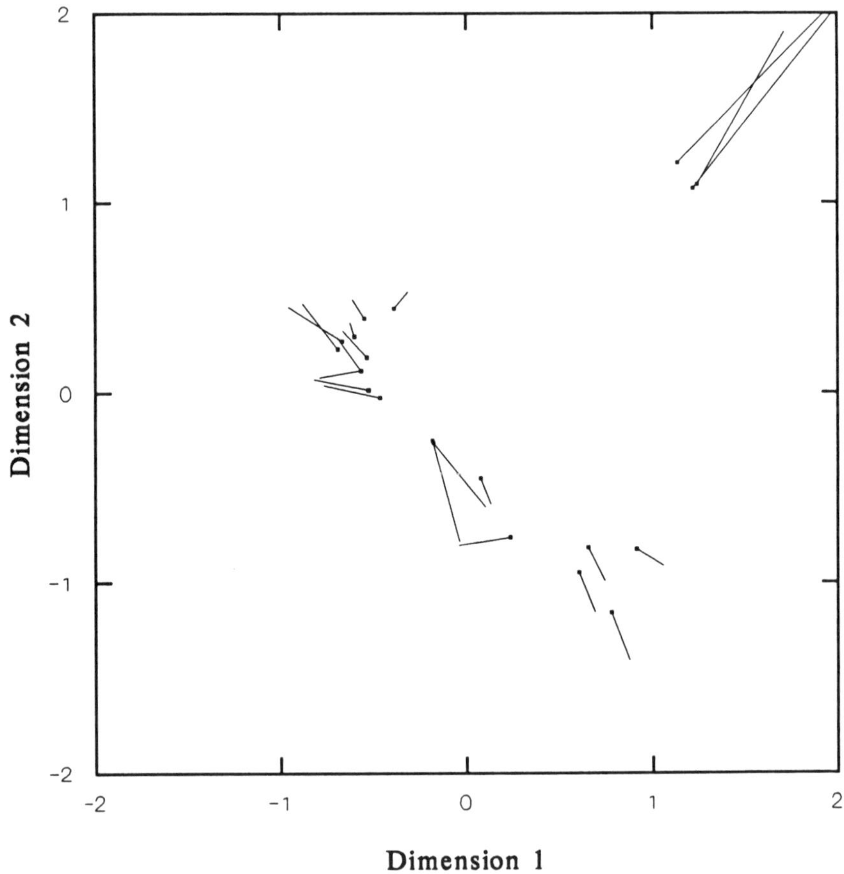

FIG. 15.11. Unstandardized plot showing Individual 3 of Pairs 1 versus aggregated picture.

tions to the aggregate picture (obtained from the means of the standardized scores). If anything, Individual 3 shows a slightly better fit. This comparison raises a standard issue in rating scales as to how many categories of comparison are optimal. In effect, Individual 3 only used 7 out of a possible 20 categories and produced a picture about as satisfactory as Individual 7, who used all 20 categories. This suggests that 20 categories are more than might be needed. Further research is necessary, but these methods of comparison should help in the solution.

We get a very different view of the two subjects if we analyze each data matrix individually and look at how well the individual representations fit the raw data.

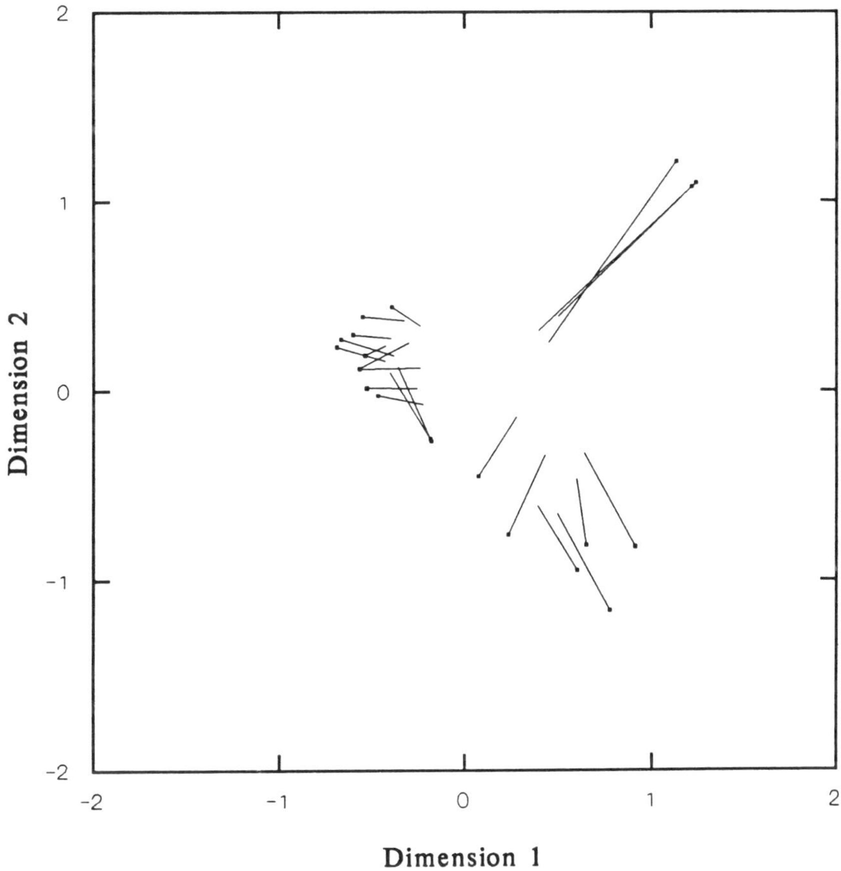

FIG. 15.12. Unstandardized plot showing Individual 7 of Pairs 1 versus aggregated picture.

Table 15.5 shows some of the possible figures taken as indicators of fit between the representation and the raw data. The singular values are better for Individual 3; however, the proportion of variance accounted for by each dimension as well as the PRE measures show that Individual 7 appears to have a much better fit than does Individual 3. Our experience with a large number of data sets would indicate that making inferences from such indicators requires a great deal of caution. The indicators may be much more a function of extraneous things such as variance, as appears to be the case here, than reflecting deeper differences in the configuration of the data.

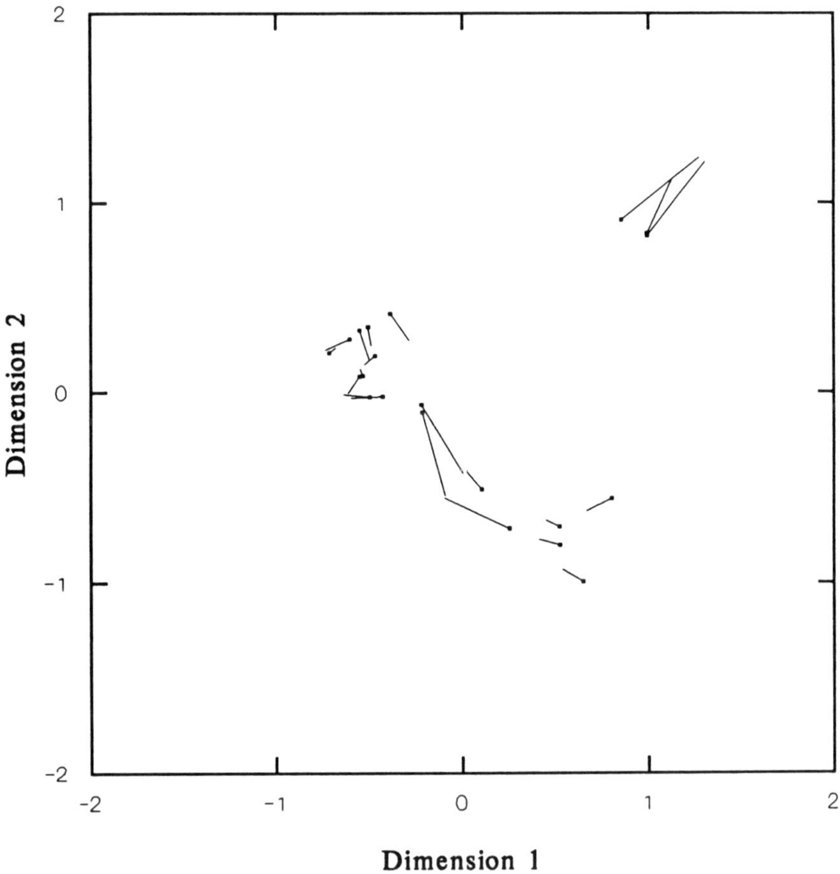

FIG. 15.13. Standardized plot showing Individual 3 of Pairs 1 versus aggregated picture.

SUMMARY AND DISCUSSION

In this chapter we have described a set of analytic procedures that provide a flexible and powerful way to scale the items of semantic domains. The means of the standardized scores provide a representation at the cultural level that is useful for the prediction of cognitive behavior. The scaling representations of each individual facilitate the comparison of individual variations resulting from any source, such as different data collection methods, membership in different demographic groups (e.g., rich versus poor or Alzheimer's versus healthy or old versus young), and differences arising from experiments.

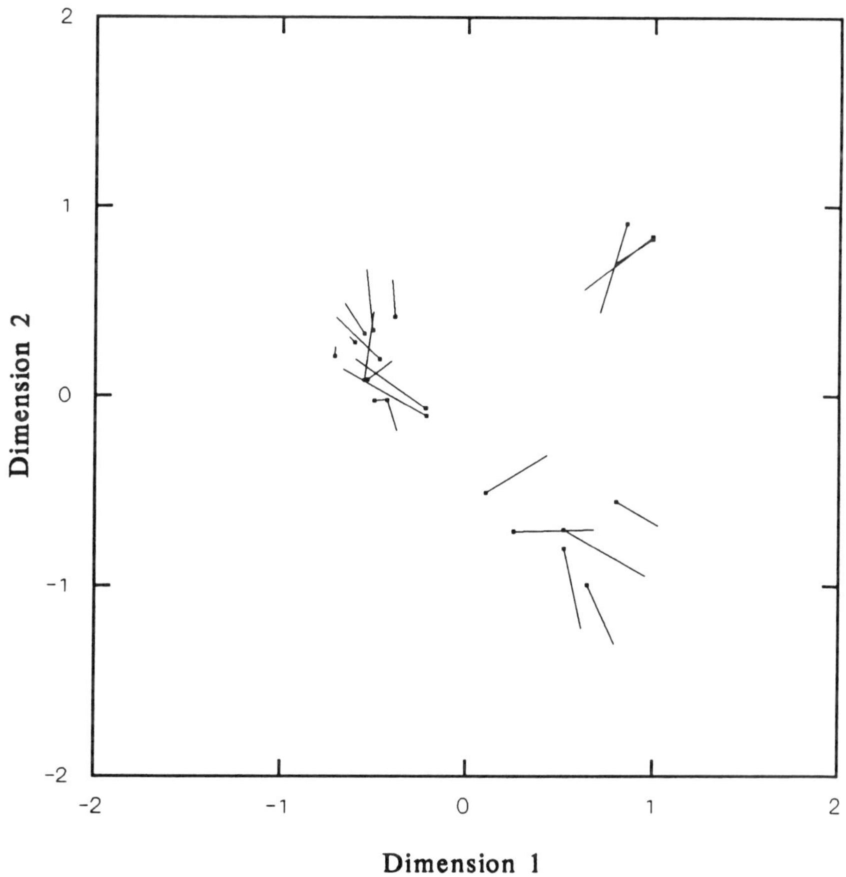

FIG. 15.14. Standardized plot showing Individual 7 of Pairs 1 versus aggregated picture.

The use of PRE to compare the paired comparison rating task with the triadic comparisons illustrates a method that should have wide application. The fact that the paired comparison rating tasks had an average PRE of .92 for the first two dimensions compared to a PRE of .72 in the triadic comparison task clearly shows a higher resolving power, which implies a more reliable and precise measurement, for the paired comparison ratings than for triadic comparisons. This method should help settle arguments about what data collection methods are most useful and appropriate in the sense of providing the highest resolving power in terms of PRE.

Further research is needed to extend and strengthen the method. Problems that need special attention include optimal aggregation techniques as well as testing

TABLE 15.5
Singular Values and PRE Statistics for Subject 3 and Subject 7 from the Paris 1 Group

Dimension	*Singular Value*	*%*	*Cum. %*	*PRE*	*Cum. PRE*
		Subject 3			
1	.708	10.7	10.7	14.3	14.3
2	.689	10.4	21.0	13.5	27.8
3	.647	9.7	30.8	11.9	39.8
4	.607	9.1	39.9	10.5	50.3
5	.526	7.9	47.8	7.9	58.1
6	.475	7.1	55.0	6.4	64.6
		Subject 7			
1	.284	28.1	28.1	62.7	62.7
2	.136	13.4	41.6	14.3	77.0
3	.083	8.2	49.7	5.3	82.3
4	.076	7.5	57.2	4.4	86.7
5	.063	6.2	63.4	3.1	89.8
6	.055	5.4	68.8	2.3	92.1

and inferential considerations. We are conducting research on the aggregation problem using, among other approaches, the consensus model (Batchelder & Romney, 1988). In the present study each subject was given equal weight in arriving at the aggregate solution. The consensus model would allow us to estimate the cultural competence of each subject. These competences could then be used to weight the contribution of the subjects on the basis of cultural knowledge.

The inferential problem should be solvable through the use of discriminant analysis and more sophisticated models of analysis of variance. One empirical problem that has been widely debated in the literature of paired comparison rating tasks is the question of how many categories to use in the rating. The most typical practice is to use a scale with five or seven values. Krantz and Tversky (1975) used 20 points. What number is optimal? The methods presented here should make it possible to design an experiment that would enable us to answer this type of question. The anecdotal evidence from our comparison of subjects 3 and 7 reported above suggests that it may not make very much difference how many categories are used.

ACKNOWLEDGMENTS

This research was supported by National Science Foundation Grant No. SES-9210009 made to A. K. Romney and W. H. Batchelder.

REFERENCES

Baker, R. F., & Young, F. W. (1975). A note on an empirical evaluation of the Isis procedure. *Psychometrika, 40,* 413–415.

Batchelder, W. H., & Romney, A. K. (1988). Test theory without an answer key. *Psychometrika, 53,* 71–92.

Borgatti, S. (1992). *ANTHROPAC 4.0.* Columbia, North Carolina: Analytic Technologies.

Caramazza, A., Hersh, H., & Torgerson, W. S. (1976). Subjective structures and operations in semantic memory. *Journal of Verbal Learning and Verbal Behavior, 15,* 103–117.

Carroll, J. D., & Chang, J. J. (1970). Analysis of individual differences in multidimensional scaling via an *N*-way generalization of "Eckart-Young" decomposition. *Psychometrika, 35,* 283–319.

Chan, A. S., Butters, N., Paulsen, J. S., Salmon, D. P., Swenson, M. R., and Maloney, L. T. (1993). An assessment of the semantic network in patients with Alzheimer's disease. *Journal of Cognitive Neuroscience, 5,* 254–261.

Chan, A. S., Butters, N., Salmon, D. P., and McGuire, K. A. (1993). Dimensionality and clustering in the semantic network of patients with Alzheimer's disease. *Psychology and Aging, 8,* 411–419.

Coombs, C. H. (1964). *A theory of data.* New York: Wiley.

Cunningham, J. P. (1978). Free trees and bidirectional trees as representations of psychological distance. *Journal of Mathematical Psychology, 17,* 165–188.

Friendly, M. (1979). Methods for finding graphic representations of associative memory structures. In C. R. Puff (Ed.), *Memory organization and structure* (pp. 85–129). New York: Academic Press.

Gifi, A. (1990). *Nonlinear multivariate analysis.* Chichester: Wiley.

Goodglass, H., Wingfield, A., Hyde, M. R., & Theurkauf, J. C. (1986). Category specific dissociations in naming and recognition by aphasic patients. *Cortex, 22,* 87–102.

Hart, J., Jr., Berndt, R. S., & Caramazza, A. (1985). Category-specific naming deficit following cerebral infarction. *Nature, 316,* 439–440.

Henley, N. M. (1969). A psychological study of the semantics of animal terms. *Journal of Verbal Learning and Verbal Behavior, 8,* 176–184.

Howard, D. V., & Howard, J. H. (1977). A multidimensional scaling analysis of the development of animal names. *Developmental Psychology, 13,* 108–113.

Hutchison, J. W., & Lockhead, G. R. (1977). Similarity as distance: A structural principle for semantic memory. *Journal of Experimental Psychology: Human Learning and Memory, 3*(6), 660–678.

Krantz, D. H., & Tversky, A. (1975). Similarity of rectangles: An analysis of subjective dimensions. *Journal of Mathematical Psychology, 12*:4–34.

Kumbasar, E., Romney, A. K., & Batchelder, W. H. (1994). Systematic biases in social perception. *Journal of Sociology, 100,* 477–505.

McCarthy, R. A., & Warrington, E. K. (1988). Evidence for modality-specific meaning systems in the brain. *Nature, 334,* 428–430.

Nosofsky, R. M. (1992). Similarity scaling and cognitive process models. *Annual Review of Psychology, 43,* 25–53.

Rips, L. J. (1975). Inductive judgments about natural categories. *Journal of Verbal Learning and Verbal Behavior, 14,* 665–681.

Rips, L. J., Shoben, E. J., & Smith, E. E. (1973). Semantic distance and the verification of semantic relations. *Journal of Verbal Learning and Verbal Behavior, 12,* 1–20.

Romney, A. K. (1989). Quantitative models, science and cumulative knowledge. *Journal of Quantitative Anthropology, 1,* 153–223.

Romney, A. K., Brewer, D. D., & Batchelder, W. H. (1993). Predicting clustering from semantic structure. *Psychological Science, 4,* 28–34.

Rumelhart, D. L., & Abrahamson, A. A. (1973). Toward a theory of analogical reasoning. *Cognitive Psychology, 5,* 1–28.

Sattath, S., & Tversky, A. (1977). Additive similarity trees. *Psychometrika, 42,* 319–345.

Sartori, G., & Job, R. (1988). The oyster with four legs: a neuropsychological study on the interaction of visual and semantic information. *Cognitive Neuropsychology, 5,* 105–132.

Shepard, R. N. (1974). Representation of structure in similarity data: Problems and prospects. *Psychometrika, 39,* 373–421.

Shoben, E. J. (1976). The verification of semantic relations in a same-different paradigm: An asymmetry in semantic memory. *Journal of Verbal Learning and Verbal Behavior, 15,* 365–379.

Silveri, M. C., & Gainotti, G. (1988). Interaction between vision and language in category-specific semantic impairment. *Cognitive Neuropsychology, 5,* 677–709.

Smith, E. E., Shoben, E. J., & Rips, L. J. (1974). Structure and process in semantic memory: A featural model for semantic decisions. *Psychological Review, 81,* 214–241.

Warrington, E. K., & McCarthy, R. A. (1987). Categories of knowledge: further fractionations and an attempted integration. *Brain, 110,* 1273–1296.

Warrington, E. K., & Shallice, T. (1984). Category specific semantic impairments. *Brain, 107,* 829–854.

Weller, S. C., & Romney, A. K. (1988). *Systematic data collection.* Newbury Park, CA: Sage.

Weller, S. C., & Romney, A. K. (1990). *Metric scaling: Correspondence analysis.* Newbury Park, CA: Sage.

Wilkinson, L. (1989). *SYGRAPH: The system for graphics.* Evanston, IL: SYSTAT, Inc.

16 A General Approach to Clustering and Multidimensional Scaling of Two-Way, Three-Way, or Higher-Way Data

J. Douglas Carroll
Rutgers University

Anil Chaturvedi
AT&T Bell Laboratories

ABSTRACT

A method, now called SINDCLUS, for fitting the ADCLUS/INDCLUS models for overlapping clustering of two-way or three-way symmetric proximity data was presented recently by Chaturvedi and Carroll (1992), utilizing a numerical procedure very similar to the Carroll and Chang CANDECOMP method used for fitting the INDSCAL model for three-way multidimensional scaling (and for multiway components analysis). SINDCLUS fits one cluster at a time to a sequence of residual data arrays, doing this by iterating among conditional least-squares estimates of parameters for each of the three ways. A separability property observed by Chaturvedi enables straightforward conditional OLS estimation of the discrete "cluster membership" parameters for the two modes corresponding to objects. Weights for the third mode (e.g., subjects) were fit by OLS regression. While the binary parameters for the two object modes are not constrained to be equal on each iteration, as in CANDECOMP applied to fitting the symmetric INDSCAL model—given data symmetric in two modes—the two sets of parameter estimates are generally equal upon convergence. CANDCLUS is a multiway generalization of this model and estimation procedure, allowing binary parameters for any p of the N modes, and continuous parameters for the remaining $N - p$. A generalized separability property allows straightforward conditional OLS estimation for each mode modeled by binary parameters. Special cases include CANDECOMP, fit "one dimension at a time," and a model in which all N modes are modeled via overlapping cluster structures. Hybrid models, in which a given mode can be modeled by a hybrid mixture of continuous (spatial) and discrete (clusterlike) dimensions, can also be fit within this framework. A generalization of CANDCLUS, called MUMCLUS (multimode clustering) is also formulated and discussed theoretically).

INTRODUCTION

There are many varieties of data that can be modeled by two or more indices or subscripts corresponding to different factors or "modes" in terms of which the data may be classified. We use the general term "way" to refer variously to such factors, facets, or modes. A common form of such data arising in the social and behavioral sciences comprises two-way data which can conveniently be displayed in a two-way table or matrix, **Y**, with general entry y_{ij}, $i = 1, 2, \ldots, I; j = 1, 2, \ldots, J$. Typical examples of such two-way data are individuals (e.g., subjects) by stimuli (or other objects) data on rating of the J stimuli by each of the I subjects on some attribute of interest (e.g., ratings of products on attributes by individual consumers, brightness of visual stimuli, or smoothness versus roughness of tactile stimuli). One very important special case of such data is that in which the attribute is some measure of degree of preference, subjective utility, or other measure of evaluation on a continuous scale. Another case of general two-way data consist of objects by rating scales (or other variables) data, where each of the I stimuli or other objects is measured on each of the J rating scales or variables. Data of these two types are often called two-way, two-mode data, since the two different *ways* (indexed by the subscripts i and j separately) correspond to different *modes* or factors (say objects and variables). Another very important class of two-way data is often called, in keeping with the taxonomy of data proposed by Carroll and Arabie (1980), two-way but one-mode data, since both "ways"—rows and columns of the square data matrix—correspond to the same factor or mode, the prototypical case being an $I \times I$ proximity matrix **S**, where I stimuli or other objects define both the rows and columns, and the general entry $s_{ii'}$ defines the similarity, dissimilarity, or other measure of *proximity* between objects i and i'.

Many proximity matrices are symmetric ($s_{ii'} = s_{i'i}$, for all i, i') since the proximity of i to i' is frequently defined to be the same as that of i' to i, but this is not always the case, since there are well-defined nonsymmetric proximity measures. In many cases, proximity data are symmetric simply because the pairs are presented in only one order, say (i, i'). The data value $s_{i'i}$ is then defined to be equal to $s_{ii'}$ by assumed symmetry.

Three-way three-mode data are indexed by three subscripts, so the three-way $I \times J \times K$ array **Y** has general entry y_{ijk}, say for the ith subject's judgment of the jth stimulus on rating scale k. Three-way but two-mode proximity data might consist of a set of $I \times I$ proximity matrices, each for a different subject, (or other source of data), indexed by k. Thus, such an array **S** might be $I \times I \times K$, where $s^k_{ii'}$ is the proximity of objects i and i' for subject or other source of data k.

Higher-way data comprise any data set indexed by more than three indices or subscripts, corresponding to an N-way $I_1 \times I_2 \times \ldots \times I_N$ array with general entry $y_{i1,i2,\ldots iN}$, where $i_n = 1, 2, \ldots, I_n$ for $n = 1, 2, \ldots, N$.

An N-way but K-mode ($K < N$) array has only K modes corresponding to the N ways indexed by the N subscripts. For example, a four-way but three-mode array **S** might consist of JK $I \times I$ proximity matrices, one for each of J subjects under K experimental conditions. Assuming the proximity relationship is a symmetric one, this $I \times I \times J \times K$ array **S** might have general entry $s_{ii'jk}$, where $s_{ii'jk} = s_{i'ijk}$, for all i, i', j k. It should be noted, however, that an $I \times J$ two-way data matrix, or submatrix of a larger array, need *not* comprise proximity data—whether symmetric or nonsymmetric. For example, this could correspond to an antisymmetric, or skew-symmetric, "dominance" matrix, **T**, where $t_{ii'}$, indicating whether i dominates i' (positive sign) or i' dominates (negative sign), while the absolute magnitude of $t_{ii'}$ indicates the degree of dominance (e.g., preference of) i over i'. As indicated, N-way dominance data will generally (but not always) be anti- or skew-symmetric; i.e., $t_{ii'} = -t_{i'i}$. Still other types of one-mode but two-, three-, or higher-way data are possible, but we will not elaborate further here. A K ($<N$) mode, N-way array will generally have subarrays that are N_1 ($\leqslant N$) way but only one mode; most typically $N_1 = 2$. Other more general such data structures are possible, but will not be discussed further here.

THE CANDCLUS MODEL

Canonical decomposition clustering (CANDCLUS) is a multilinear model for such a general multiway data array. We first state the CANDCLUS model in its most general form, for the N-way array **Y.** It might be noted that this form of the CANDCLUS model is identical to that of the Carroll and Chang CANDECOMP (canonical decomposition) model (Carroll and Chang, 1970; Carroll and Pruzansky, 1975), whose most important application to date has been to provide the computational underpinnings of the INDSCAL approach to two-mode, three-way (individual differences) multidimensional scaling (MDS). The most important difference between CANDECOMP and CANDCLUS is that, while in the former, all parameters are assumed continuous, so that the models being assumed and fitted in the concomitant data analysis are all continuous spatial models (e.g., MDS models), in CANDCLUS, some or all of the dimensions for some or all ways/modes are constrained to be discrete, typically binary (0-1) variables, which can be interpreted as class membership variables encoding whether a particular object (or other entity corresponding to a given level of a given mode/way) belongs (value = 1) or does not belong (value = 0) to a particular cluster. In CANDCLUS, the dimensions for various ways/modes may be continuous (dimenson-like) or binary (clusterlike). Other possibilities include discretely valued dimensions with k (>2) distinct possible values for a particular dimension. Specifically, the CANDCLUS model for a general N-way array **Y** can be stated in the following form:

$$y_{i_1}, \ldots, i_N \sim \sum_{p=1}^{P} a^1_{i_1,p} a^2_{i_2,p} \ldots a^N_{i_N,p}. \qquad (16.1)$$

Define $\mathbf{A}_n$ as the parameter matrix of order $I_n \times P$ with elements $a^n_{i_n,p}$ for the pth dimension and the nth way. The elements of matrices $\mathbf{A}_n$ ($n = 1, \ldots, N$) can take on any of the following values:

- Real values for all matrices $\mathbf{A}_n$, $n = 1, \ldots, N$. This results in the P-dimensional CANDECOMP model of Carroll and Chang (1970).
- Discrete integer (or finite set of real number) values for all $\mathbf{A}_n$, $n = 1, \ldots, N$.
- Mixture of real parameters for some ways and discrete values for the rest. (Other possibilities, such as those entitled "hybrid" models, will be discussed later.)

Some important special cases of the CANDCLUS model include INDCLUS (Carroll and Arabie, 1983), which is equivalent to the three-way CANDCLUS model when two of the three ways are modeled via discrete 0-1 integer parameters, while the third way is modeled via continuous parameters. ADCLUS (Shepard and Arabie, 1979), is also a special case of three-way CANDCLUS when two of the three ways are modeled via discrete 0-1 integer parameters, while the third way (with one level) is modeled via continuous parameters. Note that, in the case of the INDCLUS and ADCLUS models, the model is symmetric in two of the three ways (corresponding to stimuli or other objects on which symmetric proximity data are given). As in the case of the three-way CANDECOMP procedure (after preprocessing of data to convert it into "scalar product" form), we fit the general three-way CANDCLUS model, but find, as will be discussed, that upon convergence the **A** matrices for the two ways will be equal, as required for the symmetric model. This will be discussed in more detail.

THE PROCEDURE FOR FITTING THE CANDCLUS MODEL

The Elementary Discrete Least-Squares Procedure

The elementary discrete least-squares procedure (EDSLP) is central to much of the methodology proposed in this chapter, we shall outline this procedure in considerable detail. We start with an illustrative example. Let

$$\mathbf{L} = \begin{bmatrix} 7 & 5 & 3 & 9 \\ 8 & 6 & 5 & 1 \\ 9 & 4 & 2 & 7 \\ 5 & 3 & 4 & 6 \end{bmatrix}, \quad \mathbf{x} = \begin{bmatrix} x_1 \\ x_2 \\ x_3 \\ x_4 \end{bmatrix}, \quad \mathbf{r}' = [1 \quad 4 \quad 9 \quad 3].$$

Consider the problem of the least-squares estimate of **x**, where $\mathbf{L} = \mathbf{xr}' +$ error and elements of **x** are either 0 or 1. That is,

$$\begin{bmatrix} 7 & 5 & 3 & 9 \\ 8 & 6 & 5 & 1 \\ 9 & 4 & 2 & 7 \\ 5 & 3 & 4 & 6 \end{bmatrix} = \begin{bmatrix} 1x_1 & 4x_1 & 9x_1 & 3x_1 \\ 1x_2 & 4x_2 & 9x_2 & 3x_2 \\ 1x_3 & 4x_3 & 9x_3 & 3x_3 \\ 1x_4 & 4x_4 & 9x_4 & 3x_4 \end{bmatrix} + \text{error}.$$

If we let

$$f_1 = [7 - 1x_1]^2 + [5 - 4x_1]^2 + [3 - 9x_1]^2 + [9 - 3x_1]^2,$$

$$f_2 = [8 - 1x_2]^2 + [6 - 4x_2]^2 + [5 - 9x_2]^2 + [1 - 3x_2]^2,$$

$$f_3 = [9 - 1x_3]^2 + [4 - 4x_3]^2 + [2 - 9x_3]^2 + [7 - 3x_3]^2,$$

$$f_4 = [5 - 1x_4]^2 + [3 - 4x_4]^2 + [4 - 9x_4]^2 + [6 - 3x_4]^2,$$

then the sum of squared errors is given by

$$F = f_1 + f_2 + f_3 + f_4.$$

Note that f_1 is a function only of x_1; f_2 is a function only of x_2; f_3 is a function only of x_3, and f_4 is a function only of x_4. Thus, F is *separable* in x_1, x_2, x_3, and x_4. To minimize F, one can separately minimize f_1 w.r.t x_1, f_2 w.r.t x_2, f_3 w.r.t x_3, and f_4 w.r.t. x_4.

More generally, if we want to solve for the least squares estimate of the binary vector **x** in the equation $\mathbf{L} \sim xr'$, where $\sim$ denotes that we seek a least-squares solution for **x**, given fixed **L** and **r**. The least-squares loss function, F, will be of the form

$$F = \sum_{i=1}^{I} \sum_{j=1}^{J} [l_{ij} - x_i r_j]^2 = \sum_{i=1}^{I} f_i,$$

where $f_i = \Sigma_{j=1}^{J} [l_{ij} - x_{ir_j}]^2$ is a function only of x_i.

To minimize, say, f_1 w.r.t. x_1, one can easily evaluate f_1 at $x_1 = 1$ and $x_1 = 0$. The x_1 yielding a minimum of these two possible values is then chosen. We may minimize F as a whole by separately minimizing $f_1, f_2, \ldots, f_I$.

Thus, for I (0-1) variables, only $2I$ function evaluations and comparisons are needed, as compared to 2^I evaluations and comparisons for explicit enumeration. As already alluded to, this discretely valued vector need not be restricted to the binary case, but may be generalized to include vectors whose components are selected from k distinct integer- or real-valued numbers. Since we made no assumptions that $\mathbf{x}$ was binary above, it is clear that the separability property applies equally well to this more general case. In the case of a discrete set of k values ($k > 2$) a similar reduction in comparisons occurs, from k^I to only kI.

Application of EDLSP to Fitting of CANDCLUS

In the CANDCLUS method, parameters are estimated one "dimension" at a time. (Note that we will use the term "dimension" to denote either a continuous dimension—defined concretely as a vector of real numbers for each of a number of objects—a binary 0-1 vector defining class membership of objects in a cluster, or any of the other intermediate cases mentioned above, in which the values of each dimension take on k discrete values, for $k \geq 2$.) The following procedure describes the estimation methodology for the pth dimension, under the assumption that the current parameter estimates for all the other dimensions are known.

Let

$\mathbf{a}_p^n \equiv$ an $I_n \times 1$ parameter column vector for the pth dimension and the nth way.

$\mathbf{d}_p \equiv$ an $(I_1 \ldots I_{n-1}I_{n+1} \ldots I_N) \times 1$ vector defined as $\mathbf{d}_p = \mathbf{a}_p^1 \otimes \ldots \otimes \mathbf{a}_p^{n-1} \otimes \mathbf{a}_p^{n+1} \otimes \ldots \otimes \mathbf{a}_p^N$ where $\otimes$ is the Kronecker product.

$\mathbf{S}^n \equiv$ an $I_n \times (I_1 \ldots I_{n-1}I_{n+1} \ldots I_N)$ matrix formed from the given N-way data array by concatenating $N - 1$ ways, excluding the nth way, to form the columns, while the nth way defines the I_n rows.

$\mathbf{S}_{(-p)}^n \equiv$ an $I_n \times (I_1 \ldots I_{n-1}I_{n+1} \ldots I_N)$ matrix estimated from parameters for all the P dimensions, except the pth dimension.

If we have estimates of all but the pth dimension, then we can define $\bar{\mathbf{S}}_p^n$ as

$$\bar{\mathbf{S}}_p^n = \mathbf{S}^n - \mathbf{S}_{(-p)}^n. \tag{16.2}$$

Now, in order to estimate the parameters for the pth dimension and the nth way, we can write

$$\bar{\mathbf{S}}_p^n = f(\text{parameters for dimension } p) + \text{error}. \tag{16.3}$$

In other words, we can write Equation 16.3 as

$$\bar{\mathbf{S}}_p^n = \mathbf{a}_p^n\mathbf{d}_p' + \text{error}. \tag{16.4}$$

Assuming that the conditional estimates for the remaining $N - 1$ ways for the pth dimension are known, the only unknown in Equation 16.4 is the estimate for the

pth dimension and the nth way—$\mathbf{a}_p^n$. In the case of continuous parameters, $\mathbf{a}_p^n$ can be estimated using OLS regression. Nonnegativity constraints are imposed simply by setting all negative weights to zero, as in Carroll, De Soete, and Pruzansky (1989). In the case of discrete parameters, the elementary discrete least-squares procedure (EDLSP) can be used to find OLS estimates.

The above procedure is iterated across all N ways for the pth dimension, until convergence occurs. This iterative procedure for estimating the pth dimension for each of the N ways is embedded within another iterative cycle, consisting of what are sometimes called "major iterations," iterating over the P dimensions until (overall) convergence occurs. The iterations are continued until the improvement in fit falls below a prespecified criterion, and so at least a *local* minimum is obtained.

SINDCLUS: AN APPLICATION OF CANDCLUS TO FITTING THE ADCLUS AND INDCLUS MODELS

Assume we are given one or more symmetric proximity matrices $\mathbf{S}_k$, $k = 1, 2, \ldots, K$. Then, the INDCLUS model (Carroll, 1975; Carroll & Arabie, 1983) is written as

$$\mathbf{S}_k = \mathbf{P}\mathbf{W}_k\mathbf{P}' + \mathbf{C}_k + \text{error}, \tag{16.5}$$

where

- $\mathbf{S}_k \equiv I \times I$ similarity matrix for the kth subject (or other source of data)
- $\mathbf{W}_k \equiv R \times R$ diagonal matrix of weights for the kth subject
- $\mathbf{P} \equiv I \times R$ *binary* indicator matrix defining the (possibly overlapping) clusters
- $\mathbf{C}_k \equiv I \times I$ matrix having a constant c_k on the off-diagonals.

It should be noted that the diagonal entries in the $I \times I$ matrices $\mathbf{S}_k$ and $\mathbf{C}_k$ are not defined. This can be handled by treating these diagonals as "missing" data which can be handled very easily in the CANDCLUS approach (by simply omitting the corresponding terms from the OLS loss function being optimized). We have taken the approach of fitting the INDCLUS/ADCLUS models by fitting the more general nonsymmetric models, for a possibly nonsymmetric set of $I \times J$ matrices of the form

$$\mathbf{S}_k = \mathbf{P}\mathbf{W}_k\mathbf{Q}' + \text{error}, \tag{16.6}$$

where $\mathbf{Q}$ is a $J \times R$ matrix. In the symmetric case, of course, $\mathbf{P} = \mathbf{Q}$ and $I = J$. Following the general approach taken by Carroll and Chang in using three-way CANDECOMP to fit the symmetric INDSCAL model (once converted to scalar product form via appropriate preprocessing), we fit the symmetric INDCLUS/ADCLUS model by fitting the nonsymmetric model of Equation 16.6. This model is the three-way case of CANDCLUS, of course, with two ways (the two corresponding to stimuli) being modeled by binary parameters, and the third (indexed by k, and corresponding to one or more subjects or other sources of data) being modeled by continuous parameters, interpreted as weights for subjects or other sources (and often constrained to nonnegativity). This particular approach to fitting of INDCLUS/ADCLUS we have called SINDCLUS (Chaturvedi and Carroll, 1994), an acronym for "separability based" or "speedy INDCLUS." We have demonstrated in comparative analyses that SINDCLUS is more efficient for fitting the two-way or three-way versions of the INDCLUS/ADCLUS models than the earlier dominant algorithms for fitting these models—MAPCLUS (Arabie and Carroll, 1980) and INDCLUS (Carroll and Arabie, 1983). Mirkin (1987, 1989) has proposed a somewhat different approach for fitting a version of the two-way ADCLUS model, which, however, does not have overall least-squares (or least L_p-norm) properties; so is not strictly comparable with SINDCLUS.

Of course, as in the case of the general CANDECOMP method applied to fitting the symmetric INDSCAL model, we hope that, upon convergence, the parameters fit by this CANDCLUS approach will similarly exhibit symmetry; in particular, that upon convergence, $\hat{\mathbf{P}} = \hat{\mathbf{Q}}$. We do not explicitly constrain $\hat{\mathbf{P}} = \hat{\mathbf{Q}}$, but we find that, upon apparent attainment of a global optimum solution, that condition will generally be satisfied (as is generally the case, also, when using CANDECOMP to fit the scalar products version of the INDSCAL model, up to some arbitrary scale factor resolvable by normalization of the resulting parameters).

We then follow the approach first introduced by Carroll and Chang (1970) for fitting the symmetric INDSCAL model via CANDECOMP; i.e., we fit the nonsymmetric three-way CANDCLUS model and expect symmetric solutions at the global optimum solutions. It has been observed that, indeed, $\hat{\mathbf{P}} = \hat{\mathbf{Q}}$ so far in all cases in which a global optimum appears to have been obtained. (Local optima sometimes occur in which $\hat{\mathbf{P}} \neq \hat{\mathbf{Q}}$, but we have not yet encountered a solution where a better solution, entailing an improved least-squares loss function, does not occur in such instances in which $\hat{\mathbf{P}} = \hat{\mathbf{Q}}$.)

While the SINDCLUS algorithm is, as already discussed, a special case of that for CANDCLUS, we describe it in detail here, since SINDCLUS is one of the most important special cases of CANDCLUS, and because we feel a detailed description of this specific algorithm may further illuminate the structure of the more general CANDCLUS approach of which it is a special case. In this discussion, note that we are assuming $I = J$ throughout.

First, let us define some terms, in particular:

$\mathbf{p}_r \equiv$ an $I \times 1$ binary vector for the rth cluster

$\mathbf{w}_r \equiv$ a $K \times 1$ vector of the weights for the rth

$\mathbf{q}_r \equiv$ an $I \times 1$ binary vector for the rth cluster (not necessarily $= \mathbf{p}_r$)

$\mathbf{P}_{(-r)} \equiv$ an $I \times R$ binary matrix including the universal cluster but excluding the rth cluster

$\mathbf{W}_{(-r)} \equiv$ a $K \times R$ weight matrix including the universal constants but excluding the rth cluster

$\mathbf{Q}_{(-r)} \equiv$ an $I \times R$ matrix including the universal cluster but excluding the rth cluster

Then we can rewrite Equation 16.6 as

$$\mathbf{S}_k = \mathbf{p}_r w_{kr} \mathbf{q}_r' + \mathbf{P}_{(-r)} \mathbf{W}_{k(-r)} \mathbf{Q}_{(-r)}' + \text{error}. \tag{16.7}$$

If we have estimates of all but the rth cluster, then we can define $\mathbf{S}_k^*$ as

$$\mathbf{S}_k^* = \mathbf{S}_k - \mathbf{P}_{(-r)} \mathbf{W}_{k(-r)} \mathbf{Q}_{(-r)}' \tag{16.8}$$

to get

$$\mathbf{S}_k^* = \mathbf{p}_r w_{kr} \mathbf{q}_r' + \text{error}. \tag{16.9}$$

That is,

$$\mathbf{S}_k^* = f(\text{parameters for cluster } r) + \text{error}. \tag{16.10}$$

If we have K matrices $\mathbf{S}_k$ of order $I \times I$, we use a procedure similar to the one used in INDSCAL by Carroll and Chang (1970). Let us assume that we have the parameter estimates for all clusters, except the rth cluster. Let $\mathbf{T}_1$ be a $K \times I^2$ matrix and $\mathbf{T}_2$ and $\mathbf{T}_3$ be $I \times KI$ matrices. $\mathbf{T}_1$ has all I^2 elements of $\mathbf{S}_k$ in its kth row. $\mathbf{T}_2$ is the supermatrix that has the jth row of matrices $\mathbf{S}_1, \ldots, \mathbf{S}_K$ in the jth row. Thus,

$$\mathbf{T}_2 = [\mathbf{S}_1 | \mathbf{S}_2 | \ldots \mathbf{S}_k | \ldots | \mathbf{S}_K].$$

Similarly, we define

$$\mathbf{T}_3 = [\mathbf{S}_1' | \mathbf{S}_2' | \ldots \mathbf{S}_k' | \ldots | \mathbf{S}_K'].$$

Assuming that the estimates of $\mathbf{p}_i$, and $\mathbf{q}_i$ are known, the parameters for the rth cluster are estimated by iterating the following three steps until at least a local optimum is reached.

Step 1. Estimating $\mathbf{w}_r$ *Conditionally.* Let $\mathbf{g}_r$ be a vector of I^2 elements such that

$$\mathbf{g}_r = \mathbf{p}_r \otimes \mathbf{q}_r.$$

Then

$$\mathbf{T}_1 = \mathbf{w}_r \mathbf{g}_r' + \text{error}$$

yields a closed-form solution for $\mathbf{w}_r$. Nonnegativity constraints are imposed easily; simply set any negative weight to zero.

Step 2. Estimating $\mathbf{p}_r$ *Conditionally.* Let $\mathbf{h}_r$ be a vector of KI elements such that

$$\mathbf{h}_r = \hat{\mathbf{w}}_r \otimes \mathbf{q}_r.$$

Then, by using the elementary binary least-squares procedure in the equation

$$\mathbf{T}_2 = \mathbf{p}_r \mathbf{h}_r' + \text{error},$$

one can find $\hat{\mathbf{p}}_r$, the OLS estimates of $\mathbf{p}_r$.

Step 3. Estimating $\mathbf{q}_r$ *Conditionally.* Let $\mathbf{j}_r$ be a vector of KI elements such that

$$\mathbf{j}_r = \hat{\mathbf{w}}_r \otimes \hat{\mathbf{p}}_r.$$

Then, by using the elementary binary least-squares procedure in the equation

$$\mathbf{T}_3 = \mathbf{q}_r \mathbf{j}_r' + \text{error},$$

one can find $\hat{\mathbf{q}}_r$, the OLS estimates of $\mathbf{q}_r$. We use $\hat{\mathbf{q}}_r$ and repeat Steps 1–3 until *no* improvement in fit results. It should be noted that in Steps 2 and 3 the **p** and **q** vectors cannot be all zero. Thus, each cluster must have at least one object in it. Since the matrices $\mathbf{S}_k$ do not have diagonals, Steps 1–3 need to be modified. We simply drop the corresponding columns from the $\mathbf{T}_1$ matrix and **g** vector in Step 1 and ignore the diagonal elements in Steps 2 and 3. For the universal cluster, where **p** and **q** are fixed, we just use Step 1.

We note several features of the SINDCLUS algorithm, and features implied for CANDCLUS, that are important to understanding this method. First, the diagonals in SINDCLUS can be included, since there may well be cases in which the diagonals are present and we shall want to include them in the overall least-squares loss function, as well as the off-diagonals). The method for handling missing diagonals, in cases in which they are excluded, is simply a special case of a general feature of CANDCLUS not previously mentioned, namely the approach to handling missing data. Since, at each stage of the algorithm, we are conditionally estimating one new "dimension" (e.g., the $\mathbf{p}_r$ or $\mathbf{q}_r$ vector in SINDCLUS), through OLS regression or the use of EDLSP mentioned earlier, omission of data is accomplished quite simply by omitting the corresponding terms from the corresponding least-squares loss function. It might be noted also that generalization of SINDCLUS to *weighted* least squares, rather than OLS fitting, is quite straightforward, since each of the three conditional OLS estimation stages can simply be replaced with an appropriately weighted least-squares estimation in the usual manner. (Of course, treatment of missing data as described is simply a special case of weighted least-squares fitting, with weights of zero for missing observations and one for those that are present.)

A second comment relates to treatment of the universal cluster (or additive constant estimation) in SINDCLUS. This can be viewed as a special case of a feature that can easily be extended to the CANDCLUS approach generally, namely, that of allowing certain clusters/dimensions to be fixed a priori, rather than being reestimated on each (major) iteration. In the present case, the "universal cluster" is fixed a priori, but the weight (or set of K weights) for that cluster is reestimated on each major iteration. In other cases of CANDCLUS, it may be desired to include such a universal cluster or other a priori fixed cluster or continuous dimension; this can easily be done.

One final comment vis-à-vis SINDCLUS is that no special case need be described for the fitting of the ADCLUS model, in which $K = 1$. ADCLUS is simply fit as a special case of the INDCLUS model, in which the third way, for subjects or other sources of data, has only one level.

Applications of SINDCLUS to Some Real Data

The SINDCLUS procedure was applied to three different data sets: The kinship data of Rosenberg and Kim published in Arabie, Carroll, and DeSarbo (1987); Henley's (1969) data on animals; and data on soft drinks (Chaturvedi, 1993). While the application of SINDCLUS to these data sets is described in detail in Chaturvedi and Carroll (1994), here we just describe the application of SINDCLUS to the kinship data of Rosenberg and Kim presented in Arabie et al. (1987).

The 15 most commonly used kinship terms—aunt, brother, cousin, daughter, father, granddaughter, grandfather, grandmother, grandson, mother, nephew, niece, sister, son, and uncle—were printed on slips of paper for use in a sorting task by Rosenberg and Kim (1975). Eighty-five male and 85 female subjects were run in a condition where subjects gave (only) a single sort of the 15 terms. A different group of subjects (85 males and 85 females) were told that, after making their first sorts of the terms, they should give additional subjective partitioning(s) of these stimuli using "a different basis of meaning each time." Rosenberg and Kim (1975) used only the data from the first and second sortings for this group of subjects. Thus, we have six *conditions* which will be our subjects: females' single sort, males' single sort, females' first sort, males' first sort, females' second sort, and males' second sort. Again note that the subjects in the first two conditions were distinct from the subjects in the last four conditions.

Since the subjects' partitions of the stimuli comprise nominal scale data that do not immediately assume the form of a proximity matrix, some pre-processing is necessary to obtain such a matrix. If we form a stimuli × stimuli co-occurrence matrix for each experimental condition, with the (i, j)th entry derived as the number of subjects who placed stimuli i and j in the same group, and subtract that entry from the total number of subjects contributing to the matrix, then we have what is called the S-measure (Arabie et al., 1987). As in Arabie et al. (1987), the six matrices were reversed in scale and then analyzed using SINDCLUS via a matrix unconditional approach. A five-cluster solution explaining 81.1% variance was extracted. This optimal solution derived using the SINDCLUS

TABLE 16.1
SINDCLUS Solution for Rosenberg and Kim Data

Cluster	*Items in Cluster*	*Interpretation*
a	Brother, father, grandfather, grandson, nephew, son, uncle	Male relatives excluding cousins
b	Aunt, daughter, granddaughter, grandmother, mother, niece, sister	Female relatives excluding cousins
c	Aunt, cousin, nephew, niece, uncle	Collateral relatives
d	Brother, daughter, father, mother, sister, son	Nuclear family
e	Granddaughter, grandfather, grandmother, grandson	Grandparents and grand-children
f	All objects	Universal cluster

procedure is identical to the solution presented by Arabie et al. (1987). The five-cluster solution is presented in Table 16.1, while the importance weights are presented in Table 16.2.

The clusters are easily interpreted. In the order listed, the first two are sex defined, the third is the collateral relatives, the fourth is the nuclear family, and the fifth consists of grandparents and grandchildren. The pattern of weights also results in an interesting interpretation. The statement of Rosenberg and Kim (1975) that subjects restricted to a single sort ignore sex as a basis of organization is strongly supported by the relatively low weights for the sex-defined clusters in the first two columns (especially for female subjects) in Table 16.2. For the multiple-sort conditions, it is interesting to note that female subjects emphasized sex in the first sorting (given that the two relevant clusters have much higher weights), whereas male subjects waited until the second sorting to emphasize the salience of sex as a factor in sorting the kinship terms. Across all conditions,

TABLE 16.2
SINDCLUS Weights and VAF for the Rosenberg and Kim Data

Subject	*a*	*b*	*c*	*d*	*e*	*Universal*	*VAF (%)*
F′ single	.052	.049	.552	.478	.626	.055	78.6
M′ single	.143	.146	.397	.372	.449	.075	68.8
F′ first	.551	.554	.283	.206	.251	.132	96.3
F′ second	.241	.246	.373	.322	.385	.158	78.9
M′ first	.299	.291	.340	.241	.395	.158	82.4
M′ second	.295	.306	.237	.219	.253	.207	71.7

females' data were better fitted to the model than were males' data. Also, data from the first sort were better fitted than for the second sort, for both females and males.

NONSYMMETRIC SINDCLUS FOR FITTING OF THE NONSYMMETRIC INDCLUS AND ADCLUS MODELS

While our description of the SINDCLUS algorithm assumed $I = J$, and asymmetry of the $I \times I$ proximity matrices $\mathbf{S}_k$ ($k = 1, 2, \ldots, K$), it should be obvious that we could have easily assumed *nonsymmetric* proximity matrices $\mathbf{S}_k^*$, each $I \times J$ (where I may be different from J). In this case, the algorithm for SINDCLUS would be exactly as described for the symmetric case, except that the binary vector $\mathbf{q}_r$ will be $J \times 1$ rather than $I \times 1$ (while $\mathbf{p}_r$ will still be $I \times 1$). Also, of course, in the case of nonsymmetric proximity data, we do not necessarily expect $\mathbf{P} = \mathbf{Q}$ upon convergence. (In the cases in which $I \neq J$, this is, of course, logically impossible; in cases, such as the example below, in which $I = J$, and the rows and columns of each proximity matrix correspond to the *same* objects, whether $\mathbf{P} = \mathbf{Q}$ is simply an empirical question.)

An Example with Nonsymmetric Proximities: Nonsymmetric SINDCLUS Applied to Brand-Switching Data

DeSarbo (1982) provides brand-switching data on eight soft drinks which were collected by Bass, Pessemier, and Lehmann in 1972, which he analyzed via GENNCLUS. Bass et al. used 280 students and secretaries to select one 12-ounce can of a soft drink four days a week for three weeks. The soft drinks used in the study were Coke, 7-UP, Tab, Like, Pepsi, Sprite, Diet Pepsi, and Fresca. For the two periods measured, the asymmetric brand-switching matrix (two-way) was normalized by dividing each cell by the product of the respective row and column arithmetic mean as in DeSarbo (1982), who analyzed this by a generalized two-way clustering model called GENNCLUS. This asymmetric, 8×8 square matrix

TABLE 16.3
VAF for Bass et al. (1972) Data

# of Clusters	*VAF (%)*
1	28.33
2	47.23
3	**59.21**
4	67.58

TABLE 16.4
Solution for Bass et al. (1972) Data

Cluster	*Row Cluster*	*Column Cluster*	*Interpretation*	*Weight*
a	Coke	Coke	Coke loyalty	0.5118
b	Pepsi	Pepsi	Pepsi loyalty	0.4148
c	Sprite, 7-UP	Sprite, 7-UP	Lime/lemon loyalty	0.1650
d	All drinks	All drinks	Universal cluster	0.1002

was analyzed via the three-way CANDCLUS procedure (with only one level for the third way).

A three-cluster solution, accounting for 59.21% variance (81.25% sum of squares accounted for), was selected for reasons of interpretability. Table 16.3 presents the details of VAF. It should be noted that even though the data were nonsymmetric, the three-cluster CANDCLUS solution was symmetric. The first cluster (Table 16.4) had only Coke in the corresponding row and column clusters, and could be easily be interpreted as the "Coke loyalty" cluster, wherein subjects purchased coke in the two periods under study. The second cluster, similarly, corresponded to a Pepsi loyalty cluster, while the third cluster corresponded to the lime/lemon loyalty cluster, consisting of Sprite and 7-UP in both the row and column clusters. Thus, unlike for the leading cola drinks, Coke and Pepsi, which enjoy brand-level loyalty, the lime/lemon drinks enjoy only a category loyalty, with switching among the 2 popular lime/lemon brands in the study. It is possible that additional clusters would have captured some of the inherent nonsymmetry of the data, but, since we wanted to fit the same number of clusters as DeSarbo had in his "GENNCLUS" analysis, we restricted this analysis to only three clusters.

FITTING CANDCLUS VIA A LEAST ABSOLUTE DEVIATION CRITERION

While many possible alternate fit measures are possible (weighted least squares has already been mentioned as an option to OLS), one that is particularly attractive, and has been used in many data analytic applications is a "Least absolute deviation" (or least sum of absolute deviations rather than least sum of squared deviations) criterion, which we abbreviate as a LAD criterion. It is quite straightforward, as it happens to generalize fitting of the entire array of models we have described above to the case of optimizing a LAD rather than an OLS criterion (or, more generally a weighted LAD rather than a weighted least-squares criterion). This has been discussed by Chaturvedi, Carroll, and Lakshmi-Ratan (1993), and by Chaturvedi, Lakshmi-Ratan, and Carroll (1994). In order to accomplish this in the "one-dimension-at-a-time" approach that we have taken, it suffices to replace the EDLSP with an elementary discrete least absolute deviation procedure (EDLAD). We demonstrate how this is accomplished using the same illustrative data as before.

The Elementary Discrete LAD Procedure

Consider the following illustrative problem as before. Let

$$\mathbf{L} = \begin{bmatrix} 7 & 5 & 3 & 9 \\ 8 & 6 & 5 & 1 \\ 9 & 4 & 2 & 7 \\ 5 & 3 & 4 & 6 \end{bmatrix}, \quad \mathbf{x} = \begin{bmatrix} x_1 \\ x_2 \\ x_3 \\ x_4 \end{bmatrix}, \quad \mathbf{r}' = [1 \quad 4 \quad 9 \quad 3].$$

The estimation problem is to find the LAD estimate of **x**, where $\mathbf{L} = \mathbf{x}\mathbf{r}' + \text{error}$ and **x** is either 0 or 1. That is,

$$\begin{bmatrix} 7 & 5 & 3 & 9 \\ 8 & 6 & 5 & 1 \\ 9 & 4 & 2 & 7 \\ 5 & 3 & 4 & 6 \end{bmatrix} = \begin{bmatrix} 1x_1 & 4x_1 & 9x_1 & 3x_1 \\ 1x_2 & 4x_2 & 9x_2 & 3x_2 \\ 1x_3 & 4x_3 & 9x_3 & 3x_3 \\ 1x_4 & 4x_4 & 9x_4 & 3x_4 \end{bmatrix} + \text{error}.$$

If we let

$$f_1 = |7 - 1x_1| + |5 - 4x_1| + |3 - 9x_1| + |9 - 3x_1|,$$
$$f_2 = |8 - 1x_2| + |6 - 4x_2| + |5 - 9x_2| + |1 - 3x_2|,$$
$$f_3 = |9 - 1x_3| + |4 - 4x_3| + |2 - 9x_3| + |7 - 3x_3|,$$
$$f_4 = |5 - 1x_4| + |3 - 4x_4| + |4 - 9x_4| + |6 - 3x_4|,$$

then the sum of absolute errors is

$$F = f_1 + f_2 + f_3 + f_4.$$

Note that f_1 is a function only of x_1; f_2 is a function only of x_2; f_3 is a function only of x_3; and f_4 is a function only of x_4. Thus, F is *separable* in x_1, x_2, x_3, and x_4. To minimize F, one can separately minimize f_1 w.r.t x_1, f_2 w.r.t x_2, f_3 w.r.t x_3, and f_4 w.r.t. x_4. To minimize, say, f_1 w.r.t. x_1, one can easily evaluate f_1 at $x_1 = 1$ and $x_1 = 0$. The x_1 yielding a minimum of these two possible values is then chosen. Thus, for I 0-1 variables, only $2I$ function evaluations and comparisons are needed, as compared to 2^I evaluations and comparisons for explicit enumeration. To complete the LAD estimation scheme, we require a substitute for the OLS (or WLS) estimation procedures. We outline the approach we take, utilizing what we call the elementary continuous LAD procedure.

The Elementary Continuous LAD Procedure

Again, consider another illustrative problem. Let

$$\mathbf{L} = \begin{bmatrix} 7 & 5 & 3 & 9 \\ 8 & 6 & 5 & 1 \\ 9 & 4 & 2 & 7 \\ 5 & 3 & 4 & 6 \end{bmatrix}, \qquad \mathbf{x} = \begin{bmatrix} x_1 \\ x_2 \\ x_3 \\ x_4 \end{bmatrix}, \qquad \mathbf{r}' = [1 \quad 4 \quad 9 \quad 3].$$

The estimation problem is to find the LAD estimate of **x**, where $\mathbf{L} = \mathbf{xr}' +$ error and **x** is real. That is,

$$\begin{bmatrix} 7 & 5 & 3 & 9 \\ 8 & 6 & 5 & 1 \\ 9 & 4 & 2 & 7 \\ 5 & 3 & 4 & 6 \end{bmatrix} = \begin{bmatrix} 1x_1 & 4x_1 & 9x_1 & 3x_1 \\ 1x_2 & 4x_2 & 9x_2 & 3x_2 \\ 1x_3 & 4x_3 & 9x_3 & 3x_3 \\ 1x_4 & 4x_4 & 9x_4 & 3x_4 \end{bmatrix} + \text{error}.$$

If we let

$$f_1 = |7 - 1x_1| + |5 - 4x_1| + |3 - 9x_1| + |9 - 3x_1|,$$
$$f_2 = |8 - 1x_2| + |6 - 4x_2| + |5 - 9x_2| + |1 - 3x_2|,$$
$$f_3 = |9 - 1x_3| + |4 - 4x_3| + |2 - 9x_3| + |7 - 3x_3|,$$
$$f_4 = |5 - 1x_4| + |3 - 4x_4| + |4 - 9x_4| + |6 - 3x_4|,$$

then the sum of absolute errors is

$$F = f_1 + f_2 + f_3 + f_4.$$

Note that f_1 is a function only of x_1; f_2 is a function only of x_2; f_3 is a function only of x_3; and f_4 is a function only of x_4. Thus, F is *separable* in x_1, x_2, x_3, and x_4. To minimize F, one can separately minimize f_1 w.r.t x_1, f_2 w.r.t x_2, f_3 w.r.t x_3, and, f_4 w.r.t. x_4. In order to minimize, say, f_1 w.r.t. x_1, simply evaluate f_1 at the four points given by $x_1 = 7/1$, $5/4$, $3/9$, and $9/3$. The value of x_1 yielding the minimum is chosèn as the optimum LAD estimate of x_1. In this specific case, f_1 is minimum at $x_1 = 3/9$.

More generally, to minimize the LAD criterion as a function of a simple variable x_i, for the component f_i of the overall loss function

$$F = \sum_{i=1}^{I} f_i,$$

where $f_i = \Sigma_{j=1}^{J} |l_{ij} - x_i r_j|$ is a function only of x_i, we simply evaluate f_i at the J values

$$x_i^{(j')} = l_{ij}'/r_j'$$

and then choose the $\hat{x}_i$ to be the $x_i^{(j')}$ minimizing

$$f_i^{(j')} = \sum_{j=1}^{J} |l_{ij} - x_i^{(j')} r_j|.$$

This can easily be shown to provide the x_i minimizing f_i, thus completing the elementary continuous LAD procedure (for our "one dimension at a time" estimation scheme). While this LAD estimation scheme can be applied to any of the family of CANDCLUS models, we illustrate it here with an application to fitting the INDCLUS/ADCLUS models. The resulting approach we dub LADCLUS. (Note that this was previously called MADCLUS, for "*Minimum* Absolute Derivation Clustering," by Chaturvedi, Carroll and Lakshmi-Ratan, 1993).

Application of LADCLUS to Soft Drinks Data

Chaturvedi (1993) presents proximity data on 16 soft drinks. In the data collection task, pairwise dissimilarity ratings on 16 soft drinks were collected from 13 students of Rutgers University in 1992. The 16 soft drinks chosen in the study were Coke, Diet Coke, Cherry Coke, C&C Pineapple, Dr. Pepper, Pepsi, Diet Pepsi, Orange Slice, Sprite, Diet Sprite, Sunkist, Diet Sunkist, 7-UP, Diet 7-UP, C&C Cola, and Hawaiian Punch. The 13 proximity matrices were then analyzed via the LADCLUS procedure. A six-cluster solution was chosen for reasons of interpretability. Table 16.5 presents a summary of the fit by LADCLUS. Table 16.6 presents a summary description of the extracted clusters and their interpretation. Table 16.7 gives the weights of the 13 subjects for the six clusters. Tables 16.8–16.10 present details of the solutions attained via an application of SINDCLUS to the same data.

The six-cluster solution in Table 16.6 clearly separates the cola-based drinks (cluster *d* corresponding to Coke, Diet Coke, Cherry Coke, Dr. Pepper, Pepsi, Diet Pepsi, and C&C Cola) from the non-cola-based fruity beverages (cluster *e* corresponding to C&C Pineapple, Orange Slice, Sprite, Diet Sprite, Sunkist,

TABLE 16.5
Deviation Accounted for by LADCLUS

# of Clusters	*% Deviation Accounted For*
1	77.7
2	82.6
3	84.7
4	86.2
5	87.7
6	**87.9**

TABLE 16.6
LADCLUS Solution for Soft Drinks Data

Cluster	*Items in Cluster*	*Interpretation*
a	Coke, Diet Coke, Pepsi, Diet Pepsi, C&C Cola	Regular colas
b	Sprite, Diet Sprite, 7-UP, Diet 7-UP	Lemon/Lime
c	Orange Slice, Sunkist, Diet Sunkist	Orange
d	Coke, Diet Coke, Cherry Coke, Dr. Pepper, Pepsi, Diet Pepsi, C&C Cola	All colas
e	C&C Pineapple, Orange Slice, Sprite, Diet Sprite, Sunkist, Diet Sunkist, 7-UP, Diet 7-UP, Hawaiian Punch	All fruity beverages (noncolas)
f	Coke, Cherry Coke, Pepsi, C&C Cola	Nondiet colas

Diet Sunkist, 7-UP, Diet 7-UP, and Hawaiian Punch). The colas are further grouped into regular colas (without any fruit bases) corresponding to cluster *a* consisting of Coke, Diet Coke, Pepsi, Diet Pepsi, and C&C Cola. The fruity beverages (cluster *e*) has two distinct clusters wholly included in it: the lemon/lime-based drinks like Sprite, Diet Sprite, 7-UP, and Diet 7-UP corresponding to cluster *b*, and the Orange drinks (cluster *c*) like Orange Slice, Sunkist, and Diet Sunkist. Cluster *f* has the nondiet colas like Coke, Cherry Coke, Pepsi, and C&C Cola. It is interesting to note that only one of the six clusters derived respectively using LADCLUS and SINDCLUS is different. While

TABLE 16.7
LADCLUS Weight Matrix for Soft Drinks Data

Subject	*a* *Colas*	*b* *Lime*	*c* *Orange*	*d* *All Colas*	*e* *N-Cola Fruit*	*f* *N-Diet Cola*
1	0.22	0.22	0.67	0.00	0.00	0.00
2	1.00	1.00	1.00	0.00	0.00	0.00
3	0.33	1.00	1.00	0.00	0.00	0.00
4	0.34	0.67	0.56	0.33	0.11	0.00
5	0.11	0.56	0.56	0.33	0.22	0.22
6	0.67	0.78	0.78	0.00	0.00	0.00
7	0.51	0.89	0.00	0.27	0.00	0.11
8	0.22	0.33	0.33	0.33	0.44	0.00
9	0.67	0.78	0.78	0.00	0.00	0.00
10	0.11	0.33	0.33	0.11	0.00	0.22
11	0.22	0.00	0.56	0.33	0.00	0.33
12	0.33	0.67	0.56	0.00	0.11	0.22
13	0.12	0.56	0.67	0.44	0.00	0.01

TABLE 16.8
VAF for Soft Drinks

# of Clusters	*VAF (%)*
1	23.83
2	48.95
3	61.90
4	68.23
5	71.75
6	**73.44**
7	74.55

LADCLUS grouped all the nondiet colas in a separate cluster, SINDCLUS grouped only the fruity beverages like Cherry Coke, Dr. Pepper, and Hawaiian Punch. It is also interesting to note how much "crisper" the LADCLUS subject weights are than are those for SINDCLUS, as indicated by the large number of weights exactly equal to zero or one in the former case, while there are only a few zeros in the latter case, and no ones.

FITTING HYBRID MODELS VIA CANDCLUS

While we have assumed so far that all "dimensions" for a given mode/way of the data array are of the same form: being a k-valued discrete variable with a fixed set of k possible values, or continuous—it is also possible within this formulation to allow some dimensions to be of one type (e.g., continuous) and another subset of dimensions to be of another (e.g., discrete, clusterlike). In fact, it is even possible for some to be continuous with no constraints, others continuous with

TABLE 16.9
SINDCLUS Solution for Soft Drinks

Cluster	*Items in Cluster*	*Interpretation*
a	Coke, Diet Coke, Pepsi, Diet Pepsi, C&C Cola	Regular colas
b	Sprite, Diet Sprite, 7-UP, Diet 7-UP	Lemon/Lime
c	Orange Slice, Sunkist, Diet Sunkist	Orange
d	Coke, Diet Coke, Cherry Coke, Dr. Pepper, Pepsi, Diet Pepsi, C&C Cola	All colas
e	C&C Pineapple, Orange Slice, Sprite, Diet Sprite, Sunkist, Diet Sunkist, 7-UP, Diet 7-UP, Hawaiian Punch	All fruity beverages (noncolas)
f	Cherry Coke, Dr. Pepper, Hawaiian Punch	Fruity beverages

TABLE 16.10
Subject Weight Matrix via SINDCLUS for Soft Drinks Data

Subject	*a* *Colas*	*b* *Lime*	*c* *Orange*	*d* *All Colas*	*e* *N-Cola Fruit*	*f* *Fruit*
1	.20	.33	.52	.03	.00	.00
2	.94	.96	.96	.02	.00	.65
3	.42	.79	.77	.00	.00	.00
4	.31	.61	.57	.27	.13	.32
5	.14	.59	.35	.36	.12	.08
6	.58	.82	.84	.02	.00	.01
7	.48	.80	.30	.31	.02	.25
8	.26	.28	.33	.37	.43	.24
9	.41	.79	.84	.11	.01	.00
10	.14	.38	.49	.18	.02	.00
11	.23	.22	.50	.38	.03	.07
12	.21	.57	.33	.18	.10	.00
13	.12	.57	.72	.47	.07	.03

nonnegativity constraints, a third subset to be binary, and some other subsets to be k-ary, etc. In most cases, however, we consider here the case in which one set of R_c dimensions are continuous (spatial) dimensions, and another set of R_d are binary "cluster membership" variables. This leads to a class of what Carroll (1976) and Carroll and Pruzansky (1980) have called "hybrid" models, mixing discrete and continuous components; in the present case R_c spatial dimensions combined with R_d "discrete" clusters, defining an overlapping cluster structure. (Carroll, 1976, and Carroll & Pruzansky, 1980, discussed primarily hybrid models in which the discrete component consisted of one or more tree structures rather than an overlapping cluster structure, but since tree structure is a special case of an overlapping cluster structure in which the clusters are hierarchically nested, the cases considered in these earlier papers can be viewed as a special case of those considered here.)

We consider now three increasingly general classes of hybrid models we feel to be of special interest.

Hybrid Model Combining INDSCAL and INDCLUS for Three-Way Symmetric Proximity Data

Given K $I \times I$ symmetric "scalar products" matrices $S_{k,}$ $k = 1, \ldots, K,$ we fit a model of the form

$$\mathbf{S}_k = \mathbf{A}\mathbf{W}_k\mathbf{A}' + \mathbf{C}_k + \text{error},$$

where $\mathbf{A}$ is an $I \times R$ matrix whose columns are partitioned into R_c continuous dimensions and R_d discrete dimensions defining R_d (overlapping) clusters. $\mathbf{W}_k$ is a diagonal $R \times R$ weight matrix. In the fitting procedure we merely use the appropriate discrete least-squares (or LAD) procedure for the several R_d dimensions for the two object modes (for the i objects), while the first R_c are fit by the appropriate continuous procedure.

Nonsymmetric Hybrid Model for Three-Way Nonsymmetric Data

Given K nonsymmetric $I \times J$ matrices $\mathbf{S}_k$, we fit a model of the form

$$\mathbf{S}_k = \mathbf{A}\mathbf{W}_k\mathbf{B}' + \mathbf{C}_k + \text{error},$$

where $\mathbf{A}$ is an $I \times R$ matrix and $\mathbf{B}$ is a $J \times R$ matrix, partitioned into R_c continuous dimensions and R_d discrete dimensions defining R_d (overlapping) clusters, separately for the I row objects and the J column objects). $\mathbf{W}_k$ is a diagonal $R \times R$ weight matrix. Note that, in case of both the above models, the two-way case corresponds to the case in which $K = 1$.

General Hybrid Model for N-Way Data

The "general" hybrid model is of the same multilinear form as N-way CANDCLUS or CANDECOMP, but the matrix $\mathbf{A}_n$ ($n = 1, \ldots, N$) in the N-way decomposition has R_n^c continuous and $R_n^d = R_n - R_n^c$ discrete dimensions. R_n^c and R_n^d may differ for different values of n, and assignment of continuous and discrete dimensions within each matrix is totally arbitrary (e.g., the R_n^c continuous dimensions need not correspond to adjacent columns, nor must the R_n^d discrete dimensions. Discrete dimensions may be binary, defining cluster membership, or some or all may be k-ary for arbitrary k (possibly different for different discrete dimensions, even within same matrix $\mathbf{A}_n$). In addition to other work referred to earlier, some other work closely related to the two-way special case of CANDCLUS is reported by Mirkin (1990).

MUMCLUS (MULTIMODE CLUSTERING), GENNCLUS, GENERALIZED GENNCLUS, TUCKER'S THREE-MODE AND MULTIMODE FACTOR/COMPONENTS ANALYSIS, AND HYBRID MODELS OF GENERAL TUCKER FORM

DeSarbo's GENNCLUS Model (Two-Way and Three-Way Versions)

Given nonsymmetric $I \times J$ matrices $\mathbf{S}_k$, $k = 1, \ldots, K$, the three-way GENNCLUS model is of the form

$$\mathbf{S}_k = \mathbf{A}\mathbf{U}_k\mathbf{B}' + \mathbf{C}_k + \text{error},$$

where **A** is an $I \times R_a$ binary matrix, **B** is a $J \times R_b$ binary matrix, while $\mathbf{U}_k$ is a completely general $R_a \times R_b$ matrix.

In generalized GENNCLUS, **A** and **B** can each be continous, discrete, or a mixture of continuous and discrete dimensions (i.e., hybrid). **U** is assumed to have continuously valued entries.

Tucker's three-mode factor/components analysis (TMFA) model corresponds to the case in which both **A** and **B** (as well as **U**) are entirely continuously valued. It should be noted that two-way GENNCLUS, as proposed by DeSarbo (1982), corresponds to the case where $K = 1$.

N-Mode Factor/Components, GENNCLUS, or Other Hybrid Models

Given an $I_1 \times I_2 \times \ldots \times I_N$ array, with general entry $y_{i_1 i_2} \ldots i_N$, where $i_n = 1, 2, \ldots, I_n$ and $n = 1, 2, \ldots, N$. We fit it by a model of the general algebraic form

$$y_{i_1 i_2 \ldots i_N} \sim \sum_{t_1=1}^{T_1} \sum_{t_2=1}^{T_2} \ldots \sum_{t_n=1}^{T_N} a^1_{i_1 t_1} a^2_{i_2 t_2} \ldots a^n_{i_n t_n} u_{t_1 t_2 \ldots t_n},$$

where $\mathbf{A}_n$ is an $I_N \times T_N$ and **U** is a $T_1 \times T_2 \times \ldots \times T_N$ array (a "generalized" core array" in Tucker's terminology). Each $\mathbf{A}_n$ may be continuous, discrete, or hybrid, while **U** will have continuously valued entries. It should be noted that NMFA/GENNCLUS is the most general model, including all others as special cases. We can fit any of thee models in the MUMCLUS class of models by a generalization of the CANDCLUS algorithm, which, however, will not be discussed in detail here.

Any of these models (except GENNCLUS, generalized GENNCLUS, and others in the MUMCLUS class of models, which, at present can be fitted only by OLS or WLS procedures) may be fit via OLS, WLS (weighted least squares), LAD, WLAD (weighted least absolute deviation), or via other criteria corresponding, say, to an L_p-norm-based fit measures, or any other "additively decomposable" fit measure, i.e., one of the form

$$F = \sum_{o=1}^{O} f(y_o, \hat{y}_o)$$

where o ranges over all O observations. In case of L_p based on more general fit measures, line-search algorithms may have to be used to fit continuous parameters. Nonnegativity (or other) constraints may easily be imposed on continuous parameters.

CONCLUSIONS

We presented a very general class of models called the CANDCLUS family of models, which present an equally general approach to clustering and multidimensional scaling of N-way data. We also described in detail the estimation procedure to fit any of the models that might be a specific case of CANDCLUS, via OLS, WLS, LAD, or WLAD criteria. The fitting procedure, can, in principle, be generalized to fitting the CANDCLUS model by any L_p norm. In case of L_p norms where $p \neq 1$ or 2, the corresponding elementary continuous procedures simply have to be replaced by appropriate line-search procedures. It should also be mentioned that the CANDCLUS family of models can be fit via a generalized separability property that enables the discrete parameters for some or all of the modes/ways to be partitions rather than overlapping discrete structures.

The authors have also investigated a very interesting special case of the CANDCLUS model—a two-way CANDCLUS model where one way is modeled via continuous parameters, and the other via discrete 0-1 parameters. Even though the results of that work are being presented elsewhere, the important thing to note here is that this specific two-way case of CANDCLUS includes K-means cluster analysis as a special case when the clusters are constrained to be nonoverlapping via the generalized separability property alluded to earlier. When fit via the L_1 norm, this model results in what is called the k-medians procedure, which is also described in Hartigan (1975).[1] The unconstrained two-way CANDCLUS model, when fit via the elementary discrete procedures described earlier in this section, results in a general overlapping clustering model (Chaturvedi, Carroll, Green, & Rotondo, 1994) which extends k-means and k-medians cluster analysis to what we call overlapping k-means and overlapping k-medians clustering, respectively.

Some of this work has been done jointly with R. A. Lakshmi-Ratan and J. A. Rotondo of AT&T Bell Laboratories, and P. E. Green of the University of Pennsylvania. The two-way CANDCLUS approach, when applied in a sequential "one-cluster-at-a-time" manner also results in another approach to partitioning, via any of the L_p norms, which we dub *principal trees* because of the similarity of a component of the approach to extracting principal components from multivariate data. As mentioned earlier, the theoretical and empirical details of these approaches and further extensions of the two-way CANDCLUS model are being presented elsewhere.

REFERENCES

Arabie, P., & Carroll, J. D. (1980). MAPCLUS: A mathematical programming approach to fitting the ADCLUS model. *Psychometrika, 45,* 211–235.

Arabie, P., Carroll, J. D., & DeSarbo, W. S. (1987). *Three-way scaling and clustering*. Newbury Park, CA: Sage.

[1]We wish to thank Dr. Taskin Atilgan of AT&T Bell Laboratories for having pointed this earlier reference out to us.

Bass, F. M., Pessemier, E. A., & Lehmann, D. R. (1972). An experimental study of relationships between attitudes, brand preference, and choice. *Behavioral Science, 17,* 532–541.

Carroll, J.D. (1975). *Models for individual differences in similarities.* Paper presented at the Eighth Annual Mathematical Psychology Meeting, Purdue University, West Lafayette, IN.

Carroll, J. D. (1976). Spatial, non-spatial and hybrid models for scaling. (Presidential Address for Psychometric Society). *Psychometrika, 41,* 439–463.

Carroll, J. D., & Arabie, P. (1980). Multidimensional scaling. In M. R. Rosenzweig & L. W. Porter (Eds.), *Annual review of Psychology* (Vol. 31, pp. 607–649). Palo Alto, CA: Annual Reviews.

Carroll, J.D., & Arabie, P. (1983). An individual differences generalization of the ADCLUS model and the MAPCLUS algorithm. *Psychometrika, 48,* 157–169.

Carroll, J. D., & Chang, J. J. (1970). Analysis of individual differences in multidimensional scaling via an N-way generalization of 'Eckart-Young' decomposition. *Psychometrika,* 283–319.

Carroll, J. D., DeSoete, G., & Pruzansky, S. (1989). Fitting of the latent class model via iteratively reweighted least squares CANDECOMP with nonnegativity constraints. In R. Coppi and S. Bolasco (Eds.), *Multiway data analysis* (pp. 463–472). Amsterdam: North-Holland.

Carroll, J. D., & Pruzansky, S. (1975). Fitting of hierarchical tree structure (HTS) models, mixtures of HTS models, and hybrid models, via mathematical programming and alternating least squares. *Proceedings of the U.S.-Japan Seminar on Multidimensional Scaling,* pp. 9–19.

Carroll, J.D., & Pruzansky, S. (1980). Discrete and hybrid scaling models. In E. D. Lantermann and H. Feger (Eds.), *Similarity and choice* (pp. 108–139). Bern: Hans Huber.

Chaturvedi, A. D. (1993). *Perceived product uniqueness in product differentiation and product choice.* 9316296, University Microfilms International, Ann Arbor, Michigan.

Chaturvedi, A. D., & Carroll, J. D. (June, 1992). ALCLUS: Alternating least squares clustering. Presented at the Annual Meeting of The Classification Society of North America, East Lansing, Michigan.

Chaturvedi, A. D., & Carroll, J. D. (1994). An alternating combinatorial optimization approach to fitting the INDCLUS and generalized INDCLUS models. *Journal of Classification, 11,* 155–170.

Chaturvedi, A., Carroll, J. D., Green, P. E., & Rotondo, J. A. (1994). *A feature based approach to market segmentation via overlapping K-centroids clustering.* Manuscript submitted for publication.

Chaturvedi, A. D., Carroll, J. D., & Lakshmi-Ratan, R. A. (1993, June). *MADCLUS: Minimum absolute deviation clustering.* Presented at the Annual Meeting of the Classification Society of North America, Pittsburgh, PA.

Chaturvedi, A., Lakshmi-Ratan, R. A., & Carroll, J. D. (1994). *Two L1-norm procedures for fitting ADCLUS and INDCLUS.* Manuscript submitted for publication.

DeSarbo, W. S. (1982). GENNCLUS: New models for general nonhierarchical clustering analysis. *Psychometrika, 47,* 449–475.

Hartigan, J. A. (1975). *Clustering algorithms.* New York: Wiley.

Henley, N. M. (1969). A psychological study of the semantics of animal terms. *Journal of Verbal Learning and Verbal Behavior, 8,* 176–184.

Mirkin, B. G. (1987). Additive clustering and qualitative factor analysis methods for similarity matrices. *Journal of Classification, 4,* 7–31.

Mirkin, B. G. (1989). Erratum. *Journal of Classification, 6,* 271–272.

Mirkin, B. G. (1990). A sequential fitting procedure for linear data analysis models. *Journal of Classification, 7,* 167–195.

Rosenberg, S., & Kim, M. P. (1975). The method of sorting as a data-gathering procedure in muultivariate research. *Multivariate Behavioral Research, 10,* 489–502.

Shepard, R. N., & Arabie, P. (1979). Additive clustering representation of similarities as combinations of discrete overlapping properties. *Psychological Review, 86,* 87–123.

17 Network Models for Scaling Proximity Data

Karl Christoph Klauer
Freie Universität Berlin

J. Douglas Carroll
Rutgers University

ABSTRACT

Network models aim at representing proximity data by means of the minimum-path-length function of connected and weighted graphs. Fundamental representation and uniqueness results underlying network models as psychological representations of stimuli, given both ordinal-scale as well as interval-scale proximity measures, are discussed. In addition, computational methods for network analyses are reviewed and compared. Methods now exist to scale metric as well as nonmetric data, symmetric and nonsymmetric proximity measures, and two-way and three-way data. They are compared with respect to the factors of (a) computational cost, (b) accuracy of recovery of an underlying network, and (c) goodness of fit to the observed proximity data.

INTRODUCTION

The analysis of similarity and dissimilarity data has focused on two main approaches, the spatial and the discrete approach. In spatial models such as multidimensional scaling (Shepard, 1962; Kruskal, 1964), dissimilarities are represented by means of a continuous, low-dimensional space. Other methods derive from discrete models that yield hierarchical clusters represented as ultrametric trees (Johnson, 1967; Hartigan, 1967), more general tree structures and multiple tree structures (Carroll & Chang, 1973; Carroll, 1976; Sattath & Tversky, 1977; Cunningham, 1978; Carroll & Pruzansky, 1980; De Soete, 1983; Carroll, Clark, & DeSarbo, 1984), structures called extended trees (Corter & Tversky, 1986), additive clusters (Shepard & Arabie, 1979), and network models (Feger & Bien,

1982; Feger & Droge, 1984; Orth, 1988; Schvaneveldt, Dearholt, & Durso, 1988; Hutchinson, 1989; Klauer, 1989; Klauer & Carroll, 1989, 1991).

In recent years, much work has gone into the development of discrete models that are often found closer to psychological theories such as psychological feature models (Tversky, 1977; Tversky & Gati, 1982), network models in cognitive psychology (Collins & Loftus, 1975; Collins & Quillian, 1969), and network models in social psychology, anthropology, and sociology (Bales, 1970; Hage & Harary, 1983), and categorical theories (Anderson, 1983). Some foundations underlying multidimensional scaling as a model of the psychological representation of stimuli have been proposed and discussed by Beals, Krantz, and Tversky (1968). Although the foundations underlying the various discrete models have not in all cases been formally clarified to the same extent, the connections between discrete models and graph theory are becoming more apparent (Shepard & Arabie, 1979).

This chapter concentrates on network models that aim at representing proximities by means of the minimum-path-length function of connected and weighted graphs. Nonsymmetric proximity measures are usually scaled by means of directed graphs, whereas symmetric proximity measures give rise to undirected graphs. The first part is devoted to exploring the formal foundations underlying these network models as psychological representations of stimuli. These representation and uniqueness results (Krantz, Luce, Suppes, & Tversky, 1971) are developed for proximity measures with interval-scale as well as with ordinal-scale properties. In addition, existing algorithms for deriving network representations from observed similarities or dissimilarities are reviewed and compared.

NOTATION

A *network* G is defined as a triple (E, V, t). E is the set of nodes, V is the set of links, and t is a positive function assigning real numbers to links. For *directed* networks, links are ordered pairs (X, Y), $X \neq Y$. For undirected networks, links are unordered pairs, so that (X, Y) and (Y, X) are considered identical and analogously, $t(X, Y) = t(Y, X)$. A *path* from X to Y is a sequence of links, $l_j = (X_j, X_{j+1})$, $j = 1, \ldots, n$, with $X_1 = X$ and $X_{n+1} = Y$. The *length* $L(p)$ of the path $p = (l_1, \ldots, l_n)$ is given by

$$L(p) = \sum_{j=1}^{n} t(l_j).$$

A network is called *connected* if, for every X and Y, there is at least one path from X to Y. Given a connected network G, the minimum-path-length function $(X \neq Y)$

$$d_G(X, Y) = \min\{L(p)|p \text{ is a path from } X \text{ to } Y\},$$

is well-defined, positive, and satisfies the triangle inequality. The *triangle inequality* states that for all X, Y, and Z $d_G(X, Z) \leq d_G(X, Y) + d_G(Y, Z)$. By convention, $d_G(X, X)$ ($X \in E$) is defined separately as $d_G(X, X) = 0$. For an undirected network, d_G also satisfies the symmetry condition $d_G(X, Y) = d_G(Y, X)$. Let the number of objects in E be denoted by N and the number of ordered pairs of objects by $M_o = N(N - 1)$, whereas $M_u = M_o/2$ denotes the number of unordered pairs.

METRIC REPRESENTATION RESULTS

In the following, relevant graph-theoretical results about minimum-path-length functions and their relation to network structure are reviewed. A function d assigning real numbers to pairs (X, Y) is called *realizable* if there is a network G such that the minimum-path-length function d_G and d are identical. Graph-theoretical models usually impose severe restrictions. For example, in hierarchical clustering (Johnson, 1967), the ultrametric inequality must hold for d to be realizable by means of an ultrametric tree, whereas additive or path-length trees (Carroll, 1976; Sattath & Tversky, 1977) require a so-called four-point inequality. Thus, it is interesting to ask whether network models impose similar constraints. It turns out that they essentially do not impose any constraints over and above the triangle inequality unless the number of links in the network is constrained to be less than the maximum number possible.

> **Theorem 17.1.** The function d is realizable by a directed network if and only if it is nonnegative, satisfies $d(X, X) = 0$ if and only if $X = Y$, and satisfies the triangle inequality. It is realizable by an undirected network, if and only if in addition, it satisfies the symmetry condition $d(X, Y) = d(Y, X)$ for all $X, Y \in E$.

The proof is constructive and proceeds by considering the complete network that contains all possible pairs as links with link weights $t(X, Y) = d(X, Y)$; see for example, Hakimi and Yau (1964).

Complete networks are almost always uninteresting, however, and network representations usually strive to work with parsimonious, sparse networks that contain fewer links than the complete network. If the objective is to realize a function d with a more parsimonious network, substantial constraints are imposed on d as is easily seen. A network G is said to be *irreducible* if none of its links can be omitted without changing the minimum-path-length function. Goldman (1966), for example, proved the following theorem.

Theorem 17.2. Let G be an irreducible representation with $d_G = d$. The pair (X, Z), $X \neq Z$, is a link of G, that is $(X, Z) \in V$, if and only if

$$d(X, Z) < \min\{d(X, Y) + d(Y, Z) | Y \neq X, Z\}.$$

Since in the above inequality, the triangle inequality guarantees "less than or equal to," this means that the link (X, Z) can be omitted from the representing network, if and only if $d(X, Z) = d(X, Y) + d(Y, Z)$ for some suitable third point $Y \neq X, Z$. Thus, a triangle *additivity* allows one to drop one link. The reason is that the smallest paths connecting X and Y, and Y and Z, respectively, can be concatenated to form a path from X to Z with length given by the sum of the two component path lengths, that is by $d(X, Y) + d(Y, Z) = d(X, Z)$. By triangle inequality and the definition of the minimum-path-length function, this path connecting X and Z via Y must also be a shortest path. Any occurrence of the direct link (X, Z) in defining the minimum-path-length function can thus be replaced by the path from X to Z via Y without affecting the minimum-path-length function.

It turns out that each additional triangle additivity allows one to omit an additional link, so that the following theorem holds:

Theorem 17.3. A nonnegative function d with $d(X, Y) = 0$ if and only if $X = Y$ and satisfying the triangle inequality (and the symmetry condition) can be represented as the minimum-path-length function of a connected and directed (undirected) network with L links, if and only if there are $M_o - L$ ($M_u - L$) distinct pairs of objects (X_i, Z_i) as well as third points Y_i with $X_i \neq Y_i$, $Y_i \neq Z_i$, $X_i \neq Z_i$, such that

$$d(X_i, Z_i) = d(X_i, Y_i) + d(Y_i, Z_i),$$
$$i = 1, \ldots, M_o - L \quad (i = 1, \ldots, M_u - L).$$

Furthermore, a representing network is $G = (E, V, t)$ with V given by all pairs of objects different from the $M_o - L$ ($M_u - L$) pairs (X_i, Z_i) and with $t(X, Y) = d(X, Y)$ for $(X, Y) \in V$.

The proof is given in Klauer and Carroll (1991) and consists primarily in showing that the representing network defined in the theorem in fact reproduces the function d as its minimum-path-length function. Theorems 17.1–17.3 and their proofs amount to constructive representation theorems for ratio-scale measures of network distances (Krantz et al., 1971), and by allowing for a suitable additive constant can easily be extended to the case of interval-scale measures. As such, Theorem 17.3 in particular underlies the mathematical programming approaches to network scaling recently proposed by Klauer and Carroll (1989, 1991). Before turning to these algorithms, however, let us consider corresponding representation and uniqueness results for the case of *ordinal-scale* measures.

NONMETRIC REPRESENTATION RESULTS

For the nonmetric results, we begin with a proximity (dissimilarity) relation $\preccurlyeq$ defined upon the set of pairs of objects $E \times E$, where $(X, Y) \preccurlyeq (U, V)$ if and only if X was observed to be less than or equally dissimilar to Y than U to V. Depending upon context, the term *less than or equally dissimilar to* may of course have to be replaced by terms such as *not further from, interacted no less with, correlated no less with,* and so forth. In the following, the ordinal counterparts of representation Theorems 17.1 and 17.2 are derived. Thus, the conditions are discussed that proximity relations must satisfy in order to be realizable by a network representation as well as the conditions on proximity relations necessitating the presence of a particular link in any such representation.

Given a relation $\preccurlyeq$ upon a set $E \times E$, the *asymmetric part* $\prec$ of $\preccurlyeq$ is defined by $(a, b \in E \times E)$

$$a \prec b \text{ iff } (a \preccurlyeq b) \text{ and not } (b \preccurlyeq a),$$

whereas the *indifference relation,* $\sim$, is given by

$$a \sim b \text{ iff } (a \preccurlyeq b) \text{ and } (b \preccurlyeq a).$$

The *transitive hull* $\preccurlyeq^{+}$ of $\preccurlyeq$ is given by

$$a \preccurlyeq^{+} b \quad \text{iff there is a sequence } a_1, \ldots, a_{n+1} \text{ such that } a = a_1 \preccurlyeq \cdots \preccurlyeq a_{n+1} = b.$$

A fundamental condition for numerically representing a relation is acyclicity:

Definition 17.1. The proximity relation satisfies *acyclicity,* if

$$\text{for all } (X, Y), (U, V) \quad (X, Y) \preccurlyeq^{+} (U, V) \text{ implies not}(U, V) \prec (X, Y). \tag{17.1}$$

Note that the type of data discussed here includes triadic comparisons of paired comparisons of pairs, which may violate acyclicity. If the data are in the form of a matrix of proximities, acyclicity is necessarily satisfied.

A condition that is specific for proximity relations is *zero-minimality* corresponding to the constraint that "self-similarity" is maximal. More precisely, zero minimality requires that the similarity of different objects does not exceed self-similarity and that there are no differences in self-similarity.

Definition 17.2. The proximity relation satisfies *zero minimality,* if

$$\text{for all } X, Y, U, V \in E,\ U \neq V, \quad \text{not}(U, V) \preccurlyeq (X, X) \text{ and not}(X, X) \prec (Y, Y).$$

In the sequel, proximity relations are assumed to satisfy acyclicity and zero minimality. If the proximity relation meets additional constraints, this fact is to be reflected by the representation. A symmetry condition is the following. For X, Y, U, $V \in E$ let $\preccurlyeq^*$ be defined by

$$\begin{aligned}&(X, Y) \preccurlyeq^* (U, V)\\ \text{iff } &(X, Y) \preccurlyeq (U, V) \text{ or } (Y, X) \preccurlyeq (U, V)\\ \text{or } &(X, Y) \preccurlyeq (V, U) \text{ or } (Y, X) \preccurlyeq (V, U).\end{aligned} \tag{17.2}$$

The relation $\preccurlyeq^*$ can be interpreted as an attempt to extend $\preccurlyeq$ upon the set of unordered pairs in which (X, Y) and (Y, X) are considered identical. If the proximities are nonsymmetric, the attempt will lead to inconsistencies; if they are symmetric, the relation $\preccurlyeq^*$ itself will satisfy acyclicity.

Definition 17.3. The proximity relation $\preccurlyeq$ is *symmetric*, if $\preccurlyeq^*$ satisfies a modified acyclicity condition:

$$\begin{aligned}&\text{for all } (X, Y), (U, V)\\ &(X, Y)(\preccurlyeq^*)^+(U, V) \text{ implies not}(U, V) < (X, Y).\end{aligned}$$

Note that the symmetry condition differs from the one stated in Klauer (1989): It is a corrected version of that condition. The symmetry condition guarantees that we cannot construct closed chains of pairs of objects using the relations $\sim$ and $<$ even if we allow that (X, Y) replaces (Y, X) and vice versa in the attempt to do so.

An *ordinal network representation* (ONR) of the proximity relation is defined as a network G such that the minimum-path-length function d_G satisfies the following mapping condition:

$$\begin{aligned}&\text{for all } X, Y, U, V \in E\\ &(X, Y) < (U, V) \Rightarrow d_G(X, Y) < d_G(U, V),\\ &(X, Y) \sim (U, V) \Rightarrow d_G(X, Y) = d_G(U, V).\end{aligned}$$

The mapping condition as stated takes a strict approach to ties. A weak approach is given, if the second line in the condition is deleted. Theorem 17.4 is the ordinal equivalent of Theorem 17.1 (cf. Klauer, 1989).

Theorem 17.4. There exists an ordinal network representation G of $\preccurlyeq$, if and only if $\preccurlyeq$ satisfies acyclicity and zero minimality. G can be chosen an undirected network, if and only if $\preccurlyeq$ also satisfies the symmetry condition.

The proof is simple. If $\preccurlyeq$ satisfies acyclicity, a numerical representation satisfying the mapping condition exists, as is well known. Zero minimality ensures that all self-similarities in the numerical representation can be mapped into the same function value that is smaller than any other value. By suitable monotone transformations this value can be chosen zero, and the triangle inequality can be satisfied. Theorem 17.4 then follows from Theorem 17.1.

Theorem 17.2 stated the necessary and sufficient conditions for a given pair (X, Z), $X \neq Z$, to be present in an irreducible network representation. The conditions are that there is no third point Y, $Y \neq X, Z$, such that a triangle additivity holds for the triple (X, Y, Z): $d_G(X, Z) = d_G(X, Y) + d_G(Y, Z)$.

If G is to be an ordinal network representation, the triple (X, Y, Z) cannot satisfy a triangle additivity, if $(X, Z) \preccurlyeq^+ (X, Y)$ or $(X, Z) \preccurlyeq^+ (Y, Z)$. For, these relations imply $d_G(X, Z) \leq d_G(X, Y)$ or $d_G(X, Z) \leq d_G(Y, Z)$, so that $d_G(X, Z)$ is necessarily smaller than the sum of both (positive) function values $d_G(X, Y)$ and $d_G(Y, Z)$. This observation led Hutchinson (1989) to formulate the following Corollary 17.1 that is a partial analogue of Theorem 2 for ordinal scales.

> **Corollary 17.1.** Let G be an irreducible ordinal network representation of $\preccurlyeq$, and consider $X, Z \in E$ with $X \neq Z$. If for all distinct third points Y, $Y \neq X, Z$,
>
> $$(X, Z) \preccurlyeq^+ (X, Y) \qquad \text{or} \qquad (X, Z) \preccurlyeq^+ (Y, Z),$$
>
> then (X, Z) is a link of G.

Corollary 17.1 is only a partial nonmetric equivalent of Theorem 17.2 since it provides only sufficient conditions for the presence of links. Thus, an algorithm based on it will (in the absence of error) identify essential links, but may fail to recover all such links. To derive the necessary and sufficient conditions, it is necessary to take a second look at triples and triangle additivities. For the sake of simplicity, we assume in what follows that the proximity relation $\preccurlyeq$ is both acyclic and connected, that is, for all $a, b \in E \times E$: $a \preccurlyeq b$ or $b \preccurlyeq a$. Thus, $\preccurlyeq$ is assumed a zero-minimal weak order. In particular, $(X, Y) \sim (X, Y)$ for all $(X, Y) \in E \times E$. As was shown above, one reason for a triangle additivity to be impossible for a triple (U, V, W) is that $(U, W) \preccurlyeq (U, V)$ or $(U, W) \preccurlyeq (V, W)$. Another possible reason is that there is a second triple (X, Y, Z) that is embedded in the first in the sense that (i) $d_G(U, W) \leq d_G(X, Z)$ and (ii) $d_G(X, Y) + d_G(Y, Z) \leq d_G(U, V) + d_G(V, W)$. If one of (i) or (ii) holds strictly, then the triangle inequality for the triple (X, Y, Z) implies the strict triangle inequality for the triple (U, V, W). In terms of the ordinal restrictions imposed by the mapping condition, this can be captured in the so-called *forbidden triple* criterion first formulated by Droge (1983).

> **Definition 17.4.** The triple (U, V, W) is a *forbidden triple,* if there exists $X, Y, Z \in E$ such that
>
> (i) $(U, W) \preccurlyeq (X, Z)$ and
> {(ii) $[(X, Y) \preccurlyeq (U, V)$ and $(Y, Z) \preccurlyeq (V, W)]$, or
> (iii) $[(X, Y) \preccurlyeq (V, W)$ and $(Y, Z) \preccurlyeq (U, V)]\}$,

where one of (i), (ii) or (iii) holds *strictly*, that is, with $\preccurlyeq$ replaced by $<$ in at least one place.

Note that the condition of Corollary 17.1 is included in the forbidden-triple criterion: If $(U, W) \preccurlyeq (U, V)$, for example, it is seen that (U, V, W) is a forbidden triple by letting $(X, Y, Z) = (U, V, V)$. Forbidden triples must satisfy the triangle inequality as a strict inequality:

Theorem 17.5. If (U, V, W) is a forbidden triple, then the triangle inequality holds strictly for (U, V, W) under d_G of any ordinal network representation G of $\preccurlyeq$.

Proof. By (ii) or (iii), $d_G(X, Y) + d_G(Y, Z) \leq d_G(U, V) + d_G(V, W)$. The triangle inequality for (X, Y, Z) and (i) yield

$$d_G(U, W) \leq d_G(X, Z) \leq d_G(X, Y) + d_G(Y, Z) \leq d_G(U, V) + d_G(V, W).$$

One of these inequalities is to hold strictly, so that, in summary, $d_G(U, W) < d_G(U, V) + d_G(V, W)$.

Forbidden triples are called forbidden since they can never satisfy the triangle additivity under an ordinal network representation. Links (X, Z), for which (X, Y, Z) is a forbidden triple for every $Y \in E$, $Y \neq X, Z$, therefore cannot be dropped from an ordinal network representation (Theorem 17.2) and are called *necessary* links. Thus, the forbidden-triple criterion allows one to strengthen Corollary 17.1 by weakening the preconditions of that corollary. It so turns out that the forbidden-triple criterion provides not only sufficient but also necessary criteria for the presence of links. Again, let us look at the level of triples first, where the following uniqueness result can be shown (Klauer, 1989):

Theorem 17.6. If the triangle inequality holds strictly for the triple (U, V, W) under d_G of each ordinal network representation G of $\preccurlyeq$, then the triple is a forbidden triple.

This result, proved in Klauer (1989), immediately allows us to formulate the ordinal counterpart of Theorem 17.2.

Theorem 17.7. The link (X, Z), $X \neq Z$, is present in any ordinal network representation G of $\preccurlyeq$, if and only if (X, Z) is a necessary link.

Theorem 17.7 yields the essential links that are present in any ordinal network representation of a given proximity relation. If an ordinal network representation that uses only necessary links exists, then it is unique in the sense that any other maximally sparse ordinal network representation must use the same set of links. On the other hand, it can be shown that depending upon the proximity relation, ordinal network representations may require additional links over and above the necessary links (Klauer, 1989). These additional links will be nonunique in the sense that different ordinal network representations make use of different sets of additional links. The necessary and sufficient conditions for a given set of pairs to constitute the set of links of an ordinal network representation are currently unknown, so that an ordinal counterpart of Theorem 17.3 cannot be given here.

Droge (1983) performed a Monte Carlo study to investigate the extent of the remaining nonuniqueness. For this purpose, he randomly generated networks of varying degrees of sparseness and computed the percentage of forbidden links among the links present in the random network when only the ordinal information in the minimum-path-length metric was considered. The results suggest that the structurally unique links, that is, the necessary links, typically amount to more than 90% of the total set of links. The proportion even tends to increase with the number of nodes of the network as well as its degree of sparseness.

ALGORITHMS FOR DERIVING NETWORKS

In this section, algorithms for deriving networks from proximity data are discussed. Following the representation results described in the previous section, algorithms are grouped into those that use only the ordinal information in the data, those that use both metric and nonmetric aspects of the data, and those that rely on the metric information in the sense of presupposing at least interval-scale data.

Ordinal Network Representation

An algorithm that uses only the ordinal information in the data is the ONR algorithm proposed by Klauer (1989; different algorithms with similar objectives are discussed by Feger & Bien, 1982; Feger & Droge, 1984; Orth, 1988). The objective is to derive an ordinal network representation of a zero-minimal and acyclic proximity relation $\preccurlyeq$ that is as parsimonious as possible. According to Theorem 17.3, the number of distinct triangle additivities is a central variable governing the parsimony of the network, so that for the purpose of the ONR algorithm, parsimony is quantified as the number of triangle additivities of the representing minimum-path-length function. Theorem 17.3 then allows one to formulate the algorithm solely in terms of the representing values d without explicitly using graph-theoretical constructs.

According to Theorem 17.5, we need only look at triples (U, V, W) that are not forbidden triples in maximizing the number of triangle additivities. If the set of such triples is denoted by NF (standing for "nonforbidden"), maximizing parsimony amounts to minimizing the objective function $\Sigma_{(U,V,W)\in NF} D_{(U,V,W)}$, where $D_{(U,V,W)}$ is a dichotomous variable that assumes the value zero if the triangle additivity is satisfied for triple (U, V, W) and assumes the value one otherwise. The ONR model can be stated as the following mixed integer linear programming problem:

$$\text{Minimize} \sum_{(U,V,W)\in NF} D_{(U,V,W)}$$

subject to

$$\left.\begin{array}{ll} d(X, Y) < d(U, V) & \text{if } (X, Y) < (U, V), \\ d(X, Y) = d(U, V) & \text{if } (X, Y) \sim (U, V) \end{array}\right. \quad \text{(mapping condition)},$$

$$d(U, W) \leq d(U, V) + d(V, W), \quad \text{if } (U, V, W) \in NF \quad \text{(triangle inequalities)},$$

$$d(U, W) \geq d(U, V) + d(V, W) - cD_{(U,V,W)}, \quad \text{if } (U, V, W) \in NF.$$

A solution d of the programming problem is an ordinal network representation of $\preccurlyeq$ satisfying the triangle inequalities and the mapping condition. The values $d(X, X)$, $X \in E$, may be set equal to zero because of the zero-minimality assumption. The constant c is chosen so large that the corresponding inequalities may safely be assumed to hold if $D = 1$. If $D_{(U,V,W)} = 0$, then the corresponding inequality and the triangle inequality combine to $d(U, V) + d(V, W) \leq d(U, W) \leq d(U, V) + d(V, W)$, so that the triangle inequality is satisfied as an equality. Since minimizing the objective function is equivalent to maximizing the number of D's equal to zero, the number of triangle *equalities* is maximized. Given a solution d, the representing network $G = (E, V, t)$ with $d_G = d$ is obtained according to Theorems 17.2 and 17.3, by setting

$$V = \{(X, Z) | d(X, Z) < d(X, Y) + d(Y, Z) \text{ for all } Y \in E, Y \neq X, Z\}$$

and

$$t(X, Y) := d(X, Y) \text{ for all } (X, Y) \in V.$$

For the sake of clarity, the programming problem has been stated with redundancies. In most applications, symmetric and transitive proximity relations arise, so that many inequalities of the mapping condition are redundant. If the proximity relation satisfies acyclicity, a solution of the inequality system must exist. Since the objective function is bounded from below, maximal solutions exist and can be arrived at by mixed integer linear programming techniques (Burkhard, 1972). Due to the high cost of integer programming and the fact that in the general case the number of nonforbidden triples may still increase with N^3, N being the number of objects, the present algorithm is limited to relatively small problems with fewer than 15 objects. Programs for mixed integer linear programming are available as part of many software libraries.

Note that the algorithm is suitable both for deriving undirected networks for representing symmetric proximity relations as well as for deriving directed networks for representing nonsymmetric proximity relations. The algorithm uses only the ordinal aspects of proximity data that can be captured in a proximity relation $\preccurlyeq$. Given proximity coefficients $p(X, Y)$, $\preccurlyeq$ will usually be defined by $(X, Y) \preccurlyeq (U, V)$ iff $p(X, Y) \geq p(U, V)$.

If there are errors in the observed proximities, noise will also be introduced into the proximity relation and will in general reduce the degree of parsimony that can be obtained for the representing network. For noisy data, it has been proposed to leave open suspect comparisons such as comparisons for which $p(X, Y)$ and $p(U, V)$ are close together. Specifically, a threshold e can be used in defining a partial proximity relation $\preccurlyeq_e$ such that

$$(X, Y) \preccurlyeq_e (U, V) \text{ iff } |p(X, Y) - P(U, V)| \geq e \text{ and } p(X, Y) \geq p(U, V).$$

Thereby only those differences in the observed proximities that exceed a certain threshold e are guaranteed to be represented faithfully. Increasing the e-value will in general reduce the ordinal correspondence between the original proximites and the derived minimum-path lengths. On the other hand, it is easy to see that increasing the e-value also leads to increasingly more parsimonious network representations. In the limit, $e = \infty$, a tree representation is always possible. Thus, as always in scaling, there is the familiar trade-off between parsimony and goodness of fit that is governed, in the present case, by the value of the threshold e.

NETSCAL: An Algorithm That Uses Both Nonmetric and Metric Information

The NETSCAL algorithm proposed by Hutchinson (1989) uses both nonmetric as well as metric information. NETSCAL is formulated for nonsymmetric proximities that are to be represented by directed networks, and the objective is to recover both structural as well as quantitative aspects of a network thought to underlie the noisy data-generating process. Hutchinson reports comprehensive Monte Carlo evaluations to show that NETSCAL will in general meet this objective, with a number of qualifications discussed below.

The NETSCAL algorithm proceeds in two steps. The first step uses Corollary 17.1 to determine the links to be included in the directed network, thereby determining the network structure of the representing network. As discussed above, links identified through the use of Corollary 17.1 are essential links, but Corollary 17.1 does not identify all such links. Thus, it is not surprising that NETSCAL sometimes seriously underestimates the number of links of the network underlying the data-generating process. In a Monte Carlo study by Klauer and Carroll (1991), this tendency was most pronounced when there was only a small amount of error in the data. In this case, the forbidden-triple criteria and Theorem 17.7 can be used, in a small modification of the original NETSCAL algorithm, for a more complete identification of essential links.

Whereas the first step makes use of only the nonmetric information in the data, the second step employs the metric information to derive link weights for the links identified in the first step. For this purpose, the raw data are first subjected to a linear transformation so that the highest proximity is set equal to zero and the lowest proximity equal to 1, yielding normalized dissimilarity data values $\delta(X, Y)$. The values $\delta(X, Y)$ corresponding to links identified in the first step, are then further transformed by means of a generalized power transformation to yield link weights $t(X, Y)$ (Hutchinson, 1989, Equation 8):

$$t(X, Y) = [\gamma + \delta(X, Y)]^{\lambda}.$$

With these link weights, the minimum-path-length function is then determined and is given as a function of the coefficients γ and λ. Using an iterative direct search algorithm and linear regression, the coefficients are chosen to maximize the correlation between the minimum-path-length function and the power-transformed data values.

In a Monte Carlo study, Hutchinson (1989) investigated the ability of this algorithm to recover networks underlying the data-generating process. We postpone a discussion of these results and of applications of the NETSCAL algorithm until after an alternative mathematical programming approach to fitting directed networks has been discussed.

The Pathfinder Approach

A second approach that uses both nonmetric and metric information is the so-called Pathfinder method by Schvaneveldt et al. (1988). The Pathfinder approach takes observed dissimilarities $\delta(X, Y)$ and provides as output a connected network G. Symmetric dissimilarity data lead to an undirected network, whereas nonsymmetric data give rise to a directed network. The approach differs from the previous and the following network models in that it uses a more general definition of the path length in deriving the minimum-path-length function of the representing network. For a value $r \geq 1$, the length of path $p = (l_1, \ldots, l_n)$ is defined by

$$L_r(p) = (\Sigma t(l_j)^r)^{1/r}.$$

The case $r = 1$ corresponds to the previous definition of path lengths, and the case of $r = \infty$ can be defined as

$$L_\infty(p) = \max\{t(l_j) | j = 1, \ldots, n\}.$$

For any two values r_1 and r_2 with $r_1 \leq r_2$, it is well known that $L_{r_1}(p) \geq L_{r_2}(p)$, so that the minimum-path lengths d_{rG} of the network G become shorter the larger the value of r.

The Pathfinder method proceeds as follows. It starts out with the complete network in which the link weights are set equal to the observed dissimilarities. For a given value of r, the minimum-path-length function is then computed. As before, links (X, Y) with link weights $t(X, Y)$ that exceed the minimum-path-length function $d_{rG}(X, Y)$, that is with $t(X, Y) > d_{rG}(X, Y)$, are redundant and are dropped from the representing network in a step called "triangular reduction." The resulting network is called PFNET(r) and constitutes the representing network. Since the minimum-path lengths become shorter as r increases, more links are dropped the larger r becomes so that actually a progression of successively more parsimonious networks is defined. A second parameter with an analogous function can also be introduced and is defined as the maximum number of links in a path examined in constructing a PFNET.

It is not difficult to see that for errorless data values satisfying the triangle inequality, one has $t(X, Y) \leq d_{rG}(X, Y)$ for all links (X, Y), if $r = 1$. Thus, Pathfinder with $r = 1$ does not drop any links, and the representing network will be the complete network even if the data values are the minimum-path lengths of a highly reduced network. With increasing values of r, links with weights that equal or come close to the path length of an alternative path in the network with $r = 1$ will be dropped in the triangular reduction step.

For $r = \infty$, the approach uses only the ordinal information in the observed dissimilarities, at least during triangular reduction. For all other values of $r \geq 1$, the dissimilarities must be assumed to be measured with ratio-scale properties for the triangular reduction to yield the same network under all admissible transformations of the data.

While the Pathfinder method is undoubtedly computationally not very costly, it has some serious flaws. As outlined above, it is not capable of reproducing a latent network underlying the data-generating process even if the data are perfectly errorless. We know that for $r = 1$, Pathfinder results in the complete network, assuming the triangle inequality is satisfied, while for $r = \infty$ it results in the minimum spanning tree (or union of minimum spanning trees, if not unique) as has been proved by Carroll (1993). (See also Dearholt & Schvaneveldt, 1990.) Unfortunately, however, we have no idea whatever what happens for values of r between 1 and ∞. Pathfinder evidently provides a way to more-or-less continuously move between the complete network and the minimum spanning tree—but there are many ways to do this, and there is no reason to believe this way to be better than any other. Little is known about the capability of Pathfinder to recover underlying networks from noisy data or about its performance relative to the other algorithms discussed here. Because of the apparent problems with this algorithm, one should perhaps not expect too much from this rough-and-ready heuristic for generating a network structure from observed dissimilarities. Arabie (1993) has provided a critical review of the Pathfinder approach that points out additional problems.

MAPNET: An Algorithm for Metric Network Scaling

An algorithm that is based on interval-scale data has been proposed by Klauer and Carroll (1989, 1991). For a user-specified number of links L, the approach seeks to derive a network with L links that is a least-squares approximation of observed dissimilarity coefficients $\delta(X, Y)$. The algorithm has been developed for both symmetric and nonsymmetric dissimilarity measures. Through the use of a mathematical programming approach, both network structure and link weights are simultaneously chosen so as to optimize the goodness of fit. Based on Theorem 17.3, the MAPNET model can be stated as the following optimization problem (where c is an additive constant):

$$\text{Minimize } L(c, d) = \Sigma(\delta(X, Y) - c - d(X, Y))^2$$

subject to the following constraints on $d(X, Y)$:

- the triangle inequality, positivity, and
- $M - L$ distinct triangle additivities as specified in Theorem 17.3, where M is the number of links of the complete network.

The constrained minimization problem is solved by a series of unconstrained minimizations. For this purpose, an extended loss function $F(c, d)$ is minimized using a conjugate gradient algorithm (Powell, 1977) that is the weighted sum of the L-part defined above and additonal P- and Q_L-parts,

$$F(c, d) = L(c, d) + r_1(P(d) + aQ_L(d)).$$

The P-part is designed to move the derived values d toward satisfying the triangle inequality and constitutes a classical continuous and differentiable penalty function for imposing inequality constraints (Ryan, 1974). Measuring the deviation from triangle *equality* of triple (X, Y, Z) by $\nu(X, Y, Z) = d(X, Y) + d(Y, Z) - d(X, Z)$, $P(d)$ penalizes deviations in the direction that violates the triangle inequality; that is, with $\nu < 0$,

$$P(d) = \sum_{(X,Y,Z)} \nu^2(X, Y, Z)1_{\{\nu(X,Y,Z)<0\}}.$$

For triples (X, Y, Z) with $X = Z$, the triangle inequality amounts to nonnegativity of $d(X, Y)$. The crucial Q_L-part is a penalty function designed to push the derived values to satisfy the additivity conditions specified in Theorem 17.3. For this purpose, the function $s(X, Z)$ measures how far the pair (X, Z), $X \neq Z$, is from satisfying a triangle additivity with a suitable third point $Y \neq X, Z$:

$$s(X, Z) = \min_{Y \neq X,Z} \{\nu^2(X, Y, Z)\}.$$

A distinct third point Y such that (X, Y, Z) satisfies the triangle additivity is given iff $s(X, Z) = 0$. Considering all M values $s(X, Z)$ for all M pairs of objects ($M = M_o$ or $M = M_u$, depending on whether the symmetric or the nonsymmetric case, respectively, is treated), let $s^{(1)}$ be the smallest value arising, $s^{(2)}$ the second smallest, and, finally, $s^{(M)}$ the largest value. Setting

$$Q_L(d) = \sum_{l=1}^{M-L} s^{(l)},$$

a penalty function is defined that assumes the smallest value zero if and only if $M - L$ distinct triangle additivities involving $M - L$ distinct pairs (X, Z) have been found. $Q_L(d)$ is continuous and piecewise differentiable. The set of points at which Q_L is not differentiable has measure zero.

The overall algorithm constitutes a so-called mathematical programming approach (Arabie & Carroll, 1980; Carroll et al., 1984; Carroll & Pruzansky, 1980; Cunningham & Shepard, 1974; De Soete, 1983) to solving a partially discrete problem by embedding it in a continuous problem. The discrete solution is arrived at by a sequence of increasingly close continuous approximations. The initial values for the parameters arise from any of several sources detailed in Klauer and Carroll (1989). The loss function is minimized repeatedly with increased weight r_1 for the P- and Q_L-part until eventually the triangle inequality and the conditions of Theorem 17.3 hold essentially perfectly. Finally, the resulting network topology is fixed and a "polishing" iteration is employed to reestimate the link weights.

The actual algorithm is implemented with a number of technical modifications. In particular, positivity of $d(X, Y)$ is guaranteed by a reparametrization in terms of new parameters $e(X, Y)$ such that $d(X, Y) = r_2 + e(X, Y)$, where $X \neq Y$, and r_2 is a small positive constant that is decreased each time r_1 is increased. In addition, to avoid a certain type of degeneration, the values $s(X, Z)$ are normalized as specified in Klauer and Carroll (1989).

In MAPNET, the number of links L of the representing network is user specified. Usually, the algorithm is applied for several L-values. Decreasing L forces the algorithm to derive more parsimonious networks and simultaneously decreases the goodness-of-fit to the data that can be expected. Thus, there is the familiar trade-off between parsimony and goodness of fit that is governed, in the present case, by the parameter L. An example is given below in connection with the individual-differences generalization of MAPNET.

Extensive Monte Carlo studies were performed to investigate (a) how close MAPNET representations come to the objective of being least-squares representations, (b) how well structural aspects of networks underlying the data-generating process are recovered, and finally, (c) to compare the relative performance of NETSCAL (Hutchinson, 1989) and MAPNET in fitting proximity data and in recovering underlying networks (Klauer & Carroll, 1989, 1991). The major findings were that

- MAPNET does not always achieve optimal representations in terms of the least-squares criterion. However, the MAPNET networks were typically found to be very close to the optimal network, within a few percent of the variance accounted for by the optimal network. Thus, local minima problems exist, but do not appear to be severe. Furthermore,
- MAPNET consistently outperformed NETSCAL in obtaining a good quantitative fit to observed dissimilarities. The size of the advantage of MAPNET

over NETSCAL is a function of distributional properties of the dissimilarities and was found to vary between 5% and 50% of the variance accounted for.

- Structural aspects of an underlying network such as the set of links are recovered with high degrees of accuracy by both MAPNET and NETSCAL when the amount of errors in the data is relatively small and accounts for less than about 10% of the data variance.
- For higher amounts of error, structural recovery attains only moderate degrees of accuracy. For example, the percentage of links in the representing network that are also links of the underlying network is the *percent hit* index. In MAPNET, percent hit values decrease quickly from perfect or 100% to a level of about 70% as the amount of error mixed with the minimum-path-length function of an underlying network is increased from 0% to 10% of the total variance.

INDNET: AN INDIVIDUAL-DIFFERENCES GENERALIZATION OF MAPNET

The Monte Carlo results concerning the recovery of an underlying network were in part responsible for the recent development of an individual-differences generalization of the MAPNET approach, called INDNET: One particular expectation for INDNET is better performance in recovering underlying networks even when data are relatively noisy.

Individual-differences generalizations are based on so-called three-way data in which several data sources, typically subjects, provide proximity data. The objective is to represent the total data set using one common group structure, whereas differences between the data sources are allowed for only in the quantitative aspects of the representation. For example, in INDSCAL, the individual-differences generalization of multidimensional scaling, the same Euclidean space, with points placed in the same positions, is used for each data source (Arabie, Carroll, & DeSarbo, 1987; Carroll & Chang, 1970). The different data sources may however differentially weight the dimensions in computing the representing Euclidean distances. Similarly, in INDCLUS (Arabie et al., 1987), the three-way version of the ADCLUS model (Shepard & Arabie, 1979), the same cluster structure is used for each data source, whereas different data sources may weight the component clusters differentially. Finally, in INDTREES, an individual-differences generalization of various tree models (Carroll et al., 1984), the same tree structure is used for each data source, whereas link weights may differ between data sources.

In INDNET, the same network structure is used for each data source, and different data sources may differ in that they use different link weights. There are at least two conceptualizations of how a common network structure should be defined. The first definition, based on Theorem 17.3 and called the *same paths* structure, requires the same set of triangle additivities for each data source. Since the additivities uniquely determine the shortest paths in the network, the shortest paths connecting two points are the same for each data source, hence the name *same paths* structure. The second definition is weaker and requires the same set of links for each data set, but allows for differences in the shortest paths and the set of triangle additivities between data sources (over and above differences in link weights). This concept of a common group structure is termed the *same links* structure.

Both versions have specific advantages and disadvantages. Additivities play a central role in the psychological interpretation of networks: A dimensional interpretation requires intradimensional additivity so that additivities are imposed on triples of objects that form one-dimensional subsets. Interpretation in terms of feature sets and Restle's (1959) symmetric difference metric, on the other hand, imposes additivities on triples of objects which satisfy the betweenness relation. Both interpretations of a common network structure are possible only if the same set of additivities is guaranteed for each data source and, thus, only if the *same paths* structure is realized. On the other hand, requiring the same set of additivities or equivalently the same shortest paths for each data source, imposes inequality constraints on the link weights employed by the different data sources. From a "data-reduction" point of view, constraining the link weights is disadvantageous in that the degree of freedom introduced by each link-weight parameter is not used optimally.

In sum, the *same paths* structure has the advantage of possibly superior interpretability, but is less parsimonious a model than the *same links* version in which link weights are not constrained. Two versions of the INDNET algorithm were constructed based on the *same paths* and the *same links* structure, respectively. The algorithm is formally very similar to the MAPNET algorithm. Thus, a least-squares goodness-of-fit criterion is again employed in a mathematical programming approach that repeatedly minimizes an extended loss function involving penalty functions. Two penalty functions $P(d)$ and $Q_L(d)$ are again used. $P(d)$ is designed to impose the triangle inequality on the parameters $d_k(X, Y)$ for each data source k, $k = 1, \ldots, K$. $Q_L(d)$ is responsible for finding a set of $M - L$ additivities, where M is the total number of pairs of objects, for each data source k.

In the *same paths* structure, the additivities are the same for each data source. They are derived by means of the averaged values $d(X, Y) = (1/K)\Sigma_k d_k(X, Y)$, so that the common group structure can be considered an average structure. Specifically, $M - L$ distinct triples (X_l, Y_l, Z_l), $l = 1, \ldots, M - L$, are identified on the basis of the averaged values that are already closest to satisfying the triangle additivity in the sense that they define $Q_L(d)$ of the MAPNET algorithm used with the averaged values (see the definition of $Q_L(d)$ above). Then $Q_L(d)$ for the INDNET *same paths* model is obtained by separately penalizing deviations from triangle additivity for these $M - L$ triples for each data source k. If $v_k(X, Y, Z) = d_k(X, Y) + d_k(Y, Z) - d_k(X, Z)$, then

$$Q_L(d) = \sum_{l=1}^{L} \sum_{k=1}^{K} v_k^2(X_l, Y_l, Z_l).$$

Note that $Q_L(d) = 0$ iff the same set of triangle additivities has been imposed on each data source.

In the *same links* structure, the same $M - L$ links (X_l, Z_l) are to be dropped from the complete network, which requires, by Theorem 17.3, $M - L$ triangle additivities involving (X_l, Z_l) and suitable third points Y for each data source. In contrast to the *same paths* structure, these third points Y may differ from data source to data source, however. We can measure the deviation of a particular link from satisfying a triangle additivity in each data source via possibly different third points Y by

$$s(X, Z) = \sum_{k=1}^{K} \min_{Y \neq X,Z} \{v_k^2(X, Y, Z)\},$$

where the minimum is taken separately for each data source. If the values $s(X, Z)$ are ordered according to size as described above for the MAPNET algorithm, the appropriate penalty function $Q_L(d)$ is given by

$$Q_L(d) = \sum_{l=1}^{M-L} s^{(l)}.$$

Note that $Q_L(d) = 0$ iff $M - L$ additivities involving the same $M - L$ distinct pairs of objects and possibly different third points have been imposed upon each data source.

Monte Carlo studies show that INDNET performs about as well as MAPNET in terms of obtaining a least-squares description of observed dissimilarities. In addition, the links of underlying networks are recovered with high degrees of accuracy even when the data are relatively noisy.

As an example, the famous Rosenberg-Kim kinship data were analyzed. Rosenberg and Kim (1975) used 15 kinship terms most commonly occurring in English: aunt, brother, cousin, daughter, father, granddaughter, grandfather, grandmother, grandson, mother, nephew, niece, sister, son, uncle. The data sources in the Rosenberg-Kim study were groups that sorted kinship terms under different sorting instructions. Eighty-five males and 85 female subjects were run in a condition where subjects gave only a single sort of the terms. Members of different groups were told in advance that after making their first sort, they would be asked to give additional subjective partitionings of these stimuli using a different basis of meaning each time. The data from the first and the second sortings for these groups of subjects were used. Data were aggregated within each group and sorting condition to obtain a matrix of similarity coefficients based on how often two objects were sorted together. There are $K = 6$ data sources that can be labeled females' first sort, females' second sort, males' first sort, males' second sort, females' single sort, and males' single sort. The data matrices (converted to dissimilarities) are shown in Table 6 of Arabie et al. (1987).

INDNET analyses were performed for various numbers of links L. Figure 17.1 shows the percentages of data variance accounted for, plotted as a function of $M_u - L$, the number of links *omitted* from the complete network, for both the *same links* as well as the *same paths* structure. With 15 objects, omitting 91 links leads to a network with 14 remaining links, which is a tree and hence the smallest possible connected network. The *same links* structure consistently achieves a slightly better fit corresponding to the fact that it is the less-constrained model. The difference between the two models is small for the present data set, however. As expected, the goodness of fit decreases the more links are omitted from the representing common network structure, and noticeable drops occur after $M_u - L = 78$, $M_u - L = 81$, and $M_u - L = 88$.

The *same links* network with $M_u - L = 81$, shown in Figure 17.2, strikes a balance between the goals of parsimony and high goodness of fit. The variance accounted for (VAF) amounts to 84%. VAF is computed according to Equation 4

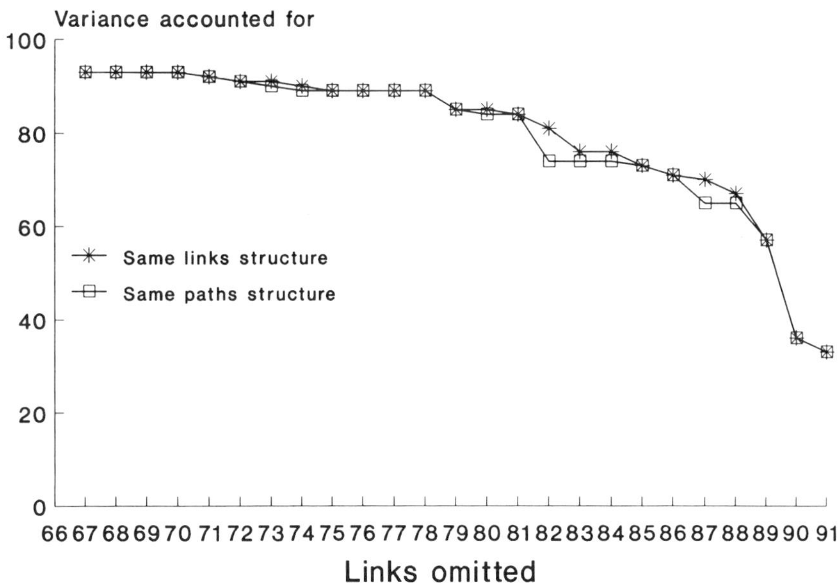

FIG. 17.1. VAF and number of links.

in Arabie et al. (1987) and thus conservatively does not include the variance accounted for by overall shifts in mean similarity between the data sources. Both the *same links* and the *same paths* network use the same set of links as well as very similar link weights. The link weights of the *same links* version are written next to the corresponding links for all six data sources in Figure 17.2. Since the

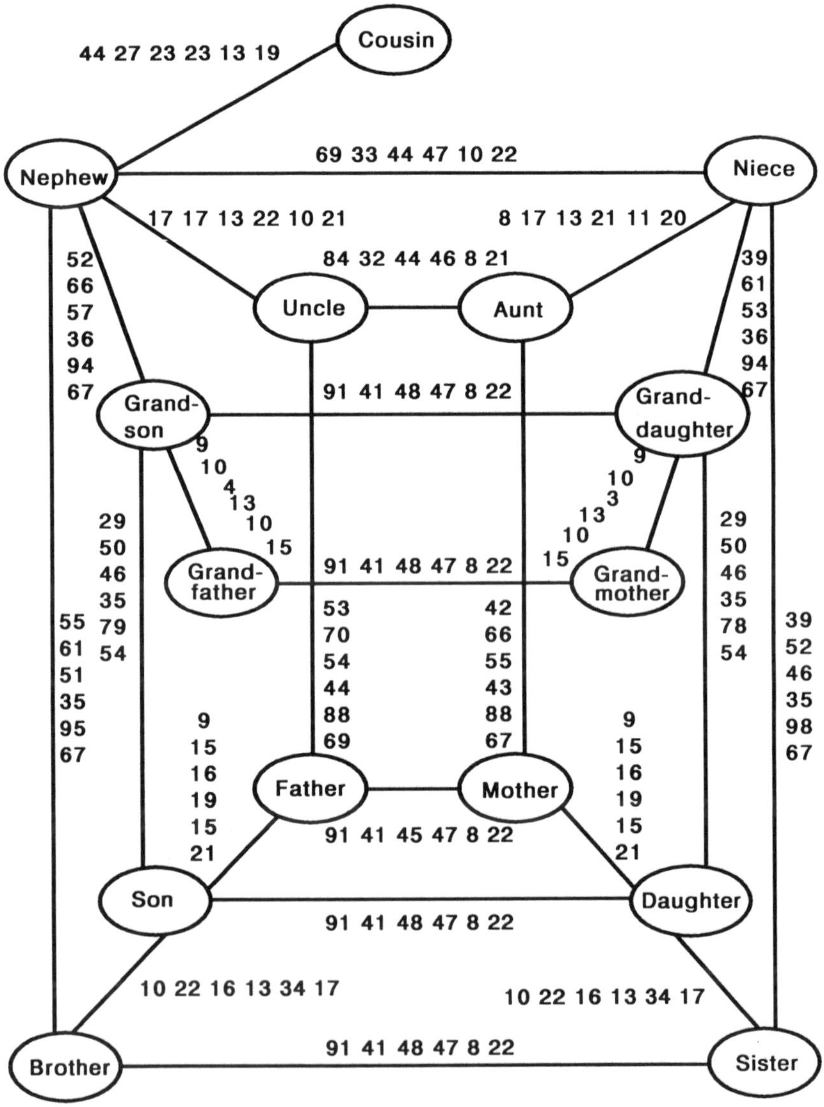

FIG. 17.2. Common network structure and link weights.

analysis is a so-called unconditional analysis (Arabie et al., 1987), their sizes can be compared between different links as well as between different data sources.

From a conceptual point of view, the kinship terms are ordered along the dimensions sex and generation. Refined conceptual analyses have been conducted in the domain of anthropology. As discussed below, so-called componential analyses in particular suggest a number of additional structural features such as a "direct kinship" versus "collateral kinship" dimension (Romney & D'Andrade, 1964). As can be seen in Figure 17.2, the sex dimension emerges clearly in that relatives who differ only in sex are always linked. In addition, the corresponding link weights are numerically similar for all such links, indicating the operation of an additive sex feature in the sense of Tversky (1977). In addition, there are a number of "generation" links, connecting relatives who differ only in one or more generation shifts. Pairs of kinship terms are said to satisfy reciprocity if the first kinship term is as far apart from the hypothetical speaker in terms of ascending generation shifts as the second is in descending generation shifts. Examples are the pairs (son, father) and (granddaughter, grandmother). All same-sex pairs of this kind, including the collateral (uncle, nephew) and (aunt, niece), are links of the network, and, as predicted by Romeny and D'Andrade (1964), consistently carry relatively small weights, indicating high proximity, as compared to other generation links that do not satisfy reciprocity. A third type of link such as (nephew, grandson) connects direct and collateral relatives of the same sex that are one generation apart. Again, the pattern of link weights is similar for this type of link over the different data sources, indicating the operation of an additive distinctive feature (Tversky, 1977). The weights for this feature are among the largest occurring in the network, so that the feature is least important in inducing similarity in the perception of the kinship terms.

Over and above these common structural aspects, there are consistent differences between the data sources in the weighting of the different features. For example, sex contributes relatively little to overall proximity in the females' first sort (large weight), but becomes more important during the second sort. Interestingly, sex is very important in the group of female students asked to perform only one sort (single sort, fifth link weight).

DISCUSSION

This chapter has reviewed and summarized fundamental representation and uniqueness results for network models as psychological theories of proximity judgments. In addition, existing methods for analyzing proximities by means of network structures have been discussed. Methods now exist that can be used to scale metric as well as nonmetric data, symmetric and nonsymmetric proximity measures, and two-way and three-way data. For the analysis of metric proximities, several alternative methods exist that differ in the dimensions of (a) computational cost, (b) accuracy of recovery of underlying networks, and (c) goodness of fit to the observed data. Methods with comparatively low computational costs are Schvaneveldt et al.'s (1988) Pathfinder and Hutchinson's (1989) NETSCAL. NETSCAL also performs satisfactorily with respect to the degree of accuracy with which the structure of an underlying latent network is recovered. It is outperformed by Klauer and Carroll's (1989, 1991) MAPNET, however, when it comes to the quantitative goodness of fit of the representing minimum-path-length function to the observed proximities. MAPNET, on the other hand, is computationally much more demanding than NETSCAL. Little is known about the Pathfinder approach with respect to the accuracy of recovery of an underlying network and the goodness-of-fit measure.

An interesting generalization of the MAPNET approach to the analysis of three-way data, termed INDNET, has been presented here for the first time. Three-way data consist of a collection of complete proximity data from each of several data sources. INDNET realizes a so-called individual-differences analysis of such data sets. Individual-differences analyses strike a balance between two extreme approaches to three-way data. One extreme possibility is to analyze the data from each data source separately. The other extreme approach is to aggregate the data over data sources and to perform a traditional two-way analysis of the aggregated data set. The latter approach completely obscures possible differences between the various data sources, the first approach yields a different structure for each data source. The individual-difference approach to network scaling uses a common network structure for each data source, whereas differences between data sources are allowed to show as differences in link weights. A particular advantage of the INDNET analysis turns out to be its relatively high degree of accuracy in recovering the structure of networks underlying the data-generating process even when there is a relatively high amount of noise in the data.

Data-driven methods for network scaling can be expected to be useful in the development and testing of psychological theories using network models. To date, construction of networks used in these approaches has usually been based upon a priori considerations (Hutchinson, 1989). The exemplary analysis of the Rosenberg-Kim kinship data clearly illustrates the usefulness of network models in revealing the underlying cognitive organization of richly organized material.

REFERENCES

Anderson, J. R. (1983). *The architecture of cognition.* Cambridge, MA: Harvard University Press.

Arabie, P. (1993). Methodology neither new nor improved [Review of Pathfinder associative networks: Studies in knowledge organization, edited by Roger W. Schvaneveldt]. *Contemporary Psychology, 38,* 66–67. (Exchanges of rejoinders: *Contemporary Psychology, 39,* 100–102.)

Arabie, P., & Carroll, J. D. (1980). MAPCLUS: A mathematical programming approach to fitting the ADCLUS model. *Psychometrika, 45,* 211–235.

Arabie, P., Carroll, J. D., & DeSarbo, W. S. (1987). *Three-way scaling and clustering*. Newbury Park, CA: Sage.

Bales, R. F. (1970). *Personality and interpersonal behavior*. New York: Holt, Rinehart, & Winston.

Beals, R., Krantz, D. H., & Tversky, A. (1968). Foundations of multidimensional scaling. *Psychological Review, 75*, 127–142.

Burkhard, R. E. (1972). *Methoden der ganzzahligen Optimierung* [Methods of integer programming]. New York: Springer.

Carroll, J. D. (1976). Spatial, non-spatial and hybrid models for scaling. *Psychometrika, 41*, 439–463.

Carroll, J. D. (1993). *Minimax length links of a dissimilarity matrix and minimum spanning trees*. Manuscript submitted for publication.

Carroll, J. D., & Chang, J. J. (1970). Analysis of individual differences in multidimensional scaling via an *N*-way decomposition of "Eckart-Young" decomposition. *Psychometrika, 35*, 283–319.

Carroll, J. D., & Chang, J. J. (1973). A method for fitting a class of hierarchical tree structure models to dissimilarities data and its applications to some 'body parts' data of Miller's. *Proceedings of the 81st Annual Convention of the American Psychological Association, 8*, 1097–1098.

Carroll, J. D., Clark, L. A., & DeSarbo, W. S. (1984). The representation of three-way proximities data by single and multiple tree structure models. *Journal of Classification, 1*, 25–74.

Carroll, J. D., & Pruzansky, S. (1980). Discrete and hybrid scaling models. In E. D. Lantermann & H. Feger (Eds.), *Similarity and choice* (pp. 108–139). Bern: Hans Huber.

Collins, A. M., & Loftus, E. F. (1975). A spreading-activation theory of semantic processing. *Psychological Review, 82*, 407–428.

Collins, A. M., & Quillian, M. R. (1969). Retrieval time from semantic memory. *Journal of Verbal Learning and Verbal Behavior, 8*, 240–247.

Corter, J. E., & Tversky, A. (1986). Extended similarity trees. *Psychometrika, 51*, 429–452.

Cunningham, J. P. (1978). Free trees and bidirectional trees as representations of psychological distance. *Journal of Mathematical Psychology, 17*, 165–188.

Cunningham, J. P., & Shepard, R. N. (1974). Monotone mapping of similarities into a general metric space. *Journal of Mathematical Psychology, 11*, 335–363.

De Soete, G. (1983). A least squares algorithm for fitting additive trees to proximity data. *Psychometrika, 48*, 621–626.

Dearholt, D. W., & Schvaneveldt, R. W. (1990). Properties of Pathfinder networks. In R. W. Schvaneveldt (Ed.), *Pathfinder associative networks: studies in knowledge organization* (1–30), Norwood, NJ: Ablex.

Droge, U. (1983). *Ordinale Netzwerkskalierung* [Ordinal network scaling]. Unpublished doctoral dissertation. University of Hamburg, Hamburg.

Feger, H., & Bien, W. (1982). Network unfolding. *Social Networks, 4*, 257–283.

Feger, H., & Droge, U.N. (1984). Ordinale Netzwerkskalierung [Ordinal network scaling]. *Kölner Zeitschrift für Soziologie und Sozialpsychologie, 3*, 417–423.

Goldman, A. J. (1966). Realizing the distance matrix of a graph. *Journal of Research of the National Bureau of Standardization, 70B*, 153–154.

Hage, P., & Harary, F. (1983). *Structural models in anthropology*. Cambridge: Cambridge University Press.

Hakimi, S. L., & Yau, S. S. (1964). Distance matrix of a graph and its realizability. *Quarterly Journal of Applied Mathematics, 22*, 305–317.

Hartigan, J. A. (1967). Representations of similarity matrices by trees. *Journal of the American Statistical Association, 62*, 1140–1158.

Hutchinson, J. W. (1989). NETSCAL: A network scaling algorithm for nonsymmetric proximity data. *Psychometrika, 54*, 25–52.

Johnson, S. C. (1967). Hierarchical clustering schemes. *Psychometrika, 32*, 241–254.

Klauer, K. C. (1989). Ordinal network representation: Representing proximities by graphs. *Psychometrika, 54*, 737–750.

Klauer, K. C., & Carroll, J. D. (1989). A mathematical programming approach to fitting general graphs. *Journal of Classification, 6,* 247–270.

Klauer, K. C., & Carroll, J. D. (1991). A comparison of two approaches to fitting directed graphs to nonsymmetric proximity measures. *Journal of Classification, 8,* 251–268.

Krantz, D. H., Luce, R. D., Suppes, P., & Tversky, A. (1971). *Foundations of measurement* (Vol. 1). New York: Academic Press.

Kruskal, J. B. (1964). Nonmetric multidimensional scaling: A numerical method. *Psychometrika, 29,* 115–129.

Orth, B. (1988). Representing similarities by distance graphs: Monotonic network analysis (MONA). In H. H. Bock (Ed.): *Classification and related methods of data analysis.* Proceedings of the First Conference of the International Federation of Classification Societies (IFCS). Amsterdam: North-Holland.

Powell, M. J. (1977). Restart procedures for the conjugate gradient method. *Mathematical Programming, 12,* 241–254.

Restle, F. (1959). A metric and an ordering on sets. *Psychometrika, 24,* 207–220.

Romney, A. K., & D'Andrade, R. G. (1964). Cognitive aspects of English kin terms. *American Anthropologist, 66,* 146–170.

Rosenberg, S., & Kim, M. P. (1975). The method of sorting as a data-gathering procedure in multivariate research. Multivariate Behavioral Research, *10,* 489–502.

Ryan, D. M. (1974). Penalty and barrier functions. In P. E. Gill & W. Murray (Eds.), *Numerical Methods for Constrained Optimization* (pp. 175–190). New York: Academic Press.

Sattath, S., & Tversky, A. (1977). Additive similarity trees. *Psychometrika, 42,* 319–344.

Schvaneveldt, R. W., Dearholt, D. W., & Durso, F. T. (1988). Graph theoretic foundations of pathfinder networks. *Computational Mathematical Applications, 15,* 337–345.

Shepard, R. N. (1962). Analysis of proximities: Multidimensional scaling with an unknown distance function: I. *Psychometrika, 27,* 125–140.

Shepard, R. N., & Arabie, P. (1979). Additive clustering: representation of similarities as combinations of discrete overlapping properties. *Psychological Review, 86,* 87–123.

Tversky, A. (1977). Features of similarity. *Psychological Review, 84,* 327–352.

Tversky, A., & Gati, I. (1982). Similarity, separability, and the triangle inequality. *Psychological Review, 89,* 123–154.

AUTHOR INDEX

A

B

T

U

V

W

SUBJECT INDEX

F

G

H

I

J, K

L

R

S